CAIRO INTERNATIONAL CONFERENCE ON HIGH ENERGY PHYSICS (CICHEP II)

To learn more about the AIP Conference Proceedings, including the Conference Proceedings Series, please visit the webpage **http://proceedings.aip.org/proceedings**

CAIRO INTERNATIONAL CONFERENCE ON HIGH ENERGY PHYSICS (CICHEP II)

Cairo, Egypt   14 – 17 January 2006

EDITOR

Shaaban Khalil

Ain Shams University & German University
Cairo, Egypt

SPONSORING ORGANIZATIONS
German University in Egypt (GUC)
International Center for Theoretical Physics (ICTP)

Melville, New York, 2007
AIP CONFERENCE PROCEEDINGS ■ VOLUME 881

Editor

Shaaban Khalil
Ain Shams University
Faculty of Science
Department of Mathematics
Cairo, 11566
Egypt
E-mail: skhalil@bue.edu.eg

L.C. Catalog Card No. 2006939107
ISBN 978-0-7354-0382-6
ISSN 0094-243X

Printed in the United States of America

CONTENTS

Preface

The second Cairo International Conference on High Energy Physics (CICHEP II) was held in the German University in Cairo (GUC) during the period January 14-17, 2006. This conference is the sequel of "The first Cairo International Conference on High Energy Physics" (CICHEP I) that was held in the period 9-14 January 2001. The conference was sponsored by the GUC and partially by the International Center for Theoretical Physics (ICTP), Trieste, Italy.

The main theme of this conference was theoretical, phenomenological and experimental aspects of the presently most active field of High Energy Physics. More than 55 given talks, that covered a wide range of these topics, were presented in the conference. I would like to express my deepest gratitude to all the lecturers for their efforts. I am also very grateful to my colleagues in the organizing committee for their help and support that made the CICHEP II possible.

Shaaban Khalil

Cairo, Egypt

October 21, 2006

The supersymmetric standard model from the Z'_6 orientifold?

David Bailin and Alex Love

Department of Physics & Astronomy, University of Sussex, Brighton, BN1 9QH, UK

Abstract. We construct $\mathcal{N} = 1$ supersymmetric fractional branes on the Z'_6 orientifold. Intersecting stacks of such branes are needed to build a supersymmetric standard model. If a, b are the stacks that generate the $SU(3)_c$ and $SU(2)_L$ gauge particles, then, in order to obtain *just* the chiral spectrum of the (supersymmetric) standard model (with non-zero Yukawa couplings to the Higgs mutiplets), it is necessary that the number of intersections $a \cap b$ of the stacks a and b, and the number of intersections $a \cap b'$ of a with the orientifold image b' of b satisfy $(a \cap b, a \cap b') = (2,1)$ or $(1,2)$. It is also necessary that there is no matter in symmetric representations of the gauge group. We have found a number of examples having these properties. Different lattices give different solutions and different physics.

Keywords: Intersecting branes, orientifold
PACS: 11.25.Wx, 12.60.Jv

1. INTRODUCTION

Intersecting D-branes provide an attractive, bottom-up route to standard-like model building [1]. In these models one starts with two stacks, a and b with $N_a = 3$ and $N_b = 2$, of D6-branes wrapping the three large spatial dimensions plus 3-cycles of the six-dimensional internal space (typically a torus T^6 or a Calabi-Yau 3-fold) on which the theory is compactified. These generate the gauge group $U(3) \times U(2) \supset SU(3)_c \times SU(2)_L$, and the non-abelian component of the standard model gauge group is immediately assured. Further, (four-dimensional) fermions in bifundamental representations $(\mathbf{N}_a, \overline{\mathbf{N}}_b) = (\mathbf{3}, \overline{\mathbf{2}})$ of the gauge group can arise at the multiple intersections of the two stacks. These are precisely the representations needed for the quark doublets Q_L of the Standard Model. In general, intersecting branes yield a non-supersymmetric spectrum, so that, to avoid the hierarchy problem, the string scale associated with such models must be low, no more than a few TeV. Then, the high energy (Planck) scale associated with gravitation does not emerge naturally. Nevertheless, it seems that these problems can be surmounted [2, 3], and indeed an attractive model having just the spectrum of the standard model has been constructed [4]. It uses D6-branes that wrap 3-cycles of an orientifold T^6/Ω, where Ω is the world-sheet parity operator. The advantage and, indeed, the necessity of using an orientifold stems from the fact that for every stack $a, b, \dots$ there is an orientifold image $a', b', \dots$. At intersections of a and b there are chiral fermions in the $(\mathbf{3}, \overline{\mathbf{2}})$ representation of $U(3) \times U(2)$, where the $\mathbf{3}$ has charge $Q_a = +1$ with respect to the $U(1)_a$ in $U(3) = SU(3)_c \times U(1)_a$, and the $\overline{\mathbf{2}}$ has charge $Q_b = -1$ with respect to the $U(1)_b$ in $U(2) = SU(2)_L \times U(1)_b$. However, at intersections of a and b' there are chiral fermions in the $(\mathbf{3}, \mathbf{2})$ representation, where the $\mathbf{2}$ has $U(1)_b$ charge $Q_b = +1$. In

CP881, *Cairo International Conference on High Energy Physics (CICHEP II)*,
edited by S. Khalil

general, besides gauge bosons, stacks of D-branes on orientifolds also have chiral matter in the symmetric $\mathbf{S}$ and antisymmetric $\mathbf{A}$ representations of the relevant gauge group; both have charge $Q = 2$ with respect to the relevant $U(1)$. For the stack a with $N_a = 3$, $\mathbf{S}_a = \mathbf{6}$ and $\mathbf{A}_a = \bar{\mathbf{3}}$. The former must be excluded on phenomenological grounds, but the latter could be quark-singlet states q_L^c. Similarly, for the stack b with $N_b = 2$, $\mathbf{S}_b = \mathbf{3}$ and $\mathbf{A}_b = \mathbf{1}$. Again, the former must be excluded on phenomenological grounds, but the latter could be lepton-singlet states ℓ_L^c. Suppose that the number of intersections $a \cap b$ of the stack a with b is p, the number of intersections $a \cap b'$ of the stack a with b' is q, and the number of copies of $\mathbf{A}_a = \bar{\mathbf{3}}$ is r. The standard model has 3 quark doublets Q_L, so that to get just the standard-model spectrum we must have $p + q = 3$. The standard model also has a total of 6 quark-singlet states. To get just the standard model spectrum we also require that $6 - r$ of the quark singlets arise from intersections of a with other stacks $c, d, \ldots$ having just a single D6-brane. These belong to the representation $(\mathbf{1}, \bar{\mathbf{3}})$ of $U(1) \times U(3)$ and each has charge $Q_a = -1$. Ramond-Ramond (RR) tadpole cancellation requires that overall Q_a sums to zero. Thus

$$2p + 2q + 2r - (6 - r) = 0 \tag{1}$$

Hence $r = 0$ and we must also exclude the representations $\mathbf{A}_a = \bar{\mathbf{3}}$. Tadpole cancellation also requires that Q_b sums to zero overall. To get just the standard model spectrum we require that there are 3 lepton doublets L arising from intersections of b with other stacks having just a single D6-brane. All have $Q_b = +1$ or $Q_b = -1$. Suppose the number of copies of $\mathbf{A}_b = \mathbf{3}$ is s. Then overall cancellation of Q_b requires that

$$-3p + 3q + 2s \pm 3 = 0 \tag{2}$$

Hence $s = 0 \bmod 3$. In the case that $s = 0$ the solutions are $(p, q) = (1, 2)$ or $(2, 1)$, whereas when $s = \pm 3$ the solutions $(p, q) = (3, 0)$ or $(0, 3)$ are also allowed [5]. (Models with $|s| > 3$ will obviously have non-standard model spectra.) However, states arising as the antisymmetric representation of $U(2)$ do not have the standard-model Yukawa couplings to the Higgs multiplet. Consequently we are only interested in models such as that in [4] with $(a \cap b, a \cap b') = (1, 2)$ or $(2, 1)$.

Despite the attractiveness of that model, there remain serious problems in the absence of supersymmetry. A generic feature of intersecting brane models is that flavour changing neutral currents are generated by four-fermion operators induced by string instantons [6]. The severe experimental limits on these processes require that the string scale is rather high, of order 10^4 TeV. This makes the fine tuning problem very severe, and the viability of such models highly questionable. Further, in non-supersymmetric theories, such as these, the cancellation of RR tadpoles does not ensure Neveu Schwarz-Neveu Schwarz (NSNS) tadpole cancellation. NSNS tadpoles are simply the first derivative of the scalar potential with respect to the scalar fields, specifically the complex structure and Kähler moduli and the dilaton. A non-vanishing derivative of the scalar potential signifies that such scalar fields are not even solutions of the equations of motion. Thus a particular consequence of the non-cancellation is that the complex structure moduli are unstable [7]. One way to stabilise these moduli is for the D6-branes to wrap 3-cycles of an orbifold T^6/P, where P is a point group, rather than a torus T^6. The FCNC problem can be solved and the complex structure moduli stabilised when the theory is supersymmetric. First, a supersymmetric theory is not obliged to have the low string scale that led to problematic FCNCs induced by string instantons. Second, in a supersymmetric theory, RR tadpole cancellation ensures cancellation of the NSNS tadpoles [8, 9]. An orientifold is then constructed by quotienting the orbifold with the world-sheet parity operator Ω. (As explained above, an orientifold is necessary to allow the possibility of obtaining just the spectrum of the supersymmetric standard model.)

Several attempts have been made to construct the MSSM [10, 11, 12, 13] using an orientifold with point group $P = \mathbf{Z}_4$, $\mathbf{Z}_4 \times \mathbf{Z}_2$ or $\mathbf{Z}_6$. The most successful attempt to date is the last of these [13, 14], which uses D6-branes intersecting on a $\mathbf{Z}_6$ orientifold to construct an $\mathcal{N} = 1$ supersymmetric standard-like model using 5 stacks of branes. We shall not discuss this beautiful model in any detail except to note that the intersection numbers for the stacks a, which generates the $SU(3)_c$ group, and b, which generates the $SU(2)_L$, are $(a \cap b, a \cap b') = (0, 3)$. In this case it is impossible to obtain lepton singlet states ℓ_L^c as antisymmetric representations of $U(2)$. Further, it was shown, quite generally, that it is impossible to find stacks a and b such that $(a \cap b, a \cap b') = (2, 1)$ or $(1, 2)$. Thus, as explained above, it is impossible to obtain exactly the spectrum of the (supersymmetric) standard model.

The question then arises as to whether the use of a different orientifold could circumvent this problem. Here we address this question for the $\mathbf{Z}_6'$ orientifold. We do not attempt to construct a standard(-like) MSSM. Instead, we merely see whether there are any stacks a, b that simultaneously satisfy the supersymmetry constraints, the absence of chiral matter in symmetric representations of the gauge groups (see below), which have not too much chiral matter in antisymmetric representations of the gauge groups, and which have $(a \cap b, a \cap b') = (2, 1)$ or $(1, 2)$. Further details of this work may be found in reference [15].

2. $\mathbf{Z}_6'$ ORIENTIFOLD

We assume that the torus T^6 factorises into three 2-tori $T_1^2 \times T_2^2 \times T_3^2$. The 2-tori T_k^2 ($k = 1, 2, 3$) are parametrised by complex coordinates z_k. The action of the generator θ of the point group $\mathbf{Z}_6'$ on the coordinates z_k is given by

$$\theta z_k = e^{2\pi i v_k} z_k \tag{3}$$

where

$$(v_1, v_2, v_3) = \frac{1}{6}(1, 2, -3) \tag{4}$$

The point group action must be an automorphism of the lattice, so in $T_{1,2}^2$ we may take an $SU(3)$ lattice. Specifically we define the basis 1-cycles by π_1 and $\pi_2 \equiv e^{i\pi/3}\pi_1$ in T_1^2, and π_3 and $\pi_4 \equiv e^{i\pi/3}\pi_3$ in T_2^2. Thus the complex structure of these tori is given by $U_1 = e^{i\pi/3} = U_2$. The orientation of $\pi_{1,3}$ relative to the real and imaginary axes of $z_{1,2}$ is arbitrary. Since θ acts as a reflection in T_3^2, the lattice, with basis 1-cycles π_5 and π_6, is

3

arbitrary. The point group action on the basis 1-cycles is then

$$\theta \pi_1 = \pi_2 \quad \text{and} \quad \theta \pi_2 = \pi_2 - \pi_1 \tag{5}$$
$$\theta \pi_3 = \pi_4 - \pi_3 \quad \text{and} \quad \theta \pi_4 = -\pi_3 \tag{6}$$
$$\theta \pi_5 = -\pi_5 \quad \text{and} \quad \theta \pi_6 = -\pi_6 \tag{7}$$

We consider "bulk" 3-cycles of T^6 which are linear combinations of the 8 3-cycles $\pi_{i,j,k} \equiv \pi_i \otimes \pi_j \otimes \pi_k$ where $i = 1,2$, $j = 3,4$, $k = 5,6$. The basis of 3-cycles that are *invariant* under the action of θ contains 4 elements $\rho_{1,3,4,6}$, where

$$\rho_1 = 2(\pi_{1,3,5} + \pi_{2,3,5} + \pi_{1,4,5} - 2\pi_{2,4,5}) \tag{8}$$
$$\rho_3 = 2(-2\pi_{1,3,5} + \pi_{2,3,5} + \pi_{1,4,5} + \pi_{2,4,5}) \tag{9}$$

and similarly for $\rho_{4,6}$ replacing π_5 by π_6 in $\rho_{1,3}$ respectively. Then the general $\mathbf{Z}'_6$-invariant bulk 3-cycle with (co-prime) wrapping numbers (n_k, m_k) of the cycles (π_{2k-1}, π_{2k}) on T_k^2 is

$$\Pi_a = A_1 \rho_1 + A_3 \rho_3 + A_4 \rho_4 + A_6 \rho_6 \tag{10}$$

where

$$A_1 = (n_1 n_2 + n_1 m_2 + m_1 n_2) n_3 \tag{11}$$
$$A_3 = (m_1 m_2 + n_1 m_2 + m_1 n_2) n_3 \tag{12}$$
$$A_4 = (n_1 n_2 + n_1 m_2 + m_1 n_2) m_3 \tag{13}$$
$$A_6 = (m_1 m_2 + n_1 m_2 + m_1 n_2) m_3 \tag{14}$$

are the "bulk coefficients". If Π_a has wrapping numbers (n_k^a, m_k^a) $(k = 1,2,3)$, and Π_b has wrapping numbers (n_k^b, m_k^b), then, in an obvious notation, the intersection number of the orbifold-invariant 3-cycles is

$$\Pi_a \cap \Pi_b = -4(A_1^a A_4^b - A_4^a A_1^b) \; + \; 2(A_1^a A_6^b - A_6^a A_1^b) + 2(A_3^a A_4^b - A_4^a A_3^b) -$$
$$- \; 4(A_3^a A_6^b - A_6^a A_3^b) \tag{15}$$

which is always even.

Besides these (untwisted) 3-cycles, there are also exceptional 3-cycles associated with (some of) the twisted sectors of the orbifold. They arise in twisted sectors in which there is a fixed torus, and consist of a collapsed 2-cycle at a fixed point times a 1-cycle in the invariant plane. We shall only be concerned with those that arise in the θ^3 sector, which has T_2^2 as the invariant plane. There is a $\mathbf{Z}_2$ symmetry acting in T_1^2 and T_3^2 and this has sixteen fixed points $f_{i,j}$ where $i, j = 1, 4, 5, 6$. There are then 32 independent exceptional cycles given by $f_{i,j} \otimes \pi_{3,4}$ from which 8 independent $\mathbf{Z}'_6$-invariant combinations may be formed. They are

$$\varepsilon_j \equiv (f_{6,j} - f_{4,j}) \otimes \pi_3 + (f_{4,j} - f_{5,j}) \otimes \pi_4 \tag{16}$$
$$\tilde{\varepsilon}_j \equiv (f_{4,j} - f_{5,j}) \otimes \pi_3 + (f_{5,j} - f_{6,j}) \otimes \pi_4 \tag{17}$$

4

TABLE 1. Relation between fixed points and exceptional 3-cycles.

Fixed point $\otimes$ 1-cycle	Invariant exceptional 3-cycle
$f_{1,j} \otimes (n_2 \pi_3 + m_2 \pi_4)$	0
$f_{4,j} \otimes (n_2 \pi_3 + m_2 \pi_4)$	$m_2 \varepsilon_j + (n_2 + m_2) \tilde{\varepsilon}_j$
$f_{5,j} \otimes (n_2 \pi_3 + m_2 \pi_4)$	$-(n_2 + m_2) \varepsilon_j - n_2 \tilde{\varepsilon}_j$
$f_{6,j} \otimes (n_2 \pi_3 + m_2 \pi_4)$	$n_2 \varepsilon_j - m_2 \tilde{\varepsilon}_j$

The non-zero intersection numbers for the invariant combinations are given by

$$\varepsilon_j \cap \tilde{\varepsilon}_k = -2\delta_{jk} \tag{18}$$

and again these are always even. The relation between the fixed points $f_{i,j}$ and the invariant exceptional cycles is given in Table 1

The embedding $\mathcal{R}$ of the world-sheet parity operator Ω may be chosen to act on the three complex coordinates z_k ($k = 1,2,3$) as complex conjugation $\mathcal{R}z_k = \bar{z}_k$, and we require that this too is an automorphism of the lattice. This fixes the orientation of the basis 1-cycles in each torus relative to the Re z_k axis. It requires them to be in one of two configurations **A** or **B**. When T_1^2 is in the **A** configuration, π_1 is aligned along the Re z_1 axis, whereas in the **B** configuration it makes an angle of $\pi/6$ below this axis. Similarly for π_3 and T_2^2. In T_3^2 the cycle π_5 is is aligned along the Re z_3 axis in both **A** and **B** configurations. The difference is that in **A** the 1-cycle π_6 aligned along the Im z_3 axis, whereas in **B** it is inclined such that its real part is one half that of π_5. In both cases the imaginary part is arbitrary, and so therefore is the imaginary part of the complex structure U_3 of T_3^2. It is then straightforward to determine the action of $\mathcal{R}$ on the bulk 3-cycles ρ_p ($p = 1,3,4,6$) and on the exceptional cycles ε_j and $\tilde{\varepsilon}_j$. In particular, requiring that a bulk 3-cycle $\Pi_a = \sum_p A_p \rho_p$ be invariant under the action of $\mathcal{R}$ gives 2 constraints on the bulk coefficients A_p, so that just 2 of the 4 independent bulk 3-cycles are $\mathcal{R}$-invariant. Which 2 depends upon the lattice.

The twist (4) ensures that the closed-string sector is supersymmetric. In order to avoid supersymmetry breaking in the open-string sector, the D6-branes must wrap special Lagrangian cycles. Then the stack Π_a with wrapping numbers (n_k^a, m_k^a) ($k = 1,2,3$) is supersymmetric if

$$\sum_{k=1}^{3} \phi_k^a = 0 \bmod 2\pi \tag{19}$$

where ϕ_k^a is the angle that the 1-cycle in T_k^2 makes with the Re z_k axis. Defining

$$Z^a \equiv \prod_{k=1}^{3} \pi_{2k-1}(n_k^a + m_k^a U_k) \equiv X^a + iY^a \tag{20}$$

where U_k is the complex structure on T_k^2, the condition (19) that Π_a is supersymmetric may be written as

$$X^a > 0, \quad Y^a = 0 \tag{21}$$

5

TABLE 2. The functions X^a and Y^a. (An overall positive factor is omitted.) A stack a of D6-branes is supersymmetric if $X^a > 0$ and $Y^a = 0$.

Lattice	X^a	Y^a
AAB	$2A_1^a - A_3^a + A_4^a - \frac{1}{2}A_6^a - A_6^a\sqrt{3}\mathrm{Im}\,U_3$	$\sqrt{3}(A_3^a + \frac{1}{2}A_6^a) + (2A_4^a - A_6^a)\mathrm{Im}\,U_3$
ABB and BAB	$\sqrt{3}(A_1^a + \frac{1}{2}A_4^a) + (A_4^a - 2A_6^a)\mathrm{Im}\,U_3$	$2A_3^a - A_1^a + A_6^a - \frac{1}{2}A_4^a + A_4^a\sqrt{3}\mathrm{Im}\,U_3$
BBB	$(A_3^a + A_1^a + \frac{1}{2}A_6^a + \frac{1}{2}A_4^a) +$	$\sqrt{3}(A_3^a - A_1^a + \frac{1}{2}A_6^a - \frac{1}{2}A_4^a)$
	$+ (A_4 - A_6)\sqrt{3}\mathrm{Im}\,U_3$	$+ (A_4 + A_6)\mathrm{Im}\,U_3$

(A stack with $Y^a = 0$ but $X^a < 0$, so that $\sum_k \phi_k^a = \pi$ mod 2π, corresponds to a (supersymmetric) stack of anti-D-branes.) In our case $T_{1,2}^2$ are $SU(3)$ lattices, and $U_1 = e^{i\pi/3} = U_2$, as already noted . Thus

$$Z^a = \pi_1\pi_3\pi_5[A_1^a - A_3^a + U_3(A_4^a - A_6^a) + e^{i\pi/3}(A_3^a + A_6^a U_3)] \tag{22}$$

It is then straightforward to evaluate X^a and Y^a for the different lattices. The results for the cases in which T_3^2 is of **B** type are given in Table 2. The (single) requirement that $Y_a = 0$ means that 3 independent combinations of the 4 invariant bulk 3-cycles may be chosen to be supersymmetric. Of these, 2 are the $\mathcal{R}$-invariant combinations. However, unlike in the case of the $\mathbf{Z}_6$ orientifold, in this case there is a third, independent, supersymmetric bulk 3-cycle that is *not* $\mathcal{R}$-invariant.

We noted earlier that the intersection numbers of both the bulk 3-cycles ρ_p ($p = 1,3,4,6$) and of the exceptional cycles $\varepsilon_j, \tilde{\varepsilon}_j$ ($j = 1,4,5,6$) are always even. However, in order to get just the (supersymmetric) standard-model spectrum, either $a \cap b$ or $a \cap b'$ must be odd. It is therefore necessary to use fractional branes of the form

$$a = \frac{1}{2}\Pi_a^{\mathrm{bulk}} + \frac{1}{2}\Pi_a^{\mathrm{ex}} \tag{23}$$

where $\Pi_a^{\mathrm{bulk}} = \sum_p A_p \rho_p$ is an invariant bulk 3-cycle, associated with wrapping numbers $(n_1^a, m_1^a)(n_2^a, m_2^a)(n_3^a, m_3^a)$, as shown in (10). The exceptional branes (in the θ^3 sector) are associated with the fixed points $f_{i,j}$, $(i,j = 1,4,5,6)$ in $T_1^2 \otimes T_3^2$, as shown in (16) and (17). If Π_a^{bulk} is a supersymmetric bulk 3-cycle, then the fractional brane a, defined in (23), preserves supersymmetry provided that the exceptional part Π_a^{ex} arises only from fixed points traversed by the bulk 3-cycle. Since the wrapping numbers (n_1^a, m_1^a) on T_1^2 are integers, the 1-cycle on T_1^2 either traverses zero fixed points or two. In the latter case we denote the fixed points by (i_1^a, i_2^a). Similarly for the 1-cycle on T_3^2, where the two fixed points are denoted by (j_1^a, j_2^a). Thus, supersymmetry requires that the exceptional part Π_a^{ex} of a derives from four fixed points, $f_{i_1^a j_1^a}, f_{i_1^a j_2^a}, f_{i_2^a j_1^a}, f_{i_2^a j_2^a}$. The choice of Wilson lines affects the relative signs with which the contributions from the four fixed points are combined to determine Π_a^{ex}. The rule is that

$$(i_1^a, i_2^a)(j_1^a, j_2^a) \to (-1)^{\tau_0^a}\left[f_{i_1^a j_1^a} + (-1)^{\tau_2^a}f_{i_1^a j_2^a} + (-1)^{\tau_1^a}f_{i_2^a j_1^a} + (-1)^{\tau_1^a + \tau_2^a}f_{i_2^a j_2^a}\right] \tag{24}$$

6

where $\tau^a_{0,1,2} = 0, 1$ with $\tau^a_1 = 1$ corresponding to a Wilson line in T^2_1 and likewise for τ^a_2 in T^2_3. The fixed point f_{i^a, j^a} with 1-cycle $n^a_2 \pi_3 + m^a_2 \pi_3$ is then associated with the orbifold invariant exceptional cycle as shown in Table 1.

In general, besides the chiral matter in bifundamental representations that occurs at the intersections of brane stacks $a, b, ...$, with each other or with their orientifold images $a', b', ...$, there is also chiral matter in the symmetric $\mathbf{S}_a$ and antisymmetric representations $\mathbf{A}_a$ of the gauge group $U(N_a)$, and likewise for $U(N_b)$. Orientifolding induces topological defects, O6-planes, which are sources of RR charge. The number of multiplets in the $\mathbf{S}_a$ and $\mathbf{A}_a$ representations is

$$\#(\mathbf{S}_a) = \frac{1}{2}(a \cap a' - a \cap \Pi_{O6}) \tag{25}$$

$$\#(\mathbf{A}_a) = \frac{1}{2}(a \cap a' + a \cap \Pi_{O6}) \tag{26}$$

where Π_{O6} is the total O6-brane homology class; it is $\mathcal{R}$-invariant. If $a \cap \Pi_{O6} = \frac{1}{2}\Pi^{bulk}_a \cap \Pi_{O6} \neq 0$, then copies of one or both representations are inevitably present. Since we require supersymmetry, Π^{bulk}_a is necessarily supersymmetric. However, we have observed above that this does not require Π^{bulk}_a to be $\mathcal{R}$-invariant, as Π_{O6} is. Thus, unlike the $\mathbf{Z}_6$ case, in this case $a \cap \Pi_{O6}$ is generally non-zero. We noted in the Introduction that we must exclude the appearance of the representations $\mathbf{S}_a$ and $\mathbf{S}_b$. Consequently, we impose the constraints

$$a \cap a' = a \cap \Pi_{O6} \tag{27}$$
$$b \cap b' = b \cap \Pi_{O6} \tag{28}$$

We also showed that demanding that the $U(1)$ charges Q_a and Q_b sum to zero overall requires that $\#(\mathbf{A}_a) = 0 = \#(\mathbf{A}_b)$, at least if we also demand standard-model Yukawa couplings. However, for the moment we proceed more conservatively. With the constraint (27) the number of multiplets in the antisymmetric representation $\mathbf{A}_a$ is $a \cap \Pi_{O6}$. For the present we require only that

$$|a \cap \Pi_{O6}| \leq 3 \tag{29}$$

since otherwise there would again be non-minimal vector-like quark singlet matter. Similarly, using just (28), we only require that

$$|b \cap \Pi_{O6}| \leq 3 \tag{30}$$

to avoid unwanted vector-like lepton singlets.

3. RESULTS AND CONCLUSIONS

We have shown [15] that, unlike the $\mathbf{Z}_6$ orientifold, at least on some lattices, the $\mathbf{Z}'_6$ orientifold *can* support supersymmetric stacks a and b of D6-branes with intersection numbers satisfying $(a \circ b, a \circ b') = (2, 1)$ or $(1, 2)$. Stacks having this property are an indispensable ingredient in any intersecting brane model that has *just* the matter content of the (supersymmetric) standard model. By construction, in all of our solutions there is no matter in symmetric representations of the gauge groups on either stack. However, some of the solutions *do* have matter, 2 quark singlets q_L^c or 2 lepton singlets ℓ_L^c, in the antisymmetric representation of gauge group on one of the stacks. This is not possible on the $\mathbf{Z}_6$ orientifold because all supersymmetric D6-branes wrap the same bulk 3-cycle as the O6-planes. In contrast, on the $\mathbf{Z}_6'$ orientifold there exist supersymmetric 3-cycles that do not wrap the O6-planes. Thus, there is more latitude in this case, and the solutions with antisymmetric matter exploit this feature. Unfortunately, however, none of the solutions of this nature that we have found can be enlarged to give just the standard-model spectrum, since the overall cancellation of the relevant $U(1)$ charge cannot be achieved with this matter content. Nevertheless, some of our solutions have no antisymmetric (or symmetric) matter on either stack. We shall attempt in a future work to construct a realistic (supersymmetric) standard model using one of these solutions.

The presence of singlet matter on the branes in some, but not all, of our solutions is an important feature of our results. It is clear that different orbifold point groups produce different physics, as indeed, for the reasons just given, our results also illustrate. The point group must act as an automorphism of the lattice used, but it is less clear that realising a given point group symmetry on different lattices produces different physics. Our results show that different lattices can produce different physics. The observation that the lattice does affect the physics suggests that other lattices are worth investigating in both the $\mathbf{Z}_6$ and $\mathbf{Z}_6'$ orientifolds. In particular, since Z_6 can be realised on a G_2 lattice, as well as on an $SU(3)$ lattice, one or more of all three $SU(3)$ lattices in the $\mathbf{Z}_6$ case, and of the two on $T_{1,2}^2$ in the $\mathbf{Z}_6'$ case, could be replaced by a G_2 lattice. We shall explore this avenue too in future work.

The construction of a realistic model will, of course, entail adding further stacks of D6-branes $c, d, ..$, with just a single brane in each stack, arranging that the matter content is just that of the supersymmetric standard model, the whole set satisfying the condition for RR tadpole cancellation. In a supersymmetric orientifold, RR tadpole cancellation ensures that NSNS tadpoles are also cancelled, but some moduli, (some of) of the complex structure moduli, the Kähler moduli and the dilaton, remain unstabilised. Recent developments have shown how such moduli may be stabilised using RR, NSNS and metric fluxes [16, 17, 18, 19, 20], and indeed Cámara, Font & Ibáñez [21, 22] have shown how models similar to the ones we have been discussing can be uplifted into ones with stabilised Kähler moduli using a "rigid corset". In general, such fluxes contribute to tadpole cancellation conditions and might make them easier to satisfy. In which case, it may be that one or other of our solutions with antisymmetric matter could be used to obtain just the standard-model spectrum. In contrast, the rigid corset can be added to any RR tadpole-free assembly of D6-branes in order to stabilise all moduli. Thus our results represent an important first step to obtaining a supersymmetric standard model from intersecting branes with all moduli stabilised.

REFERENCES

1. For a review, see D. Lüst, Intersecting brane worlds: A path to the standard model?, Class. Quant. Grav. **21** (2004) S1399 [hep-th/0401156].
2. R. Blumenhagen, V. Braun, B. Körs and D. Lüst, The standard model on the quintic, hep-th/0210083.
3. A. M. Uranga, Local models for intersecting brane worlds, JHEP **0212** (2002) 058 [hep-th/0208014].
4. L. E. Ibáñez, F. Marchesano and R. Rabadán, Getting just the standard model at intersecting branes, JHEP **0111** (2001) 002 [hep-th/0105155].
5. R. Blumenhagen, B. Kors, D. Lust and T. Ott, "The standard model from stable intersecting brane world orbifolds," Nucl. Phys. B **616** (2001) 3 [arXiv:hep-th/0107138].
6. S. A. Abel, O. Lebedev and J. Santiago, Flavour in intersecting brane models and bounds on the string scale, Nucl. Phys. B **696** (2004) 141 [hep-ph/0312157].
7. R. Blumenhagen, B. Körs, D. Lüst and T. Ott, Intersecting brane worlds on tori and orbifolds, Fortsch. Phys. **50** (2002) 843 [hep-th/0112015].
8. M. Cvetič, G. Shiu and A. M. Uranga, Three-family supersymmetric standard like models from intersecting brane worlds, Phys. Rev. Lett. **87** (2001) 201801 [hep-th/0107143].
9. M. Cvetič, G. Shiu and A. M. Uranga, Chiral four-dimensional N = 1 supersymmetric type IIA orientifolds from intersecting D6-branes, Nucl. Phys. B **615** (2001) 3 [hep-th/0107166].
10. R. Blumenhagen, L. Görlich and T. Ott, Supersymmetric intersecting branes on the type IIA T**6/Z(4) orientifold, JHEP **0301** (2003) 021 [hep-th/0211059].
11. G. Honecker, Chiral supersymmetric models on an orientifold of Z(4) x Z(2) with intersecting D6-branes, Nucl. Phys. B **666** (2003) 175 [hep-th/0303015].
12. G. Honecker, Chiral N = 1 4D orientifolds with D-branes at angles, Mod. Phys. Lett. A **19** (2004) 1863 [hep-th/0407181].
13. G. Honecker and T. Ott, Getting just the supersymmetric standard model at intersecting branes on the Z(6)-orientifold, Phys. Rev. D **70** (2004) 126010 [Erratum-ibid. D **71** (2005) 069902] [hep-th/0404055].
14. T. Ott, Catching the phantom: The MSSM on the Z6-orientifold, Fortsch. Phys. **53** (2005) 955 [hep-th/0505274].
15. D. Bailin and A. Love, arXiv:hep-th/0603172.
16. J. P. Derendinger, C. Kounnas, P. M. Petropoulos and F. Zwirner, Superpotentials in IIA compactifications with general fluxes, Nucl. Phys. B **715** (2005) 211 [hep-th/0411276].
17. S. Kachru and A. K. Kashani-Poor, Moduli potentials in type IIA compactifications with RR and NS flux, JHEP **0503** (2005) 066 [hep-th/0411279].
18. T. W. Grimm and J. Louis, The effective action of type IIA Calabi-Yau orientifolds, Nucl. Phys. B **718** (2005) 153 [hep-th/0412277].
19. G. Villadoro and F. Zwirner, N = 1 effective potential from dual type-IIA D6/O6 orientifolds with JHEP **0506** (2005) 047 [hep-th/0503169].
20. O. DeWolfe, A. Giryavets, S. Kachru and W. Taylor, Type IIA moduli stabilization, JHEP **0507** (2005) 066 [hep-th/0505160].
21. P. G. Cámara, A. Font and L. E. Ibáñez, Fluxes, moduli fixing and MSSM-like vacua in a simple IIA orientifold, JHEP **0509** (2005) 013 [hep-th/0506066].
22. G. Aldazabal, P. G. Cámara, A. Font and L. E. Ibáñez, More dual fluxes and moduli fixing, hep-th/0602089.

Split Supersymmetry in Brane Models

Ignatios Antoniadis

Department of Physics, CERN - Theory Division, 1211 Geneva 23, Switzerland[1]

Abstract. Type I string theory in the presence of internal magnetic fields provides a concrete realization of split supersymmetry. To lowest order, gauginos are massless while squarks and sleptons are superheavy. For weak magnetic fields, the correct Standard Model spectrum guarantees gauge coupling unification with $\sin^2 \theta_W = 3/8$ at the compactification scale of $M_{GUT} \simeq 2 \times 10^{16}$ GeV. I discuss mechanisms for generating gaugino and higgsino masses at the TeV scale, as well as generalizations to models with split extended supersymmetry in the gauge sector.

Keywords: supersymmetry, D-branes, magnetic fluxes, gaugino masses

INTRODUCTION

During the last decades, physics beyond the Standard Model (SM) was guided from the stabilization of mass hierarchy. For instance, compositeness, supersymmetry, extra dimensions, low string scale and little Higgs are different approaches to address the hierarchy. However, the actual precision tests, implying the absence of any deviation from the SM to a great accuracy, suggest that any new physics at a TeV needs to be fine-tuned at the per-cent level. Thus, either the underlying theory beyond the SM is very special, or our notion of naturalness should be reconsidered. The latter is also motivated from the recent evidence for the presence of a tiny non-vanishing cosmological constant that raises another more severe hierarchy problem. This raises the possibility that the same mechanism may solve both problems and casts some doubts on all previous proposals.

On the other hand, the necessity of a Dark Matter (DM) candidate and the fact that LEP data favor the unification of the three SM gauge couplings are smoking guns for the presence of new physics at high energies. Supersymmetry is then a nice candidate offering both properties. Moreover, it arises naturally in string theory, which provides a framework for incorporating the gravitational interaction in our quantum picture of the universe. It was then proposed to consider that supersymmetry might be broken at high energies without solving the gauge hierarchy problem. More precisely, making squarks and sleptons heavy does not spoil unification and the existence of a DM candidate while at the same time it gets rid of all unwanted features of the supersymmetric SM related to its complicated scalar sector. On the other hand, experimental hints to the existence of supersymmetry persist since there are still gauginos and higgsinos at the electroweak scale. This is the so-called split supersymmetry framework [1, 2].

Split supersymmetry has a natural realization in type I string theory with magnetized

---

[1] On leave from CPHT (UMR CNRS 7644) Ecole Polytechnique, F-91128 Palaiseau

D9-branes, or equivalently with branes at angles [3]. We first show that the general spectrum has the required properties and then discuss the conditions for gauge coupling unification near the string scale. It turns out that equality of the two non-abelian couplings is a consequence of the correct SM spectrum for weak magnetic fields, while the value for the weak angle $\sin^2\theta_W = 3/8$ is easily obtained even in simple constructions. Indeed, we perform a general study of SM embedding in three brane stacks and find a simple model realizing the conditions for unification [3]. We then discuss mass scales and in particular a mechanism generating light gaugino and higgsino masses in the TeV region, while scalars are superheavy, of order 10^{13} GeV [4]. Finally, it was shown how splitting supersymmetry reconciles toroidal models of intersecting branes with unification [5]. The gauge sector in these models arises in multiplets of extended supersymmetry while matter states are in $\mathcal{N} = 1$ representations. In general, split supersymmetry offers new possibilities for realistic string model building, that were previously unavailable because they were mainly restricted in the context of large dimensions and low string scale [6, 7].

GENERAL FRAMEWORK

We start with type I string theory, or equivalently type IIB with orientifold 9-planes and D9-branes [8]. Upon compactification in four dimensions on a Calabi-Yau manifold, one gets $\mathcal{N} = 2$ supersymmetry in the bulk and $\mathcal{N} = 1$ on the branes. Moreover, various fluxes can be turned on, to stabilize part or all of the closed string moduli. We then turn on internal magnetic fields [9, 10], which, in the T-dual picture, amounts to intersecting branes [11, 12]. For generic angles, or equivalently for arbitrary magnetic fields, supersymmetry is spontaneously broken and described by effective D-terms in the four-dimensional (4d) theory [9]. In the weak field limit, $|H|\alpha' < 1$ with α' the string Regge slope, the resulting mass shifts are given by:

$$\delta M^2 = (2k+1)|qH| + 2qH\Sigma \quad ; \quad k = 0, 1, 2, \ldots, \tag{1}$$

where H is the magnetic field of an abelian gauge symmetry, corresponding to a Cartan generator of the higher dimensional gauge group, on a non-contractible 2-cycle of the internal manifold. Σ is the corresponding projection of the spin operator, k is the Landau level and $q = q_L + q_R$ is the charge of the state, given by the sum of the left and right charges of the endpoints of the associated open string. We recall that the exact string mass formula has the same form as (1) with qH replaced by:

$$qH \longrightarrow \theta_L + \theta_R \quad ; \quad \theta_{L,R} = \arctan(q_{L,R}H\alpha'). \tag{2}$$

Obviously, the field theory expression (1) is reproduced in the weak field limit.

The Gauss law for the magnetic flux implies that the field H is quantized in terms of the area of the corresponding 2-cycle A:

$$H = \frac{m}{nA}, \tag{3}$$

where the integers m, n correspond to the respective magnetic and electric charges; m is the quantized flux and n is the wrapping number of the higher dimensional brane around the corresponding internal 2-cycle.

For simplicity, we consider below the case where the internal manifold is a product of three factorized tori $\prod_{I=1}^{3} T_{(I)}^2$. Then, the mass formula (1) becomes:

$$\delta M^2 = \sum_I (2k_I + 1)|qH_I| + 2qH_I\Sigma_I, \tag{4}$$

where Σ_I is the projection of the internal helicity along the I-th plane. For a ten-dimensional (10d) spinor, its eigenvalues are $\Sigma_I = \pm 1/2$, while for a 10d vector $\Sigma_I = \pm 1$ in one of the planes $I = I_0$ and zero in the other two $(I \neq I_0)$. Thus, charged higher dimensional scalars become massive, fermions lead to chiral 4d zero modes if all $H_I \neq 0$, while the lightest scalars coming from 10d vectors have masses

$$M_0^2 = \begin{cases} |qH_1| + |qH_2| - |qH_3| \\ |qH_1| - |qH_2| + |qH_3| \\ -|qH_1| + |qH_2| + |qH_3| \end{cases} \tag{5}$$

Note that all of them can be made positive definite, avoiding the Nielsen-Olesen instability, if all $H_I \neq 0$. Moreover, one can easily show that if a scalar mass vanishes, some supersymmetry remains unbroken [10, 11].

GENERIC SPECTRUM

We turn on now several abelian magnetic fields H_I^a of different Cartan generators $U(1)_a$, so that the gauge group is a product of unitary factors $\prod_a U(N_a)$ with $U(N_a) = SU(N_a) \times U(1)_a$. In an appropriate T-dual representation, it amounts to consider several stacks of D6-branes intersecting in the three internal tori at angles. An open string with one end on the a-th stack has charge ± 1 under the $U(1)_a$, depending on its orientation, and is neutral with respect to all others. Using the results described above, the massless spectrum of the theory falls into three sectors [12, 10]:

1. Neutral open strings ending on the same stack, giving rise to $\mathcal{N} = 1$ gauge supermultiplets of gauge bosons and gauginos.
2. Doubled charged open strings from a single stack, with charges ± 2 under the corresponding $U(1)$, giving rise to massless fermions transforming in the antisymmetric or symmetric representation of the associated $SU(N)$ factor. Their bosonic superpartners become massive. The multiplicities of chiral fermions are given by:

$$\text{Antisymmetric}: \frac{1}{2}\left(\prod_I 2m_I^a\right)\left(\prod_J n_J^a + 1\right)$$

$$\text{Symmetric}: \frac{1}{2}\left(\prod_I 2m_I^a\right)\left(\prod_J n_J^a - 1\right) \tag{6}$$

where m_I^a, n_I^a are the integers entering in the expression of the magnetic field (3). For orbifolds or more general Calabi-Yau spaces, the above multiplicities may be further reduced by the corresponding supersymmetry projection down to $\mathcal{N} = 1$.

In the degenerate case where a magnetic field vanishes, say, along one of the tori ($m_I^a = 0$ for some I), there are no chiral fermions in $d = 4$ dimensions, but the same formula with the products extending over the other two magnetized tori gives the multiplicities of chiral fermions in $d = 6$. In this case, chirality in four dimensions may arise only when the last T^2 compactification is combined with some additional orbifold-type projection.

3. Open strings stretched between two different brane stacks, with charges ± 1 under each of the corresponding $U(1)$'s. They give rise to chiral fermions transforming in the bifundamental representation of the two associated unitary group factors. Their multiplicities, for toroidal compactifications, are given by:

$$(N_a, N_b) \quad : \quad \prod_I (m_I^a n_I^b + n_I^a m_I^b)$$

$$(N_a, \overline{N}_b) \quad : \quad \prod_I (m_I^a n_I^b - n_I^a m_I^b). \tag{7}$$

As in the previous case, when a factor in the products of the above multiplicities vanishes, there are no 4d chiral fermions, but the same formula with the product restricted over the other two magnetized tori gives the corresponding multiplicity of chiral fermions in $d = 6$.

As mentioned already above, all charged bosons are massive. Massless scalars can appear only when some supersymmetry remains unbroken. It is now clear that this framework leads to models with a tree-level spectrum realizing the idea of split supersymmetry. Embedding the Standard Model (SM) in an appropriate configuration of D-brane stacks, one obtains tree-level massless gauginos while all scalar superpartners of quarks and leptons typically get masses at the scale of the magnetic fields, whose magnitude is set by the compactification scale of the corresponding internal space. On the other hand, the condition to obtain a (tree-level) massless Higgs in the spectrum implies that supersymmetry remains unbroken in the Higgs sector, leading to a pair of massless higgsinos, as required by anomaly cancellation.

GAUGE COUPLING UNIFICATION

On general grounds, there are two conditions to obtain unification of SM gauge interactions, consistently with extrapolation of gauge couplings from low-energy data using the minimal supersymmetric SM spectrum. (i) Equality of the $SU(3)$ color and weak $SU(2)$ non-abelian gauge couplings and (ii) the correct prediction for the weak mixing angle $\sin^2 \theta_W = 3/8$ at the grand unification (GUT) scale. On the other hand, a generic D-brane model using several stacks, as described in the framework of the previous section, does not satisfy either of the two conditions. Indeed, this framework was developed in connection to the idea of low-scale strings [7], where the concept of unification is radically different from conventional GUTs. In this section, we study precisely the general requirements for satisfying the first of the above two conditions, namely natural unification of non-abelian gauge couplings.

The 4d non-abelian gauge coupling α_{N_a} of the a-th brane stack is given by:

$$\frac{1}{\alpha_{N_a}} = \frac{V}{g_s} \prod_I |n_I^a| \sqrt{1 + (H_I^a \alpha')^2}, \tag{8}$$

where g_s is the string coupling and V the compactification volume in string units. The presence of the wrapping numbers $|n_I^a|$ can be understood from the fact that $|n_I^a| V_I$ is the effective area of the 2-torus $T_{(I)}^2$ wrapped n_I^a times by the D9-brane, and $V = \prod_I V_I$. The additional factor in the square root follows from the non-linear Dirac-Born-Infeld (DBI) action of the abelian gauge field, $\sqrt{\det(\delta_{ij} + F_{ij}\alpha')}$, which in the case of two dimensions with $F_{ij} = \varepsilon_{ij}H$, it is reduced to $\sqrt{1 + (H\alpha')^2}$. Obviously, the expression (8) holds at the compactification scale, since above it gauge couplings receive important corrections and become higher dimensional. Finally, the gauge couplings of the associated abelian factors, in our convention of $U(1)$ charges, are given by

$$\alpha_{U(1)a} = \frac{\alpha_{N_a}}{2N_a}. \tag{9}$$

Here, non-abelian generators are normalized according to $\mathrm{Tr} T^a T^b = \delta^{ab}/2$.

From equation (8), it follows that unification of non-abelian gauge couplings holds if (i) $\prod_I |n_I^a|$ are independent of a, and (ii) the magnetic fields are either a-independent as well, or they are much smaller than the string scale.

Condition (i) follows from eq. (6), by requiring the absence of chiral fermions transforming in the symmetric representations of the non-abelian groups, i.e. no chiral $SU(3)$ color sextets and no weak $SU(2)$ triplets.

Condition (ii) of weak magnetic fields is more quantitative. Allowing for 1% error in the unification condition at high scale, one should have $|H_I^a| \alpha' \lesssim 0.1$. From the quantization condition (3), this implies that the volume $V \gtrsim 10^3$ for three magnetized tori, which is rather high to keep the theory weakly coupled above the compactification scale. Indeed, eq. (8) gives a string coupling g_s of order $\mathscr{O}(10)$ for gauge couplings $\alpha_{N_a} \simeq 1/25$ at the unification scale. On the other hand, for one or two magnetized tori one obtains $V \gtrsim 10 - 10^2$, which is compatible with a string weak coupling regime ($g_s \sim 0.1 - 1$). Fortunately, this condition can be partly relaxed in some direction, by requiring the absence of chiral antiquark doublets in the spectrum. Indeed eq. (7), for open strings stretched between the strong $SU(3)$ and weak $SU(2)$ interactions brane stacks, implies the vanishing of one of the factors in the product. This leads to the equality of the ratio m_I^a/n_I^a for the two stacks and for some I, and thus, to the equality of the two corresponding magnetic fields via eq. (3).[2] As a result, the condition of perturbativity is weakened and becomes possible even in the case of three factorized magnetized tori.

The above analysis concerns the non-abelian couplings α_3 and α_2 of strong and weak interactions. The case of hypercharge is more subtle since it can be in general a linear

---

[2] This argument is true only when the $U(1)$ accompanying the weak interactions brane stack participates in the hypercharge combination. Otherwise, quark anti-doublets are equivalent to quark doublets.

combination of several $U(1)$'s coming from different brane stacks. In the following section, we present an explicit example with the correct prediction of the weak mixing angle. It is based on a minimal SM embedding in three brane stacks with the hypercharge being a linear combination of two abelian factors. This provides an existence proof that can be generalized in different constructions. We notice for instance that in a class of supersymmetric models with four brane stacks, the equality of the two non-abelian couplings $\alpha_2 = \alpha_3$ implies the value $3/8$ for $\sin^2 \theta_W$ at the unification scale [13].

MINIMAL STANDARD MODEL EMBEDDING

In this section, we perform a general study of SM embedding in three brane stacks with gauge group $U(3) \times U(2) \times U(1)$ [14], and present an explicit example having realistic particle content and satisfying gauge coupling unification.

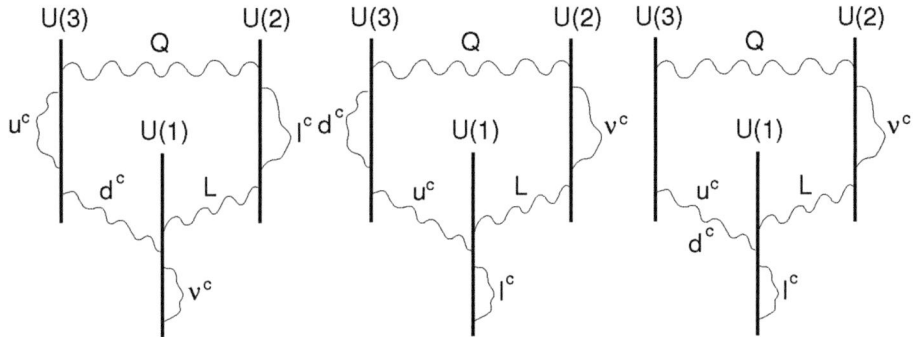

FIGURE 1.   Pictorial representation of models A, B and C

The quark and lepton doublets (Q and L) correspond to open strings stretched between the weak and the color or $U(1)$ branes, respectively. On the other hand, the u^c and d^c antiquarks can come from strings that are either stretched between the color and $U(1)$ branes, or that have both ends on the color branes and transform in the antisymmetric representation of $U(3)$ (which is an anti-triplet). There are therefore three possible models, depending on whether it is the u^c (model A), or the d^c (model B), or none of them (model C), the state coming from the antisymmetric representation of color branes. It follows that the antilepton l^c comes in a similar way from open strings with both ends either on the weak brane stack and transforming in the antisymmetric representation of $U(2)$ which is an $SU(2)$ singlet (in model A), or on the abelian brane and transforming in the "symmetric" representation of $U(1)$ (in models B and C). The three models are presented pictorially in Fig. 1.

Thus, the members of a family of quarks and leptons have the following quantum numbers:

	Model A	Model B	Model C
Q	$(\mathbf{3},\mathbf{2};1,1,0)_{1/6}$	$(\mathbf{3},\mathbf{2};1,\varepsilon_Q,0)_{1/6}$	$(\mathbf{3},\mathbf{2};1,\varepsilon_Q,0)_{1/6}$
u^c	$(\bar{\mathbf{3}},\mathbf{1};2,0,0)_{-2/3}$	$(\bar{\mathbf{3}},\mathbf{1};-1,0,1)_{-2/3}$	$(\bar{\mathbf{3}},\mathbf{1};-1,0,1)_{-2/3}$

$$d^c \quad (\overline{\mathbf{3}},\mathbf{1};-1,0,\varepsilon_d)_{1/3} \qquad (\overline{\mathbf{3}},\mathbf{1};2,0,0)_{1/3} \qquad (\overline{\mathbf{3}},\mathbf{1};-1,0,-1)_{1/3} \qquad (10)$$

$$L \quad (\mathbf{1},\mathbf{2};0,-1,\varepsilon_L)_{-1/2} \qquad (\mathbf{1},\mathbf{2};0,\varepsilon_L,1)_{-1/2} \qquad (\mathbf{1},\mathbf{2};0,\varepsilon_L,1)_{-1/2}$$

$$l^c \quad (\mathbf{1},\mathbf{1};0,2,0)_1 \qquad (\mathbf{1},\mathbf{1};0,0,-2)_1 \qquad (\mathbf{1},\mathbf{1};0,0,-2)_1$$

$$v^c \quad (\mathbf{1},\mathbf{1};0,0,2\varepsilon_v)_0 \qquad (\mathbf{1},\mathbf{1};0,2\varepsilon_v,0)_0 \qquad (\mathbf{1},\mathbf{1};0,2\varepsilon_v,0)_0$$

where the last three digits after the semi-column in the brackets are the charges under the three abelian factors $U(1)_3 \times U(1)_2 \times U(1)$, that we will call Q_3, Q_2 and Q_1 in the following, while the subscripts denote the corresponding hypercharges. The various sign ambiguities $\varepsilon_i = \pm 1$ are due to the fact that the corresponding abelian factor does not participate in the hypercharge combination (see below). In the last lines, we also give the quantum numbers of a possible right-handed neutrino in each of the three models. These are in fact all possible ways of embedding the SM spectrum in three sets of branes.

The hypercharge combination is:

$$\text{Model A} \quad : \quad Y = -\frac{1}{3}Q_3 + \frac{1}{2}Q_2 \tag{11}$$

$$\text{Model B, C} \quad : \quad Y = \frac{1}{6}Q_3 - \frac{1}{2}Q_1$$

leading to the following expressions for the weak angle:

$$\text{Model A} \quad : \quad \sin^2 \theta_W = \frac{1}{2 + 2\alpha_2/3\alpha_3} = \left.\frac{3}{8}\right|_{\alpha_2=\alpha_3} \tag{12}$$

$$\text{Model B, C} \quad : \quad \sin^2 \theta_W = \frac{1}{1 + \alpha_2/2\alpha_1 + \alpha_2/6\alpha_3}$$

$$= \left.\frac{6}{7 + 3\alpha_2/\alpha_1}\right|_{\alpha_2=\alpha_3}$$

In the second part of the above equalities, we used the unification relation $\alpha_2 = \alpha_3$, that can be naturally imposed as described in the previous section. It follows that model A admits natural gauge coupling unification of strong and weak interactions, and predicts the correct value for $\sin^2 \theta_W = 3/8$ at the unification scale M_{GUT}.

Besides the hypercharge combination, there are two additional $U(1)$'s. It is easy to check that one of the two can be identified with $B-L$. For instance, in model A choosing the signs $\varepsilon_d = \varepsilon_L = -\varepsilon_v = -\varepsilon_H = \varepsilon_{H'}$, it is given by:

$$B - L = -\frac{1}{6}Q_3 + \frac{1}{2}Q_2 - \frac{\varepsilon_d}{2}Q_1. \tag{13}$$

Finally, the above spectrum can be easily implemented with a Higgs sector, since the Higgs field H has the same quantum numbers as the lepton doublet or its complex conjugate:

	Model A	Model B, C	
H	$(\mathbf{1},\mathbf{2};0,-1,\varepsilon_H)_{-1/2}$	$(\mathbf{1},\mathbf{2};0,\varepsilon_H,1)_{-1/2}$	(14)
H'	$(\mathbf{1},\mathbf{2};0,1,\varepsilon_{H'})_{1/2}$	$(\mathbf{1},\mathbf{2};0,\varepsilon_{H'},-1)_{1/2}$	

MASS SCALES

String scale

To preserve gauge coupling unification, the compactification scale (actually the smallest, if there are several) must be of order of the unification scale $M_{GUT} \simeq 10^{16}$ GeV. Above this energy, gauge interactions acquire a higher dimensional behavior. Moreover, to keep the theory weakly coupled, the string scale $M_s \equiv \alpha'^{-1/2}$ should be close to the compactification scale and therefore to M_{GUT}. On the other hand, as we discussed above, to ensure that corrections to the unification of gauge couplings are within 1%, the magnetic fields should be weak, $|H_I^a|\alpha' \lesssim 0.1$. From the quantization condition (3), it follows that the string scale should be roughly a factor of 3 higher than the compactification scale,

$$M_s \simeq 3 M_{GUT}. \tag{15}$$

Scalar masses

The supersymmetry breaking scale m_0 is given by the heaviest charged scalar mass (5): $m_0^2 \sim \delta H^a \equiv \sum_{I=1}^{3} \varepsilon_I H_I^a$ on brane stacks, and $m_0^2 \sim \delta H^a - \delta H^b$ on brane intersections. Here, ε_I are signs: two positive and one negative. Thus, even for strong magnetic fields, of order of the string scale, m_0 can be much smaller and corresponds to an arbitrary parameter. Although values much lower than M_{GUT} require an apparent fine tuning of radii, such a tuning is technically natural since the supersymmetric point $m_0 = 0$ is radiatively stable.

All scalar masses are of the order of the supersymmetry breaking scale m_0, which is assumed to be very high in split supersymmetry, except for those coming from supersymmetric sectors, which are vanishing to lowest order, such as the higgses. The latter are expected to acquire masses from one loop corrections, proportional to m_0 but suppressed by a loop factor. Note that off diagonal elements of the 2×2 Higgs mass matrix, usually denoted by $B\mu$, should also be generated at the same order as the diagonal elements, in the absence of a Peccei-Quinn (PQ) symmetry. For high m_0, a fine tuning between $B\mu$ and the diagonal elements is then required to ensure a light Higgs.

Gaugino masses

It remains to discuss the corrections to gaugino and higgsino masses, $m_{1/2}$ and μ, which are vanishing at the tree-level. In the absence of gravity, they are both protected by an R-symmetry. Actually, higgsino masses are protected in addition by a PQ symmetry which must be broken in order to generate a $B\mu$ mixing term in the Higgs mass matrix, as we argued above. Then, a μ-term can be generated via $B\mu$, or directly using the PQ symmetry breaking, if R-symmetry is broken. Indeed, R-symmetry is in general broken in the gravitational sector by the gravitino mass $m_{3/2}$ and thus, in the presence of gravity, $m_{1/2}$ and μ are not anymore protected. Since supersymmetry breaking in the gravity sector is model dependent and brings more uncertainties, here we will assume that gravitational corrections are negligible. For instance, if supergravity breaking occurs via a Scherk-Schwarz compactification on an interval transverse to our braneworld [15], using the usual $\mathbb{Z}_2$ fermion number in the bulk, the gravitino acquires Dirac mass together with its Kaluza-Klein modes and R-symmetry remains unbroken [3]. One can therefore discuss other sources of R-symmetry breaking within only global supersymmetry.

As discussed previously, supersymmmetry breaking via internal magnetic fields is described in the 4d effective field theory by vacuum expectation values (VEVs) of D-term auxiliaries for all magnetic $U(1)$'s. In the low energy limit, one has:

$$\langle D \rangle \simeq m_0^2, \tag{16}$$

and thus R-symmetry remains unbroken. However, it is broken by α'-string corrections, that modify for instance the gauge kinetic terms to the DBI form. In particular, gaugino masses can be induced by a dimension-seven effective operator which is the chiral F-term [4]:

$$F_{(0,3)} \int d^2\theta \, \mathcal{W}^2 \text{Tr} W^2 \quad \Rightarrow \quad m_{1/2} \sim \frac{m_0^4}{M_s^3}, \tag{17}$$

where $\mathcal{W}$ and W denote the magnetic $U(1)$ and non-abelian gauge superfield, respectively. The coefficient $F_{(0,3)}$ is a moduli dependent function given by the topological partition function on a world-sheet with no handles and three boundaries. It is non-vanishing when the three brane stacks associated to the boundaries do not intersect at a point in any of the three internal torii. From the effective field theory point of view, it corresponds to a two-loop correction involving massive open string states. Upon a VEV $\langle \mathcal{W} \rangle = \theta \langle D \rangle$, the above F-term generates gaugino masses given in eq. (17). They are in the TeV region for scalar masses at intermediate energies, $m_0 \sim \mathcal{O}(10^{13})$ GeV.

ACKNOWLEDGMENTS

This work was supported in part by the European Commission under the RTN contract MRTN-CT-2004-503369, and in part by the INTAS contract 03-51-6346.

REFERENCES

1. N. Arkani-Hamed and S. Dimopoulos, JHEP **0506**, 073 (2005) [arXiv:hep-th/0405159].
2. G. F. Giudice and A. Romanino, Nucl. Phys. B **699** (2004) 65 [Erratum-ibid. B **706** (2005) 65] [arXiv:hep-ph/0406088].
3. I. Antoniadis and S. Dimopoulos, Nucl. Phys. B **715** (2005) 120 [arXiv:hep-th/0411032].
4. I. Antoniadis, K. S. Narain and T. R. Taylor, Nucl. Phys. B **729** (2005) 235 [arXiv:hep-th/0507244].
5. I. Antoniadis, A. Delgado, K. Benakli, M. Quiros and M. Tuckmantel, Phys. Lett. B **634** (2006) 302 [arXiv:hep-ph/0507192].
6. I. Antoniadis, Phys. Lett. B **246** (1990) 377; J. D. Lykken, Phys. Rev. D **54** (1996) 3693 [arXiv:hep-th/9603133].
7. N. Arkani-Hamed, S. Dimopoulos and G. R. Dvali, Phys. Lett. B **429** (1998) 263 [arXiv:hep-ph/9803315]; I. Antoniadis, N. Arkani-Hamed, S. Dimopoulos and G. R. Dvali, Phys. Lett. B **436** (1998) 257 [arXiv:hep-ph/9804398].

8. C. Angelantonj and A. Sagnotti, Phys. Rept. **371** (2002) 1 [Erratum-ibid. **376** (2003) 339] [arXiv:hep-th/0204089].
9. C. Bachas, arXiv:hep-th/9503030.
10. C. Angelantonj, I. Antoniadis, E. Dudas and A. Sagnotti, Phys. Lett. B **489** (2000) 223 [arXiv:hep-th/0007090].
11. M. Berkooz, M. R. Douglas and R. G. Leigh, Nucl. Phys. B **480** (1996) 265 [arXiv:hep-th/9606139].
12. R. Blumenhagen, L. Goerlich, B. Kors and D. Lust, JHEP **0010** (2000) 006 [arXiv:hep-th/0007024]; G. Aldazabal, S. Franco, L. E. Ibanez, R. Rabadan and A. M. Uranga, J. Math. Phys. **42** (2001) 3103 [arXiv:hep-th/0011073].
13. R. Blumenhagen, D. Lust and S. Stieberger, JHEP **0307** (2003) 036 [arXiv:hep-th/0305146].
14. I. Antoniadis, E. Kiritsis and T. N. Tomaras, Phys. Lett. B **486** (2000) 186 [arXiv:hep-ph/0004214]; I. Antoniadis, E. Kiritsis, J. Rizos and T. N. Tomaras, Nucl. Phys. B **660** (2003) 81 [arXiv:hep-th/0210263]; R. Blumenhagen, B. Kors, D. Lust and T. Ott, Nucl. Phys. B **616** (2001) 3 [arXiv:hep-th/0107138]; I. Antoniadis and J. Rizos, 2003 unpublished work.
15. I. Antoniadis, E. Dudas and A. Sagnotti, Nucl. Phys. B **544** (1999) 469 [arXiv:hep-th/9807011].

Dynamical Gauge-Higgs Unification

Yutaka Hosotani

Department of Physics, Osaka University, Toyonaka, Osaka 560-0043, Japan

Abstract. In the dynamical gauge-Higgs unification the 4D Higgs field is unified with gauge fields and the electroweak symmetry is dynamically broken by the Hosotani mechanism. Interesting phenomenology is obtained in the Randall-Sundrum warped spacetime. (i) The Higgs boson mass is predicted at the LHC energies. (ii) The hierarchy in the fermion mass spectrum is naturally explained. (iii) Tiny violation of the universality in the charged current interactions is predicted. (iv) Yukawa couplings of quarks and leptons are suppressed compared with those in the standard model. (v) WWZ, WWH, ZZH couplings are suppressed compared with those in the standard model.

Keywords: gauge-Higgs unification, dynamical gauge symmetry breaking
PACS: 11.10.Kk, 11.15.Ex, 12.60.-i

INTRODUCTION

In the gauge-Higgs unification 4D Higgs scalar fields are unified with 4D gauge fields within the framework of higher dimensional gauge theory. Low energy modes of extra-dimensional components of gauge potentials are 4D Higgs fields. The scenario works remarkably well when the extra-dimensional space is non-simply-connected.[1, 2] There arise Yang-Mills AB (Aharonov-Bohm) phases along the extra dimension, whose fluctuations in four dimensions are nothing but the 4D Higgs fields. The most notable feature is that quantum dynamics generate non-trivial, finite effective potential for the Higgs fields, inducing dynamical gauge symmetry breaking and generating finite masses for the Higgs fields at the same time. Even though the theory is non-renormalizable, many properties of the Higgs fields can be deduced irrespective of unknown dynamics at the cutoff scale. The Higgs boson mass turns out to be much smaller than the Kaluza-Klein mass scale. This is contrasted to the earlier proposal of the gauge-Higgs unification based on the ad hoc symmetry ansatz.[3, 4]

In the last ten years the scenario of the gauge-Higgs unification has been applied to the electroweak interactions and grand unified theories with the aid of orbifolds as extra dimensions.[5]-[21] In this article we focus on applications to the electroweak interactions where, besides the Higgs boson mass, many illuminating predictions are made for LHC and linear colliders.

To achieve the gauge-Higgs unification in the electroweak interactions, there are a few requirements to be fulfilled. First of all, the electroweak gauge symmetry is $SU(2)_L \times U(1)_Y$ which breaks down to $U(1)_{\text{EM}}$ triggered by non-vanishing VEV of an $SU(2)_L$ doublet Higgs field. In order for the 4D Higgs field be a part of gauge potentials the original gauge group must be larger than $SU(2)_L \times U(1)_Y$. Secondly, fermion content must be chiral. The second requirement is restrictive, as fermions in higher dimensions tend to lead to vectorlike theory in the effective 4D theory at low energies, unless the extra-dimensional space has nontrivial topology or there exists nonvanishing flux in the

CP881, *Cairo International Conference on High Energy Physics (CICHEP II)*,
edited by S. Khalil

extra dimensions. These requirements can be naturally and easily fulfilled when the extra dimensional space is an orbifold.

Gauge theory on an orbifold

Consider S^1 with a coordinate y where y and $y + 2\pi R$ are identified. Further we identify y and $-y$, which gives an orbifold S^1/Z_2. There appear two fixed points under parity; $y_0 = 0$ and $y_1 = \pi R$. Let us analyse gauge theory on $M^4 \times (S^1/Z_2)$, which is first defined on a covering space M^5, supplemented with restrictions appropriate to preserve the nature of S^1/Z_2. Although $y_j + y$ and $y_j - y$ represent the same physical point, gauge potentials need not be the same. They may differ from each other up to a gauge transformation. The orbifold structure is respected if

$$\begin{pmatrix} A_\mu \\ A_y \end{pmatrix} (x, y_j - y) = P_j \begin{pmatrix} A_\mu \\ -A_y \end{pmatrix} (x, y_j + y) P_j^\dagger \tag{1}$$

where P_j is an element of the gauge group satisfying $P_j^2 = I$. Similarly for fermions in the spinor representation in an $SU(N)$ group or in the vector representation in an $SO(N)$ group

$$\psi(x, y_j - y) = \pm P_j \gamma^5 \psi(x, y_j + y) . \tag{2}$$

If $P_j \not\propto I$, the gauge symmetry G apparently breaks down to a smaller subgroup H_{BC}. $\{P_0, P_1\}$ defines the symmetry of boundary condition. H_{BC} is not necessarily the physical symmetry H_{phys} which survives at the end. H_{phys} can be either smaller or larger than H_{BC}. Put it differently, two distinct sets of boundary conditions, $\{P_0, P_1\}$ and $\{P_0', P_1'\}$ can be equivalent to each other in physics content. All of these are due to dynamics of Yang-Mills AB phases. It is called the Hosotani mechanism.[1, 2, 10] In the application to the electroweak interactions we would like to have $H_{\mathrm{BC}} = SU(2)_L \times U(1)_Y$ and $H_{\mathrm{phys}} = U(1)_{\mathrm{EM}}$.

In the $SU(3)$ model, $P_0 = P_1 = \mathrm{diag}\,(-1, -1, 1)$ gives $H_{\mathrm{BC}} = SU(2) \times U(1)$. Zero modes exist for the H_{BC} part of A_μ and for the G/H_{BC} part of A_y which forms an $SU(2)$ doublet and idetified with the 4D Higgs field. Although this model gives an incorrect Weinberg angle, it gives a nice working ground to investigate physics of the W boson and fermions. Another model of interest is the $SO(5) \times U(1)_{B-L}$ model proposed by Agashe et al.[16] For the $SO(5)$ part we take $P_0 = P_1 = \mathrm{diag}\,(-1, -1, -1, -1, 1)$, which gives $H_{\mathrm{BC}}' = SO(4) \times U(1)_{B-L} = SU(2)_L \times SU(2)_R \times U(1)_{B-L}$. With additional dynamics on the one of the branes at $y = 0$, the symmetry of boundary conditions is reduced to $H_{\mathrm{BC}} = SU(2)_L \times U(1)_Y$. Zero modes of A_y are located at

$$A_y \sim \begin{pmatrix} & & & & -i\phi_1 \\ & & & & -i\phi_2 \\ & & & & -i\phi_3 \\ & & & & -i\phi_4 \\ i\phi_1 & i\phi_2 & i\phi_3 & i\phi_4 & \end{pmatrix} , \quad \Phi = \begin{pmatrix} \phi_1 + i\phi_2 \\ \phi_4 - i\phi_3 \end{pmatrix} . \tag{3}$$

Φ is the 4D Higgs doublet in the standard model. We note that with the given $\{P_0, P_1\}$, A_M becomes periodic; $A_M(x, y + 2\pi R) = A_M(x, y)$.

Yang-Mills AB phase θ_H

The zero modes of A_y lead to non-Abelian generalization of the Aharonov-Bohm phases (Yang-Mills AB phases).[1] The configuration gives vanishing field strengths $F_{MN} = 0$, but gives nontrivial phases

$$e^{i\Theta_H/2} = P\exp\left\{ig_A \int_0^{\pi R} dy\, A_y\right\} . \tag{4}$$

The spectrum of gauge fields and fermions depends on Θ_H. The phase Θ_H is a physical quantity. As seen in eq. (3), the 4D Higgs fields are four-dimensional fluctuations of the Yang-Mills AB phases. This property leads to the finiteness of the Higgs boson mass.[1], [5], [22]-[25]

In the $SO(5) \times U(1)_{B-L}$ model one can suppose with the use of the residual $SU(2)_L$ symmetry that only the ϕ_4 component of A_y is nonvanishing in the vacuum. The Yang-Mills AB phase θ_H is given by

$$\Theta_H = \theta_H \cdot \Lambda \quad , \quad \Lambda = \begin{pmatrix} 0 & & & \\ & 0 & & \\ & & 0 & \\ & & & -i \\ & & i & \end{pmatrix} \tag{5}$$

There exist large gauge transformations which shift θ_H by multiples of 2π, while preserving the boundary conditions;

$$A'_M = \Omega A_M \Omega^\dagger + \frac{i}{g}\Omega \partial_M \Omega^\dagger \quad , \quad \Omega = e^{iny/R\cdot\Lambda}$$

$$\theta'_H = \theta_H + 2\pi n . \tag{6}$$

It is seen that the phase nature of θ_H is a consequence of the large gauge invariance.

DIFFICULTIES IN FLAT SPACE

Before going into detailed discussions in the Randall-Sundrum warped spacetime, we briefly summarize difficulties one encounters in gauge-Higgs unification in flat spacetime. The value of θ_H is dynamically determined once the matter content is specified. In typical situation the global minimum of the effective potential $V_{\text{eff}}(\theta_H)$ is located either at $\theta_H = 0$ or at $\theta_H = O(1)$. In the former case the gauge symmetry H_{BC} is unbroken, whereas in the latter case the symmetry breaks down to $U(1)_{\text{EM}}$.

The W boson mass, m_W, becomes non-vanishing for $\theta_H \neq 0$ at the tree level. In flat space

$$m_W \sim \frac{|\sin\theta_H|}{\pi R} \tag{7}$$

---

[1] In the literature they are often called Wilson line phases.

which implies that the Kaluza-Klein mass scale M_{KK} is too low. Since the 4D Higgs boson corresponds to four-dimensional fluctuations of θ_H, its mass arises as radiative corrections. It turns out finite, but is given by

$$m_H \sim \sqrt{\frac{\alpha_W}{30}} \frac{1}{R} \sim \sqrt{\frac{\alpha_W}{30}} \frac{\pi m_W}{|\sin \theta_H|} \tag{8}$$

which typically gives too small $m_H \sim 10\,\text{GeV}$. Of course, θ_H can be small as a result of cancellations among contributions from various matter fields. However, it requires tuning of the matter content. We argue that natural resolution of the problem can be found once the gauge-Higgs unification is achieved in the Randall-Sundrum spacetime.

THE RANDALL-SUNDRUM WARPED SPACETIME

The Randall-Sundrum (RS) warped spacetime is given by

$$\begin{aligned} ds^2 &= e^{2\sigma(y)}(\eta_{\mu\nu}dx^\mu dx^\nu - dy^2) \\ \sigma(y + 2\pi R) &= \sigma(y) = \sigma(-y) \ , \\ \sigma(y) &= k|y| \quad \text{for } |y| \le \pi R \ , \end{aligned} \tag{9}$$

where $\eta_{\mu\nu} = \text{diag}\,(1, -1, -1, -1)$. It has topology of S^1/Z_2. In the bulk five-dimensional spacetime $0 < y < \pi R$, the cosmological constant is given by $-k^2$. In other words, the RS spacetime is the 5D anti-de Sitter space sandwiched by the Planck brane (at $y = 0$) and the TeV brane (at $y = \pi R$). The warp factor $e^{\pi k R}$ provides natural explanation of the large hierarchy factor $M_{\text{Pl}}/m_W \sim 10^{17}$, as was originally pointed out by Randall and Sundrum.[26] We examine gauge theory defined on the RS spacetime. Many surprises are hidden there.[19, 21]

The spectrum of various fields in the RS spacetime has been analyzed by many authors.[27, 28] Each field has a Kaluza-Klein tower, which has a spectrum $m_n \sim M_{KK} n$ for large n with the Kaluza-Klein mass scale M_{KK} given by

$$M_{KK} \sim \frac{\pi k}{e^{\pi k R} - 1} = \begin{cases} 1/R & \text{as } k \to 0, \\ \pi k e^{-\pi k R} & \text{for } kR > 2. \end{cases} \tag{10}$$

It is legitimate to suppose that $k = O(M_{\text{Pl}})$. For $kR \sim 12$, M_{KK} comes out in the TeV range. In the RS spacetime, the spectrum is not with an equal spacing for small n.

W BOSON AND Z BOSON

As in the flat spacetime, W bosons and Z bosons acquire finite masses as the Yang-Mills AB phase θ_H becomes nonvanishing. In the $SU(3)$ model the eigenstate of the W boson becomes a mixture of A_μ^{21} and A_μ^{31}. Its mass is given by

$$m_W \sim \sqrt{\frac{2k}{\pi R}} \, e^{-\pi k R} \, \sin \frac{\theta_H}{2} \ . \tag{11}$$

The neutral current sector is not realistic at all.

In the $SO(5) \times U(1)_{B-L}$ model we denote $SO(5)$ gauge fields in the $SU(2)_L$, $SU(2)_R$, and $SO(5)/SU(2)_L \times SU(2)_R$ parts by A_μ^{jL}, A_μ^{jR}, and $A_\mu^{\hat{a}}$ ($j = 1,2,3$, $a = 1,2,3,4$), and $U(1)_{B-L}$ gauge fields by B_μ, respectively. The W boson becomes a mixture of $A_\mu^{1L} + iA_\mu^{2L}$, $A_\mu^{1R} + iA_\mu^{2R}$, and $A_\mu^{\hat{1}} + iA_\mu^{\hat{2}}$ with a mass

$$m_W \sim \sqrt{\frac{k}{\pi R}} \, e^{-\pi k R} \, \sin \theta_H \ . \tag{12}$$

The Z boson becomes a mixture of A_μ^{3L}, B_μ, A_μ^{3R} and $A_\mu^{\hat{3}}$ with a mass

$$m_Z \sim \frac{m_W}{\cos \theta_W} \ ,$$

$$\sin \theta_W = \frac{g_B}{\sqrt{g_A^2 + 2g_B^2}} = \frac{g_Y}{\sqrt{g_A^2 + g_Y^2}} \tag{13}$$

where g_A and g_B are the $SO(5)$ and $U(1)$ gauge coupling constants, respectively. g_Y is the weak hypercharge gauge coupling constant.

When $\sin \theta_H$ is $O(1)$, or unless $\theta_H \ll 1$, the relation (12) implies that $kR \sim 12$ (6) for $k \sim O(M_{\text{Pl}})$ (10^{12} GeV). With $kR = 12$, M_{KK} turns out $1.8 \sim 3.5$ TeV for $\theta_H = (0.4 \sim 0.2)\pi$.

HIGGS BOSON

The 4D Higgs field corresponds to four-dimensional fluctuations of the Yang-Mills AB phase θ_H so that its mass and self-couplings can be obtained from the effective potential for θ_H. In the $SO(5) \times U(1)_{B-L}$ model the zero mode of A_y is related to the 4D neutral Higgs field ϕ^0 by

$$A_y = \sqrt{\frac{k}{2(e^{2\pi kR} - 1)}} \, e^{2ky} \, \phi^0(x) \cdot \Lambda \ . \tag{14}$$

The effective potential $V_{\text{eff}}(\theta_H)$ at one loop is given by

$$V_{\text{eff}}(\theta_H) = \sum \mp \frac{i}{2} \int \frac{d^4 p}{(2\pi)^4} \sum_n \ln \left\{ p^2 + m_n^2(\theta_H) \right\} \ . \tag{15}$$

It is shown that the θ_H-dependent part of $V_{\text{eff}}(\theta_H)$ is finite, independent of the cutoff scale.[1, 2, 17] In a model with standard matter content it is

$$V_{\text{eff}}(\theta_H) \sim \frac{3}{128\pi^6} \, M_{KK}^4 \, f(\theta_H) \ , \tag{16}$$

where the amplitude of $f(\theta_H) = f(\theta_H + 2\pi)$ is $O(1)$ as confirmed in various models.

Suppose that the global minimum of $V_{\text{eff}}(\theta_H)$ is located at $\theta_H \neq 0 \ (mod \ \pi)$ so that the electroweak symmetry breaks down. By expanding the effective potential around the global minimum one can determine the mass of the Higgs boson and its self-couplings. The mass and quartic coupling are found, in the $SO(5) \times U(1)_{B-L}$ model, to be

$$m_H \sim \sqrt{\frac{3\alpha_W}{32\pi} f^{(2)}(\theta_H)} \ \frac{\pi k R}{2} \ \frac{\sqrt{2} m_W}{\sin \theta_H} \ ,$$

$$\lambda \sim \frac{\alpha_W^2}{16} f^{(4)}(\theta_H) \left(\frac{\pi k R}{2} \right)^2 . \tag{17}$$

Notice the presence of the enhancement factor $\pi k R/2 \sim 20$, which is absent when evaluated in the flat spacetime. For $\theta_H = (0.2 \sim 0.5)\pi$, one finds that $m_H = (120 - 210)$ GeV and $\lambda \sim 0.3$. Note that in flat spacetime the values are $m_H \sim 10$ GeV and $\lambda \sim 0.0008$, which already contradicts with observation.

It is surprising that the Higgs mass turns out to be at LHC energies, though there exists ambiguity in $f^{(2)}(\theta_H)$. The enhancement factor originating from the curved space is essential.

QUARKS AND LEPTONS

The Lagrangian density for quarks and leptons in a generic form is given by

$$\mathcal{L}_f = \overline{\psi} \, i \Gamma^a e_a{}^M \left\{ \partial_M + \frac{1}{8} \omega_{bcM} [\Gamma^b, \Gamma^c] - i g_A A_M - \frac{i}{2} g_B q_{B-L} B_M \right\} \psi - c k \varepsilon(y) \overline{\psi} \, \psi \tag{18}$$

where $\varepsilon(y + 2\pi R) = \varepsilon(y) = -\varepsilon(-y)$ and $\varepsilon(y) = 1$ for $0 < y < \pi R$. The last term is called a bulk kink mass.[27] The dimensionless parameter c plays an important role in determining wave functions of fermions.

Let us consider a fermion multiplet in the spinor representation of $SO(5)$ in the $SO(5) \times U(1)_{B-L}$ model. The boundary condition matrices in (2) are given by $P_0 = P_1 = \text{diag}\,(1,1,-1,-1)$. ψ contains

$$\psi = \begin{pmatrix} q_L & q_R \\ Q_L & Q_R \end{pmatrix} \quad : \quad \begin{bmatrix} (+,+) & (-,-) \\ (-,-) & (+,+) \end{bmatrix} \tag{19}$$

where q and Q belong to $(\mathbf{2}, \mathbf{1})$ and $(\mathbf{1}, \mathbf{2})$ of $SU(2)_L \times SU(2)_R$, respectively. q_L and Q_R are even under parity, and have zero modes in the absence of A_M irrespective of the value of c.

When $\theta_H \neq 0$, the gauge coupling $g_A \overline{\psi} \Gamma^5 e_5{}^y A_y \psi$ mixes q and Q. Further $A_y(y)$ has nontrivial y-dependence in the RS spacetime so that the mixing with KK excited states also results. The fermion mass in four dimensions is determined by finding eigenstates under such mixing.

To good accuracy the lightest mass eigenvalue is given by

$$m_f \sim k \left(\frac{c^2 - \frac{1}{4}}{e^{\pi k R} \sinh \left[(c + \frac{1}{2}) k \pi R \right] \sinh \left[(c - \frac{1}{2}) k \pi R \right]} \right)^{1/2} |\sin \tfrac{1}{2} \theta_H| . \tag{20}$$

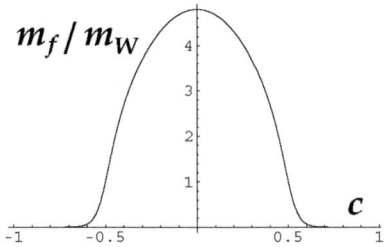

FIGURE 1. The lightest mass eigenvalue m_f as a function of c at $\theta_W = \frac{1}{2}\pi$. The vertical axis is in the unit of m_W, or the fermion mass value at $c = \pm\frac{1}{2}$.

TABLE 1. The values of c for leptons and quarks

	m_e	m_μ	m_τ	m_u	m_c	m_t
c	0.87	0.71	0.63	0.81	0.64	0.43

The result is depicted in fig. 1 for $\theta_H = \frac{1}{2}\pi$. $c = 1$ corresponds to $m_f = m_W$. Given the fermion mass m_f, the value of c is determined.

The remarkable fact is that the values of c are distributed in the range $0.43 < c < 0.87$. The huge hierarchy in the quark-lepton masses is explained by standard distribution of c, which is a natural quantity in the RS spacetime. The mass becomes exponentially small for $c > 0.6$.

Suppressed Yukawa couplings

By inserting the wave functions of the 4D Higgs field and fermions into $g_A \overline{\psi} \Gamma^5 e_5{}^y A_y \psi$ and integrating over y, one finds the Yukawa coupling in four dimensions. In the standard model the Yukawa coupling is proportional to the mass of the fermion. The relation in the dynamical gauge-Higgs unification is modified, becoming

$$y_\psi \sim \frac{g_A \sqrt{k(c^2 - \frac{1}{4})}}{2e^{\pi kR(c - \frac{1}{2})}} \cdot \cos\frac{\theta_H}{2} = \frac{g m_f}{2m_W} \cdot \cos^2\frac{\theta_H}{2} \ . \tag{21}$$

It is suppressed by a factor $\cos^2\frac{1}{2}\theta_H$.

GAUGE COUPLINGS OF QUARKS AND LEPTONS

The electric charge is conserved and the electromagnetic coupling is universal. It is the same to all charged particles. The weak coupling constants, however, may not be universal once $SU(2)_L \times U(1)_Y$ breaks down to $U(1)_{EM}$. In the standard model those

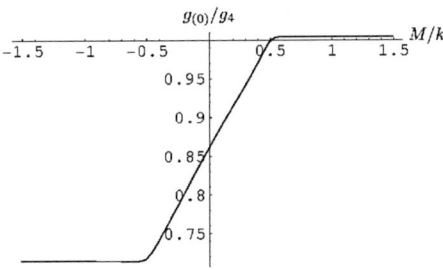

FIGURE 2. The 4D gauge coupling $g_{(0)}/g_4$ as a function of $c = M/k$ for $\theta_W = \frac{1}{2}\pi$ and $kR = 12$ in the $SU(3)$ model.

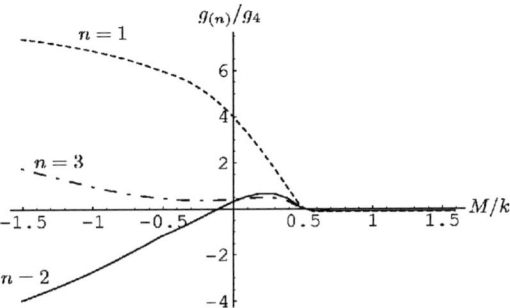

FIGURE 3. The gauge couplings to the n-th KK excited states of W, $g_{(n)}$ $(n = 1, 2, 3)$ as functions of $c = M/k$ at $\theta_W = \frac{1}{2}\pi$ in the $SU(3)$ model.

weak couplings are universal at least at the tree level. In the dynamical gauge-Higgs unification small deviation results.

Each fermion multiplet couples to the W boson with $g_{(0)}(\theta_H, c)$ obtained by integrating over y with wave functions of W and fermions inserted in the gauge interaction term in (18). $g_{(0)}$ depends on both θ_H and c. It is depicted in fig. 2 as a function of $c = M/k$ at $\theta_H = \frac{1}{2}\pi$. For $c > \frac{1}{2}$ the deviation is very small. The asymptotic value for $c < \frac{1}{2}$ is $\cos\theta_H$ in the $SU(3)$ model.

For the values of $c(> 0.43)$ for quarks and leptons in table 1, the dependence of $g_{(0)}$ on c is small. The violation of the μ-e, τ-e, and t-e universality in the charged current interactions is of order of 10^{-8}, 2×10^{-6} and 2×10^{-2}, respectively.

Each quarks and leptons couples to the KK excited states of gauge bosons as well. It was noticed that those couplings can be large at $c = 0$, which gives rise to contradition with observation unless M_{KK} is sufficiently large.

The coupling $g_{(n)}$ of a fermion to the n-th KK excited state of W is depicted in fig. 3 at $\theta_H = \frac{1}{2}$. It is seen that the couplings are very small for $c > \frac{1}{2}$ as noted by Gherghetta and Pomarol so that the earlier constgraint on M_{KK} is evaded.

WWZ, *WWH*, AND *ZZH* COUPLINGS

At $\theta_H \neq 0$, an eigenstate of each field becomes mixture of various components of the multiplet which the field belong to. All of the W, Z, and H fields in four dimensions are parts of the gauge field multiplets. The mixing pattern is not identical among these fields so that the effective 4D couplings necessarily depend on θ_H. This would provide critical tests for the dynamical gauge-Higgs unification particularly in the *WWZ*, *WWH*, and *ZZH* couplings.

The *WWZ* coupling is found to be

$$g_{WWZ} \simeq g \cos\theta_W \cdot \frac{1 + \cos^2\theta_H}{2} . \tag{22}$$

The coupling is suppressed by a factor of $(1 + \cos^2\theta_H)/2$ compared with the value in the standard model. The experiment at LEP2 indicates the validity of the standard model, and therefore gives a strong constraint on $\cos\theta_H$. We remark that in the process $e^+e^- \to W^+W^-$ there are important contributions from KK states of Z which have to be included in the analysis.

Similarly the *WWH* and *ZZH* couplings are

$$\lambda_{WWH} \simeq g m_W \cdot p_H \cos\theta_H ,$$
$$\lambda_{ZZH} \simeq \frac{g m_Z}{\cos\theta_W} \cdot p_H \cos\theta_H , \tag{23}$$

where $p_H = \mathrm{sign}(\sin\theta_H)$. The couplings are suppressed by a factor $\cos\theta_H$. Note that these couplings are important in drawing a constraint for the Higgs boson mass from the LEP data as well.

ACKNOWLEDGMENTS

The author would like to thank the Aspen Center for Physics for its hospitality where a part of this work was performed. This work was supported in part by Scientific Grants from the Ministry of Education and Science, Grant No. 17540257, Grant No. 13135215 and Grant No. 18204024.

References

1. Y. Hosotani, *Phys. Lett.* **B126** (1983) 309.
2. Y. Hosotani, *Ann. Phys. (N.Y.)* **190** (1989) 233.
3. D.B. Fairlie, *Phys. Lett.* **B82** (1979) 97; *J. Phys.* G5 (1979) L55.
4. N. Manton, *Nucl. Phys.* **B158** (1979) 141.
5. H. Hatanaka, T. Inami and C.S. Lim, *Mod. Phys. Lett.* **A13** (1998) 2601.
6. I. Antoniadis, K. Benakli and M. Quiros, *New. J. Phys.* **3** (2001) 20.
7. M. Kubo, C.S. Lim and H. Yamashita, *Mod. Phys. Lett.* **A17** (2002) 2249.

8. C. Csaki, C. Grojean and H. Murayama, *Phys. Rev.* D**67** (2003) 085012; C.A. Scrucca, M. Serone and L. Silverstrini, *Nucl. Phys.* B**669** (2003) 128.
9. L.J. Hall, Y. Nomura and D. Smith, *Nucl. Phys.* B**639** (2002) 307; L. Hall, H. Murayama, and Y. Nomura, *Nucl. Phys.* B**645** (2002) 85; G. Burdman and Y. Nomura, *Nucl. Phys.* B**656** (2003) 3; C.A. Scrucca, M. Serone, L. Silvestrini and A. Wulzer, *JHEP* **0402** (2004) 49.
10. N. Haba, M. Harada, Y. Hosotani and Y. Kawamura, *Nucl. Phys.* B**657** (2003) 169; *Erratum, ibid.* B**669** (2003) 381.
11. N. Haba, Y. Hosotani, Y. Kawamura and T. Yamashita, *Phys. Rev.* D**70** (2004) 015010; N. Haba, K. Takenaga, and T. Yamashita, *Phys. Lett.* B**615** (2005) 247.
12. Y. Hosotani, S. Noda and K. Takenaga, *Phys. Lett.* B**607** (2005) 276.
13. G. Cacciapaglia, C. Csaki and S.C. Park, *JHEP* **0603** (2006) 099.
14. G. Panico, M. Serone and A. Wulzer, *Nucl. Phys.* B**739** (2006) 186.
15. R. Contino, Y. Nomura and A. Pomarol, *Nucl. Phys.* B**671** (2003) 148.
16. K. Agashe, R. Contino and A. Pomarol, *Nucl. Phys.* B**719** (2005) 165.
17. K. Oda and A. Weiler, *Phys. Lett.* B**606** (2005) 408.
18. Y. Hosotani and M. Mabe, *Phys. Lett.* B**615** (2005) 257.
19. Y. Hosotani, S. Noda, Y. Sakamura and S. Shimasaki, *Phys. Rev.* D**73** (2006) 096006.
20. M. Carena, E. Ponton, J. Santiago and C.E.M. Wagner, hep-ph/0607106.
21. Y. Sakamura and Y. Hosotani, hep-ph/0607236.
22. Y. Hosotani, in the Proceedings of *"Dynamical Symmetry Breaking"*, ed. M. Harada and K. Yamawaki (Nagoya University, 2004), p. 17. (hep-ph/0504272).
23. N. Irges and F. Knechtli, *Nucl. Phys.* B**719** (2005) 121; hep-lat/0604006.
24. N. Maru and T. Yamashita, hep-ph/0603237.
25. Y. Hosotani, hep-ph/0607064.
26. L. Randall and R. Sundrum, *Phys. Rev. Lett.* **83** (1999) 3370.
27. T. Gherghetta and A. Pomarol, *Nucl. Phys.* B**586** (2000) 141.
28. S. Chang, J. Hisano, H. Nakano, N. Okada and M. Yamaguchi, *Phys. Rev.* D**62** (2000) 084025.

What can we learn from a Higher-Dimensional Decaying Black Hole?

Panagiota Kanti

Department of Mathematical Sciences, University of Durham, Science Site, South Road, Durham DH1 3LE, United Kingdom

Abstract. In the context of the scenario with Large Extra Dimensions, we investigate the emission spectra for Hawking radiation on the brane from a variety of higher-dimensional black holes: Schwarzschild, Kerr and Schwarzschild - de Sitter. The energy emission spectra in each case are presented, and their dependence on the spin of the emitted particle, dimension of spacetime, angular momentum of the black hole and cosmological constant of spacetime is discussed.

Keywords: Large Extra Dimensions, Black Holes, Hawking Radiation
PACS: 04.50.+h, 04.62.+v, 04.70.Dy

INTRODUCTION

The introduction of the idea of the existence of additional, spacelike dimensions in nature, by Kaluza and Klein almost a century ago, has radically changed our notion of the universe. The very same idea was used, only slightly changed, in the formulation of the superstring theory, as well as, more recently, in that of the scenario with Large Extra Dimensions [1]. The latter scenario was proposed in an attempt to resolve the hierarchy problem, and its basic constituents was a 3-brane, on which all Standard Model fields are localized, and the bulk – the spacetime transverse to the brane – where only gravity can freely propagate. The scenario postulated the existence of $n \geq 2$ extra, spacelike, compact dimensions with size $R \leq 1$ mm, and of a fundamental $(4 + n)$-dimensional Planck mass, M_*, related to the traditional 4-dimensional one by $M_P^2 \sim M_*^{2+n} R^n$. For $R \gg l_P$, M_* can be much lower than M_P, thus enhancing the Newton's constant and the strength of the gravitational forces by orders of magnitude.

The above opens the way for exciting low-energy quantum gravitational effects: particles with energies larger than M_* may trigger the appearance of strong gravitational phenomena and lead, during their collisions, to the creation of extended objects (p-branes, string balls, string states, etc), rather than ordinary particles [2]. Tiny small black holes with $l_P < r_H < R$ may also be produced during these transplanckian collisions, when any two partons of the colliding particles pass within the horizon radius corresponding to their centre-of-mass energy. These miniature black holes can, nevertheless, be treated as classical objects as long as $M_{BH} \gg M_*$. In addition, these higher-dimensional black holes are expected to go through the same stages in their life as their 4-dimensional counterparts: a short *balding* phase, during which the black hole will shed all additional quantum numbers apart its mass, angular momentum and charge, the more familiar *spin-down* phase, during which the black hole will lose mainly its angular momentum, then the *Schwarzschild* phase, when the black hole gradually loses its actual mass, and

CP881, *Cairo International Conference on High Energy Physics (CICHEP II)*,
edited by S. Khalil

finally the *Planck* phase, where the black hole has reduced to a quantum object with unknown properties.

During the two intermediate phases, the axially symmetric spin-down and spherically-symmetric Schwarzschild, the black hole loses energy (either kinetic or rest mass) through the emission of Hawking radiation [3]. This is realised by the creation of a virtual pair of particles just outside the horizon of the black hole: the antiparticle falls inside the black hole whereas the particle can now propagate away from the black hole and escape to infinity. Nevertheless, this does not happen always successfully since the strong gravitational field that surrounds the black hole forces many particles to reverse their motion and fall back into the black hole. For that reason, although the spectrum of the resulting Hawking radiation is very close to that of a black body, it is not exactly the same. The main modification is that the area of the emitting body that appears in the expression of the thermal spectrum of a black body in a flat spacetime is now replaced by the so-called greybody factor, the absorption cross-section for a particle to overcome the gravitational barrier and escape to infinity. This factor turns out to depend on many properties of the propagating particle (energy, spin, angular momentum numbers) but also on properties of the gravitational background (dimensionality of spacetime, angular momentum of the black hole, etc).

These miniature black holes, being created by ordinary particles localised on our brane, will be, at least initially, centered on the brane. However, being higher-dimensional objects, they will extend off our brane as well, and their decay will take place through the emission of Hawking radiation (elementary particles) both in the bulk and on the brane, that will be characterized by a very distinct thermal spectrum. For phenomenological reasons, we are mostly interested in the emission of particles directly on our brane since, being ourselves made of ordinary matter, we have no access to the bulk. It is this brane emission therefore that upon successful detection will provide the most convincing evidence not only for the creation of microscopic black holes but of the very existence of extra spacelike dimensions. In addition, the information about the gravitational background hidden inside the spectrum of Hawking radiation might lead to valuable information about topological properties of our universe such as the number of additional spacelike dimensions.

In the next two sections, I will discuss the emission spectra for Hawking radiation on the brane obtained during the Schwarzschild and spin-down phases in the life of a black hole. The emission channels for all Standard Model particles (scalars, gauge bosons and fermions) will be presented and discussed. In the following section, I will also present results for the "scalar channel" of the emission from a higher-dimensional Schwarzschild-de Sitter black hole, and see what information about the bulk cosmological constant we can also obtain from the corresponding spectrum. Our conclusions are briefly presented in the last section.

SCHWARZSCHILD PHASE

During the Schwarzschild phase, the black hole is believed to have lost all of its angular momentum and to have settled down to a rather simple spherically-symmetric geometry. Although the properties of these higher-dimensional black holes are not thoroughly known, it is generally believed that this phase will be the longest one in the life of the black hole and will account for the greatest proportion of the mass loss through the emission of Hawking radiation, in analogy to the $4D$ case. For the above reasons, the Schwarzschild phase was the first one to undergo a systematic study both analytically [4, 5] and numerically [6]. Here, we briefly discuss the brane emission channels of scalars, fermions and gauge bosons, and the dependence of the derived emission spectra on the number of additional spacelike dimensions in nature.

The line-element that describes the spacetime around a higher-dimensional, spherically-symmetric neutral black hole is given by the Tangherlini solution [7]

$$ds^2 = -\left[1 - \left(\frac{r_H}{r}\right)^{n+1}\right] dt^2 + \left[1 - \left(\frac{r_H}{r}\right)^{n+1}\right]^{-1} dr^2 + r^2 d\Omega^2_{2+n}, \tag{1}$$

where

$$d\Omega^2_{2+n} = d\theta^2_{n+1} + \sin^2\theta_{n+1}\left(d\theta^2_n + \sin^2\theta_n\left(... + \sin^2\theta_2\left(d\theta^2_1 + \sin^2\theta_1\, d\varphi^2\right)...\right)\right). \tag{2}$$

In the above, $0 < \varphi < 2\pi$ and $0 < \theta_i < \pi$, for $i = 1,...,n+1$. The horizon radius r_H of such a black hole is related to its mass M_{BH} through the following relation

$$r_H = \frac{1}{\sqrt{\pi}M_*}\left(\frac{M_{BH}}{M_*}\right)^{\frac{1}{n+1}}\left(\frac{8\Gamma\left(\frac{n+3}{2}\right)}{n+2}\right)^{\frac{1}{n+1}}, \tag{3}$$

while its temperature, given in terms of its surface gravity, comes out to be

$$T_H = \frac{n+1}{4\pi r_H}. \tag{4}$$

The above black hole, characterized by a non-vanishing temperature, emits Hawking radiation [3] both in the bulk and on the brane. As mentioned in the introduction, here we focus on the more phenomenologically interesting case of emission of Standard Model particles directly on our brane. These particles are restricted to propagate in a 4-dimensional background which is the projection of the line-element (1) onto our brane. This follows by setting $\theta_i = \pi/2$, for $i > 1$, in which case the line-element of the unit sphere $d\Omega^2_{2+n}$ reduces to the usual $d\Omega^2_2$. As the rest of the metric (1) remains unchanged, the horizon and temperature of the "projected" black hole are still given by the above expressions. The emission rate of brane-localised modes is then given by

$$\frac{dE(\omega)}{dt} = \sum_j \frac{\sigma_j(\omega)\,\omega}{\exp(\omega/T_H) \mp 1}\frac{d^3k}{(2\pi)^3}. \tag{5}$$

In the above, ω is the energy of the emitted particle, j the total angular momentum number, and the statistics factor in the denominator is -1 for bosons and $+1$ for fermions. The ω-dependent factor σ_j is the *greybody factor* that causes the black-hole spectrum to deviate from the one of a pure blackbody. The absorption cross-section

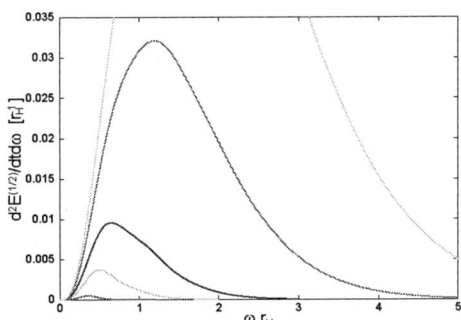

FIGURE 1. Energy emission rates for fermions on the brane from a $(4+n)D$ Schwarzschild BH [6].

$\sigma_j(\omega)$ is related to the corresponding absorption probability $|\mathscr{A}_{j,n}|^2$, for emission on the brane, through the relation

$$\sigma_j(\omega) = \frac{\pi}{\omega^2}(2j+1)|\mathscr{A}_{j,n}|^2. \tag{6}$$

In terms of the latter, the energy emission rate can take the following simplified form

$$\frac{d^2E(\omega)}{d\omega\,dt} = \sum_j \frac{(2j+1)|\mathscr{A}_{j,n}(\omega)|^2}{\exp(\omega/T_H)\mp 1}\frac{\omega}{(2\pi)}. \tag{7}$$

To find the absorption probability, we must solve the corresponding equation of motion of the particular type of field whose emission from the black hole we study. By using the Newman-Penrose formalism and the factorisation $\Psi_s = e^{-i\omega t}e^{im\varphi}\Delta^{-s}P_s(r)S_{s,j}^m(\theta)$, a brane *master* equation may be derived, that describes the propagation of an arbitrary field with spin s on the projected-on-the-brane background. This has the form [8, 9]

$$\Delta^s \frac{d}{dr}\left(\Delta^{1-s}\frac{dP_s}{dr}\right) + \left[\frac{\omega^2 r^2}{h} + 2i\omega sr - \frac{is\omega r^2 h'}{h} - \Lambda_{sj}\right]P_s(r) = 0, \tag{8}$$

where

$$\Delta(r) = r^2 h(r) = r^2\left[1 - \left(\frac{r_H}{r}\right)^{n+1}\right], \qquad \Lambda_{sj} = j(j+1) - s(s-1). \tag{9}$$

Although an equation for propagation of a field in a 4-dimensional background, the above equation has an explicit dependence, through the expression of the metric function $\Delta(r)$, on the transverse-to-the-brane spacelike dimensions. As a result, $\mathscr{A}_{j,n}$ and consequently the radiation spectrum will bear a similar dependence.

The above equation was solved for all types of Standard Model fields (scalars, fermions and gauge bosons) both analytically [4, 5] and numerically [6]. Here, we focus on the exact, complete radiation spectrum following by using numerical analysis. In Fig. 1, we present, as an indicative case, the radiation spectrum for the emission of fermions

TABLE 1. Total emissivities for emission of SM fields on the brane [6].

n	0	1	2	3	4	5	6	7
Scalars	1.0	8.94	36.0	99.8	222	429	749	1220
Fermions	1.0	14.2	59.5	162	352	664	1140	1830
G. Bosons	1.0	27.1	144	441	1020	2000	3530	5740

on the brane. The different curves correspond to a different number of extra spacelike dimensions, with this number increasing from bottom to top. From these curves, it becomes obvious that the amount of emitted energy per unit time strongly depends on the number of spacelike dimensions that exist transverse to the brane, and it is enhanced as n increases. For quantitative purposes, in Table 1, we display the total emissivities – i.e. the emission rates integrated over the whole frequency regime – for all three types of fields living on the brane, and for different values of n. The emissivities are normalised to the ones for $n = 0$, and they quickly increase, as n increases, eventually reaching enhancements of orders of magnitude.

Let us, finally, briefly comment on the dependence of the relative emission rates on the number of extra dimensions n. In Figs. 2(a,b), we display the emission rates for scalars, fermions and gauge bosons for $n = 0$ and $n = 6$, respectively. In the first case, it is obvious that a 4-dimensional black hole prefers to emit mainly scalar particles, then fermions and finally gauge bosons. In terms of total emissivity, if the one for scalars is normalised to 1, the one for fermions comes out to be 0.55 and the one for gauge bosons 0.23. In the presence of 6 additional spacelike dimensions, though, the situation changes radically: now the dominant modes are the gauge bosons, then come the scalars and finally the fermions (with total emissivities 1.06:1:0.84, respectively). We may therefore conclude that not only the rate but also the type of emitted radiation depends strongly on the dimensionality of spacetime.

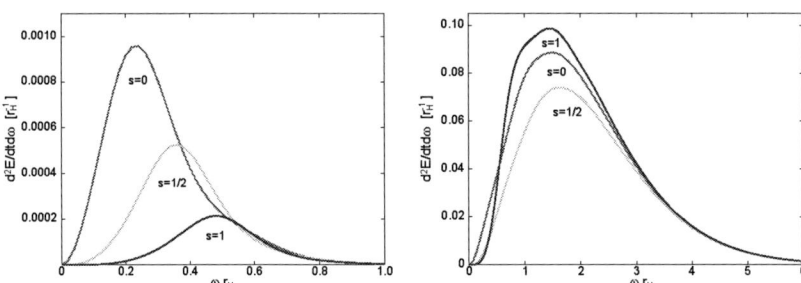

FIGURE 2. Emission rates for scalars, fermions and gauge bosons for **(a)** $n = 0$, and **(b)** $n = 6$ [6].

SPIN-DOWN PHASE

Going backwards in time, the phase that precedes the spherically-symmetric Schwarzschild one in the life of the black hole is the spin-down phase, during which the black hole resembles a Kerr one living in a higher-dimensional spacetime. The emission of

Hawking radiation by such a black hole was studied in two early analytical works [9, 10]. During the last year, these were supplemented by a number of works that offered complete, exact numerical results for the emission of Hawking radiation from a $(4+n)$-dimensional rotating black hole on the brane [11, 12, 13, 14].

The gravitational background around a higher-dimensional, rotating, neutral black hole is given by the Myers-Perry solution [15]. Again, what we would be interested in here is the line-element that describes the background on a brane which is placed in the vicinity of such a higher-dimensional black hole. This follows again by fixing the values of all additional azimuthal coordinates to $\pi/2$, that brings the projected-on-the-brane line-element to the form [8]

$$
ds^2 = \left(1 - \frac{\mu}{\Sigma r^{n-1}}\right) dt^2 + \frac{2a\mu \sin^2 \theta}{\Sigma r^{n-1}} dt\, d\varphi - \frac{\Sigma}{\Delta} dr^2
$$
$$
- \Sigma d\theta^2 - \left(r^2 + a^2 + \frac{a^2 \mu \sin^2 \theta}{\Sigma r^{n-1}}\right) \sin^2 \theta\, d\varphi^2 ,
\tag{10}
$$

where

$$
\Delta = r^2 + a^2 - \frac{\mu}{r^{n-1}} , \qquad \Sigma = r^2 + a^2 \cos^2 \theta .
\tag{11}
$$

The mass and angular momentum (transverse to the $r\varphi$-plane) of the black hole are then given by

$$
M_{BH} = \frac{(n+2)\, \pi^{(n+1)/2}}{8G\, \Gamma[(n+3)/2]} \mu , \qquad J = \frac{2}{n+2} M_{BH}\, a ,
\tag{12}
$$

with G being the $(4 + n)$-dimensional Newton's constant. By applying again the Newman-Penrose formalism for the above brane background, and using the factorisation $\Psi_s = e^{-i\omega t}\, e^{im\varphi}\, R_s(r)\, \hat{S}_{s,j}^m(\theta)$, the master equation now takes the form [8]

$$
\Delta^{-s} \frac{d}{dr} \left(\Delta^{s+1} \frac{dR_s}{dr}\right) + \left[\frac{K^2 - iKs\Delta'}{\Delta} + 4is\omega r + s\left(\Delta'' - 2\right) - \Lambda_j^m\right] R_s = 0 ,
\tag{13}
$$

where $K = (r^2 + a^2)\, \omega - am$. The above equation describes again the propagation of an arbitrary field with spin $s = 0, 1/2$ and 1 in the brane background (10). The exact value of the angular eigenvalue, Λ_j^m, that couples the angular and radial part of the equation of motion of the field, does not exist now in closed form, and a numerical analysis must be used to determine its value before the radial equation (13) can be itself integrated.

The differential energy emission rate for Hawking radiation emitted from a rotating black hole is given by the expression

$$
\frac{d^2 E(\omega)}{d\omega\, dt} = \sum_{j,m} |\mathscr{A}_{j,n}^m|^2 \frac{\omega}{\exp\left[(\omega - m\Omega)/T_H\right] - 1} \frac{1}{2\pi} ,
\tag{14}
$$

with the Hawking temperature and rotation velocity of this brane black hole given by

$$
T_H = \frac{(n+1)r_H^2 + (n-1)a^2}{4\pi(r_H^2 + a^2)r_H} , \qquad \Omega = \frac{a}{(r_H^2 + a^2)} .
\tag{15}
$$

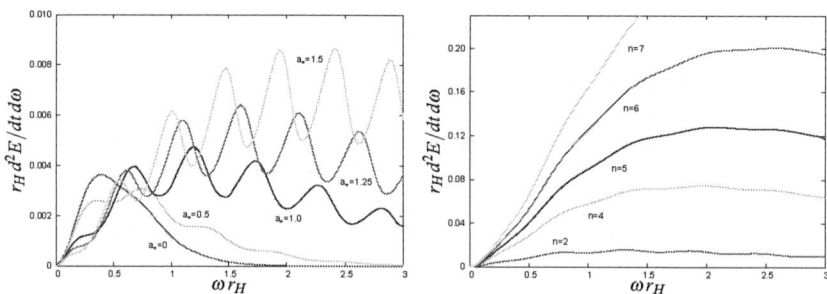

FIGURE 3. Energy emission rates for scalar fields on the brane from a $(4+n)D$ Kerr black hole, for **(a)** variable a $(n=1)$, and **(b)** variable n $(a=1)$.

We start the presentation of our exact, numerical results with those obtained for the case of scalar fields emitted on our brane [11, 13]. In Figs. 3(a,b), we present the energy emission rates for scalar fields for variable angular momentum parameter a and number of extra dimensions n, respectively. As it is clear from the plots, the energy emission rates are strongly enhanced with both parameters: in terms of a, the enhancement is more obvious in the high-energy regime, while, in terms of n, the enhancement is present in all energy regimes and is of a larger scale. The work of [11, 13] was soon supplemented by another comprehensive study of the emission of gauge bosons on the brane by a rotating black hole [14]. In Figs. 4(a,b), we present the corresponding graphs for the energy emission rates for gauge bosons, in terms of a and n. As it is again obvious, the emission rates for gauge fields on the brane are also greatly enhanced by both the dimension of spacetime and the angular momentum of the black hole.

In the case of a spherically-symmetric black hole, the emitted radiation is uniformly distributed over the whole solid angle $\Omega_2 = 4\pi$, since there is no preferred direction in space. In the case of a rotating black hole, though, the situation changes as the rotation axis of the black hole breaks the symmetry and provides a preferred direction. It is found that the angular distribution of emitted particles in this case is not uniform, and is affected by a number of factors. In Figs. 5(a,b,c), we display the angular distribution of

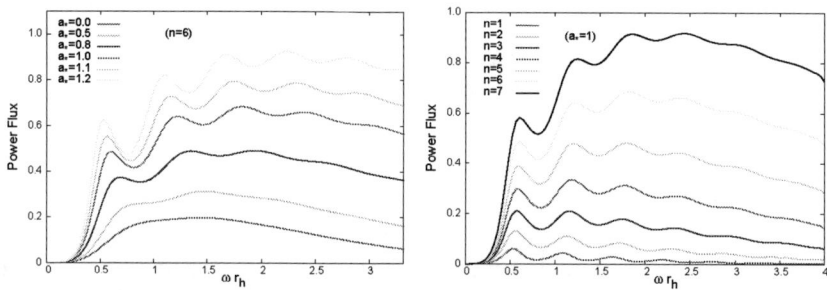

FIGURE 4. Energy emission rates for gauge bosons on the brane from a $(4+n)D$ Kerr black hole, for **(a)** variable a $(n=6)$, and **(b)** variable n $(a=1)$.

36

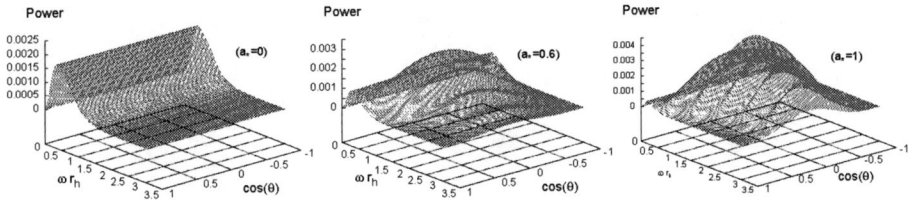

FIGURE 5. Angular distribution of the power spectra for scalars emitted on the brane by a rotating black hole, for $n = 1$ and $a = (0, 0.6, 1)$.

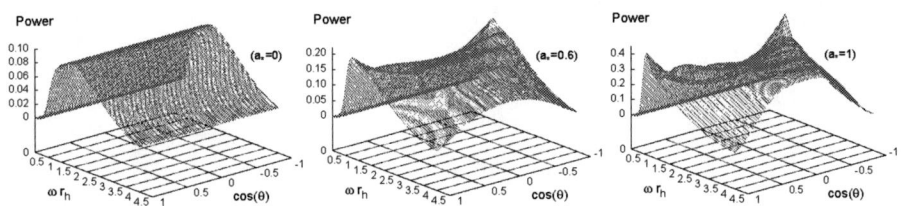

FIGURE 6. Angular distribution of the power spectra for gauge bosons emitted on the brane by a rotating black hole, for $n = 6$ and $a = (0, 0.6, 1)$.

scalar particles emitted on the brane, for three different values of the angular momentum parameter: $a = (0, 0.6, 1)$, and for $n = 1$. While for $a = 0$, there is no angular variation as expected, as soon as a takes a non-zero value, most of the emitted particles are concentrated near the equatorial plane ($\theta = \pi/2$). The latter effect is due to the centrifugal potential that becomes stronger as the angular momentum of the black hole increases. In Figs. 6(a,b,c), we display the corresponding graphs for the emission of gauge bosons on the brane, for the same values of a and for $n = 6$. In this case, there is an additional factor affecting the angular variation of the spectrum: the spin-rotation coupling, absent for the spin-0 scalar particles, that polarises the gauge bosons along the rotation axis ($\theta = 0, \pi$). This factor is found to be most effective for particles with low-energy, and it becomes subdominant in the high-energy regime.

SCHWARZSCHILD - DE SITTER BLACK HOLES

There have been several studies in the literature that address the question of the Hawking radiation from other types of higher-dimensional black-hole backgrounds. One particular case is the one of a black hole created in a spacetime with a positive cosmological constant Λ. For simplicity, we ignore here the angular momentum of such a black hole, in which case the line-element of the resulting Schwarzschild - de Sitter black hole is given again by the Tangherlini solution [7]. Projecting this line-element onto the brane,

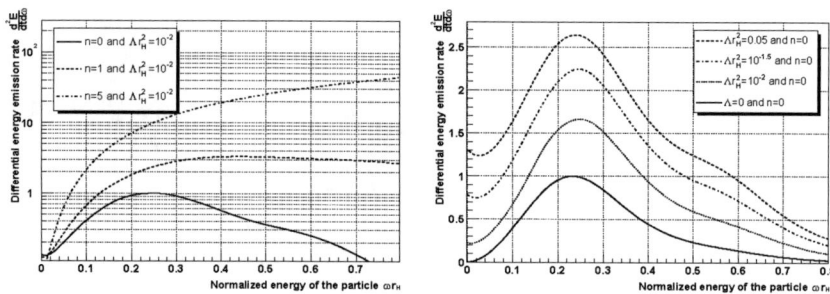

FIGURE 7. Energy emission rates for scalar fields on the brane from a $(4+n)D$ Schwarzschild - de Sitter black hole, for **(a)** variable n, and **(b)** variable Λ, respectively.

we obtain the following 4-dimensional background

$$ds^2 = -h(r)\,dt^2 + \frac{dr^2}{h(r)} + r^2\,d\Omega_2^2, \tag{16}$$

where

$$h(r) = 1 - \frac{\mu}{r^{n+1}} - \frac{2\kappa_D^2 \Lambda r^2}{(n+3)(n+2)}, \tag{17}$$

with $\kappa_D = 8\pi G$. The temperature of the above black hole is now given by the expression

$$T_H = \frac{1}{\sqrt{h(r_0)}} \frac{1}{4\pi r_H} \left[(n+1) - \frac{2\kappa_D^2 \Lambda}{(n+2)} r_H^2 \right], \tag{18}$$

where r_0 is the point along the radial coordinate where $|g_{tt}|$ reaches its maximum value.

The differential energy emission rate for the emission of particles on the brane by a higher-dimensional Schwarzschild - de Sitter black hole is again given by Eq. (7). The particular case of the emission of scalar particles, both in the bulk and on the brane, was studied in [16]. In Figs. 7(a,b), we present the energy emission rates for scalar fields on the brane from a higher-dimensional Schwarzschild - de Sitter black hole, for variable dimension of spacetime n and bulk cosmological constant Λ, respectively. The effect of the number of dimensions was found to be the same as in the case of a higher-dimensional Schwarzschild-like or Kerr-like black hole, with an increase in n causing a significant enhancement in the emission rate. A similar, but of a smaller magnitude, enhancement is observed as the cosmological constant increases, too. However, there is now an additional feature in the spectrum characteristic of the existence of a positive cosmological constant in the spacetime: the energy emission rate asymptotes to a non-vanishing value as the energy goes to zero – this asymptotic value, and thus the number of soft quanta emitted, is increasing as Λ increases.

CONCLUSIONS

Until today, we have not yet detected the Hawking radiation emitted from a decaying black hole. For astrophysical black holes, this is due to their large mass resulting into an extremely low temperature and thus to an undetectable radiation. For primordial black holes, whose spectrum is centered on frequencies more accessible to us, our failure might be due to the fact that no such black holes exist in our neighbourhood.

If Extra Dimensions exist, however, then the creation of tiny black holes might be possible. The presence of extra dimensions lowers the fundamental scale of gravity [1], and enhances the black-hole production cross-section [2]. In addition, quite conveniently, the black holes that might be created in the present or next-generation accelerators will have a radiation spectrum that can be easily probed with existing detectors [8].

Upon detection, the emission spectra can help us determine the *dimensionality* of spacetime, with possible signatures being both the *rate* and the *type* of the emitted radiation. In addition, the emission spectra strongly depend on additional features of the gravitational background, such as the angular momentum of the black hole and the cosmological constant, thus making the Hawking radiation an extremely rich source of information for the topological properties of the higher-dimensional background.

ACKNOWLEDGMENTS

I would like to thank A. Barrau, M. Casals, G. Duffy, J. Grain, C. Harris and E. Winstanley for our enjoyable and fruitful collaborations. My research activities are funded by the UK PPARC Research Grant PPA/A/S/2002/00350.

REFERENCES

1. N. Arkani-Hamed, S. Dimopoulos and G. R. Dvali, *Phys. Lett.* B **429**, 263 (1998); *Phys. Rev.* D **59**, 086004 (1999);
   I. Antoniadis, N. Arkani-Hamed, S. Dimopoulos and G. R. Dvali, *Phys. Lett.* B **436**, 257 (1998).
2. T. Banks and W. Fischler, hep-th/9906038;
   S. B. Giddings and S. Thomas, *Phys. Rev.* D **65**, 056010 (2002);
   S. Dimopoulos and G. Landsberg, *Phys. Rev. Lett.* **87**, 161602 (2001).
3. S. W. Hawking, *Commun. Math. Phys.* **43**, 199 (1975).
4. P. Kanti and J. March-Russell, *Phys. Rev.* D **66**, 024023 (2002); *Phys. Rev.* D **67**, 104019 (2003).
5. V. P. Frolov and D. Stojkovic, *Phys. Rev.* D **66**, 084002 (2002).
6. C. M. Harris and P. Kanti, *JHEP* **0310**, 014 (2003).
7. F. R. Tangherlini, *Nuovo Cim.* **27**, 636 (1963).
8. P. Kanti, *Int. J. Mod. Phys.* A **19**, 4899 (2004).
9. D. Ida, K. y. Oda and S. C. Park, *Phys. Rev.* D **67**, 064025 (2003) [Erratum-ibid. D **69**, 049901 (2004)].
10. V. P. Frolov and D. Stojkovic, *Phys. Rev.* D **67**, 084004 (2003).
11. C. M. Harris and P. Kanti, *Phys. Lett.* B **633**, 106 (2006).
12. D. Ida, K. y. Oda and S. C. Park, *Phys. Rev.* D **71**, 124039 (2005).
13. G. Duffy, C. Harris, P. Kanti and E. Winstanley, *JHEP* **0509**, 049 (2005).
14. M. Casals, P. Kanti and E. Winstanley, *JHEP* **0602**, 051 (2006).
15. R. C. Myers and M. J. Perry, *Ann. Phys. (NY)* **172**, 304 (1986).
16. P. Kanti, J. Grain and A. Barrau, *Phys. Rev.* D **71**, 104002 (2005).

D-branes in Type IIB plane wave background

Bum-Hoon Lee

Center for Quantum Spacetime and Department of Physics
Sogang University, Seoul 121-742, Korea

Abstract. We classify and summarize the intersecting supersymmetric D-branes in the type IIB plane wave background, based on the Green-Schwarz superstring formulation. Many new configurations appears if we turn on the electric or magnetic background fields or boost the D-branes. Applications to the phenomelogical models are left for further study.

Keywords: D-branes, pp-wave background
PACS: 11.25.Uv

INTRODUCTION

The discovery of the D-branes in the string theory has given deep influence to various areas, including the phenomenological model building of the particle physics. Various D-brane configurations especially with some unbroken supersymmetries provide many interesting gauge theory models. Recently, the geometry corresponding to the Penrose limit of the $AdS_5 \times S^5$ background was shown to be a maximally supersymmetric type IIB string background, [1] - [4],

$$ds^2 = -2dx^+dx^- - \mu^2 x_I^2 (dx^+)^2 + dx_I{}^2, \qquad (1)$$
$$F_{+1234} = F_{+5678} = 2\mu.$$

The presence of D-branes will break some of these supersymmetries. We systematically classify *static* D-branes in the maximally supersymmetric type IIB plane wave background (1) [5] - [18] using the Green-Schwarz superstring theory based on the light-cone open string theory. [16, 17, 18] We restrict such D-branes with the light-cone worldvolume coordinates $X^{\pm}$ satisfying the Neumann boundary condition hence the instantonic branes and branes with only one light-cone coordinate along the worldvolume in [5] will be outside of our classification. The dual gauge theory descriptions will not be considered here.

FLAT D-BRANES IN A PLANE WAVE BACKGROUND

The Green-Schwarz action with the light-cone gauge $X^+ = \tau$ in the plane wave background (1) describes eight free massive bosons and fermions [2] and is given by

$$S = \frac{1}{2\pi\alpha' p^+} \int d\tau \int_0^{2\pi\alpha' p^+} d\sigma \left[\frac{1}{2}\partial_+ X_I \partial_- X_I - \frac{1}{2}\mu^2 X_I^2 - i\bar{S}(\rho^A \partial_A - \mu\Pi)S \right] \qquad (2)$$

CP881, *Cairo International Conference on High Energy Physics (CICHEP II)*,
edited by S. Khalil

where $\partial_\pm = \partial_\tau \pm \partial_\sigma$. The equations of motion from the action (2) take the form

$$\partial_+\partial_- X^I + \mu^2 X^I = 0, \tag{3}$$

$$\partial_+ S^1 - \mu\Pi S^2 = 0, \qquad \partial_- S^2 + \mu\Pi S^1 = 0. \tag{4}$$

The open string action is described by the action (2) with string length $\alpha = 2\alpha' p^+$ and with appropriate boundary conditions on each end of the string. We will use the notation and the convention in [16] with more refined indices. Neumann coordinates X^r are decomposed into oblique directions $X^{\hat{r}}$ and usual parallel directions $X^{\dot{r}} : r = (\hat{r}, \dot{r})$. Similarly, Dirichlet coordinates $X^{r'}$ are also decomposed into oblique directions $X^{\hat{r}'}$ and usual parallel directions $X^{\dot{r}'} : r' = (\hat{r}', \dot{r}')$. For longitudinal coordinates X^r on D-branes without any worldvolume flux, we impose the Neumann boundary condition

$$\partial_\sigma X^r \, |_{\partial\Sigma} = 0, \tag{5}$$

while for transverse coordinates $X^{r'}$ we have the Dirichlet boundary condition

$$\partial_\tau X^{r'} \, |_{\partial\Sigma} = 0. \tag{6}$$

In the case to include gauge field excitations considered later, some Neumann boundary conditions have to be modified as follow [5, 11, 13, 16, 19]

$$(\partial_\sigma X^r \pm \mu X^r) \, |_{\partial\Sigma} = 0 \tag{7}$$

for some $r \in N$. The fermionic coordinates also have to satisfy the following boundary condition at each end of the open string [20]

$$(S^1 - \Omega S^2) \, |_{\partial\Sigma} = 0, \tag{8}$$

For D-branes with the flux F_{+I} and the angular momentum L_{IJ} only, the gluing matrix Ω is exactly the same as the trivial backgrounds and is simply given by the product of γ-matrices along the Neumann directions:

$$\Omega = \prod_{r \in N} \gamma^r. \tag{9}$$

So, in this case,

$$\Omega^2 = \pm 1, \tag{10}$$

$$\gamma^r \Omega = -\Omega\gamma^r, \quad \forall r \in N, \tag{11}$$

$$\gamma^{r'} \Omega = \Omega\gamma^{r'}, \quad \forall r' \in D. \tag{12}$$

For D-branes with the flux F_{rs}, however, Ω has the following form [21, 22]:

$$\Omega = \tilde{\Omega} \exp^{\frac{1}{4}\Theta_{rs}\gamma^{rs}}, \tag{13}$$

TABLE 1. Flat D-branes with $\Gamma^2 = 1$

D-brane type	Γ	Ω
$D_\pm$	± 1	$\Omega_{D_\pm}$
$OD3$	$\pm\gamma^{1256}$	$\frac{1}{2}(\gamma^1-\gamma^6)(\gamma^2\pm\gamma^5)$
$OD5$	$\pm\gamma^{1256}$	$\frac{1}{2}(\gamma^1-\gamma^6)(\gamma^2\mp\gamma^5)\gamma^{34}$ $\frac{1}{2}(\gamma^1-\gamma^6)(\gamma^2\mp\gamma^5)\gamma^{78}$ $\frac{1}{2}(\gamma^1-\gamma^6)(\gamma^2\pm\gamma^5)\gamma^{37}$
$OD7$	$\pm\gamma^{1256}$	$\frac{1}{2}(\gamma^1-\gamma^6)(\gamma^2\pm\gamma^5)\gamma^{3478}$
$OD_\pm 5$	$\pm\gamma$	$\frac{1}{4}(\gamma^1-\gamma^6)(\gamma^2\pm\gamma^5)(\gamma^3-\gamma^8)(\gamma^4+\gamma^7)$

where $\widetilde{\Omega}$ is the gluing matrix of the type (9) for the Neumann directions without flux and the parameters Θ_{rs} depend on the flux F_{rs}. In this case, the nice properties, eqs.(10) and (11), no longer hold due to the additional exponential factor. However the property (12) is still true since the flux F_{rs} extends only along the Neumann directions. If $\Theta_{rs} \neq 0$, e.g., with rank 2, the gluing matrix Ω continuously interpolates among codimension 2 D-branes. When $\Theta_{rs} \to 0$ or π, we have to recover the case (9) [22].

The boundary condition (8) has to be compatible with the fermionic equation of motion (4) and thus the possible type of D-branes shall be characterized by the matrix Γ defined by

$$\Gamma \equiv \Pi\Omega\Pi\Omega. \tag{14}$$

The simplest type of D-branes, $D_\pm$-branes [5]-[10], are satisfying $\Gamma = \pm 1$.

Table 1 shows possible flat D-branes with particular polarizations. Other flat D-branes with different polarizations can be generated by $SO(4) \times SO(4)'$ rotations of those in Table 1.

For D_--branes, there are the following possibilities [5, 8, 10, 11]:

$$
\begin{aligned}
D3 &: (m,n) = (2,0),\ (0,2), \\
D5 &: (m,n) = (3,1),\ (1,3), \\
D7 &: (m,n) = (4,2),\ (2,4).
\end{aligned}
\tag{15}
$$

For D_+-branes, there are the following possibilities [5, 8, 11]:

$$
\begin{aligned}
D1 &: (m,n) = (0,0), \\
D3 &: (m,n) = (1,1), \\
D5 &: (m,n) = (4,0),\ (2,2),\ (0,4), \\
D7 &: (m,n) = (3,3), \\
D9 &: (m,n) = (4,4).
\end{aligned}
\tag{16}
$$

The D-branes discussed in [13, 14] correspond to the $OD3$-brane with $\Gamma = -\gamma^{1256}$ and OD_-5-brane in Table 1.

The mode expansions for the bosonic and fermionic fields can be done in a straightforward way [18].

INTERSECTING D-BRANES

In this section, we will generalize the previous analysis to the case of intersecting D-branes using the formalism in [17]. In particular, the fermionic coordinates have to satisfy the following boundary condition at each end of the open string

$$(S^1 - \Omega_0 S^2)\,|_{\sigma=0} = 0, \qquad (S^1 - \Omega_\pi S^2)\,|_{\sigma=\pi\alpha} = 0, \qquad (17)$$

with the matrix $\Omega_\theta = (\Omega_0, \Omega_\pi)$ satisfying

$$\Pi\Omega_0\Pi\Omega_0 = \Gamma_0, \qquad \Pi\Omega_\pi\Pi\Omega_\pi = \Gamma_\pi. \qquad (18)$$

Here the D-brane is either a $D_\pm$-brane or an OD-brane.

The coordinates $X^I(\tau, \sigma)$ of a $p - q$ string can be partitioned into four sets, NN, DD, ND, and DN, according to whether the coordinate X^I has Neumann (N) or Dirichlet (D) boundary condition at each end. For intersecting D-branes, we will use indices $(r, s, \cdots) = (\hat{r}, \hat{s}, \cdots; \dot{r}, \dot{s}, \cdots)$ $(r', s', \cdots) = (\hat{r}', \hat{s}', \cdots; \dot{r}', \dot{s}', \cdots)$, $(i, j, \cdots) = (\hat{i}, \hat{j}, \cdots; \dot{i}, \dot{j}, \cdots)$, and $(i', j', \cdots) = (\hat{i}', \hat{j}', \cdots; \dot{i}', \dot{j}', \cdots)$ for NN, DD, ND, and DN coordinates, respectively, with a distinction between hatted indices for oblique directions and dotted indices for parallel directions.

The mode expansion of the spinor field can be determined [18]. We take an appropriate combination of spinor fields $\xi^A(\tau, \sigma)$ with integer modes and $\eta^A(\tau, \sigma)$ with half-integer modes or with **R**-modes to be compatible with supersymmetry:

$$S^1(\tau, \sigma) = \begin{cases} I_+\xi^1(\tau, \sigma) + I_-\eta^1(\tau, \sigma), & \text{for A-type;} \\ I_-\xi^1(\tau, \sigma) + I_+\eta^1(\tau, \sigma), & \text{for B-type,} \end{cases}$$

$$S^2(\tau, \sigma) = I_+\xi^2(\tau, \sigma) + I_-\eta^2(\tau, \sigma), \qquad (19)$$

where 16×16 matrices I_+ and I_- are defined by

$$I_+ = \frac{1}{2}(1 + \Omega_0^T\Omega_\pi), \qquad I_- = \frac{1}{2}(1 - \Omega_0^T\Omega_\pi). \qquad (20)$$

The spinors $\xi^A(\tau, \sigma)$ and $\eta^A(\tau, \sigma)$ are taken as the solution of the equations of motion (4) satisfying the boundary condition (17) at $\sigma = 0$.

We now require the spinors $S^A(\tau, \sigma)$ in Eq. (19) to satisfy the equations of motion (4) and then we need the following condition on $I_\pm$:

$$\Pi I_\pm = \begin{cases} I_\pm\Pi, & \text{for A-type;} \\ I_\mp\Pi, & \text{for B-type.} \end{cases} \qquad (21)$$

One can see that the condition (21) is equivalent to the following constraint

$$\Gamma_0\Gamma_\pi = \begin{cases} 1 \text{ or } \gamma, & \text{for } \Gamma_\theta^T = \Gamma_\theta; \\ -1 \text{ or } -\gamma, & \text{for } \Gamma_\theta^T = -\Gamma_\theta. \end{cases} \qquad (22)$$

The condition (22) clearly explains why a D_--brane cannot have a supersymmetric intersection with a D_+-brane, as was shown in [17], since $\Gamma_0 = -1$ and $\Gamma_\pi = 1$ for

43

TABLE 2. Supersymmetry of flat D-branes. A D-brane with a gauge field condensate is denoted by the boldface. $q_{D_-}^+$ ($q_{D_+}^+$) is the number of unbroken kinematical supersymmetry of D_--type (D_+-type).

D-brane type	Γ	Ω	$q_{D_-}^+$	$q_{D_+}^+$	q^-
D_5	-1	$(3,1),\ (1,3)$	8	0	8
D$_+$(2n+1)	1	$(n,n),\ n=1,2,3,4$	0	8	4
OD3	$\pm\gamma^{1256}$	$\frac{1}{2}(\gamma^1-\gamma^6)(\gamma^2\pm\gamma^5)$	4	4	4
OD5	$\pm\gamma^{1256}$	$\frac{1}{2}(\gamma^1-\gamma^6)(\gamma^2\mp\gamma^5)\gamma^{34}$ $\frac{1}{2}(\gamma^1-\gamma^6)(\gamma^2\mp\gamma^5)\gamma^{78}$	4	4	4
OD5	$\pm\gamma^{1256}$	$\frac{1}{2}(\gamma^1-\gamma^6)(\gamma^2\pm\gamma^5)\gamma^{37}$	4	4	2
OD7	$\pm\gamma^{1256}$	$\frac{1}{2}(\gamma^1-\gamma^6)(\gamma^2\pm\gamma^5)\gamma^{3478}$	4	4	2
OD_+5	γ	$\frac{1}{4}(\gamma^1-\gamma^6)(\gamma^2+\gamma^5)(\gamma^3-\gamma^8)(\gamma^4+\gamma^7)$	0	8	0
OD_-5	$-\gamma$	$\frac{1}{4}(\gamma^1-\gamma^6)(\gamma^2-\gamma^5)(\gamma^3-\gamma^8)(\gamma^4+\gamma^7)$	8	0	8

this kind of intersection. In addition, the condition (22) implies that there may be a supersymmetric intersection between different classes of OD-brane or an $OD_\pm5$-brane and a $D_\pm p$-brane only if they satisfy $\Gamma_0\Gamma_\pi = \gamma$. This case preserves only kinematical supersymmetries.

The case $\Omega_0^T\Omega_\pi = 1$ corresponds to parallel Dp-branes while the case $\Omega_0^T\Omega_\pi = -1$ corresponds to Dp-anti-Dp branes, but the cases $\Omega_0^T\Omega_\pi = \pm\gamma$ and $\Omega_0^T\Omega_\pi = \pm\Xi$ correspond to $Dp - Dq$ or Dp-anti-Dq branes with $\sharp_{ND} = 8$ and $\sharp_{ND} = 4$, respectively. Note that the B-type branes allow only the $\sharp_{ND} = 4$ case.

SUPERSYMMETRY OF FLAT D-BRANES

In a light-cone gauge, the 32 components of the supersymmetries for a closed string decompose into kinematical supercharges, Q_a^{+A}, and dynamical supercharges, $Q_{\dot{a}}^{-A}$. [2]

An open string on a D-brane satisfying $\Gamma^T = \Gamma$ preserves 8 kinematical supersymmetries of which 4 supersymmetries are generated by S_0^- and another 4 supersymmetries are generated by S_0^+. On the other hand, an open string on a D-brane satisfying $\Gamma^T = -\Gamma$ preserves no kinematical supersymmetry.

Now we investigate the dynamical supersymmetry preserved by an open string on a D-brane characterized by Γ in Eq. (14). The dynamical supercharge of an open string is given by a combination of those of a closed string compatible with the open string boundary conditions. Due to the boundary condition (8), it turns out that the conserved dynamical supercharge is given by (a subset of)

$$q^- = Q^{-1} - \Omega Q^{-2}. \tag{23}$$

Using the similar recipe used in the kinematical supersymmetry, it is not difficult to show that the dynamical supercharge density q_τ^- in Eq. (23) also satisfies the conservation law

$$\frac{\partial q_\tau^-}{\partial \tau} + \frac{\partial q_\sigma^-}{\partial \sigma} = 0, \tag{24}$$

where

$$\begin{aligned}
q_\sigma^- = \sqrt{\frac{1}{2p^+}} \Big(& (\partial_\tau X^r \gamma^r - \partial_\sigma X^{r'} \gamma^{r'})(S^1 - \Omega S^2) \\
& + (\partial_\tau X^{r'} \gamma^{r'} - \partial_\sigma X^r \gamma^r)(S^1 + \Omega S^2) \\
& + \mu X^r \gamma^r \Omega \Pi (S^1 + \Gamma \Omega S^2) - \mu X^{r'} \gamma^{r'} \Omega \Pi (S^1 - \Gamma \Omega S^2) \Big).
\end{aligned} \tag{25}$$

If $\Gamma = \pm 1$, we definitely recover the $D_\pm$-brane case [16]. It was shown in [5] that D_--branes of type $(+, -, 3, 1)$ or $(+, -, 1, 3)$ with a constant worldvolume flux also preserve 16 supersymmetries whose possibility was not discussed in [16].

When $\Gamma = \Omega \Pi \Omega \Pi = 1$, there is also a new possibility for D_+-branes of type $(+, -, n, n)$ with $n = 1, 2, 3, 4$ to preserve dynamical supersymmetries by introducing a gauge field excitation, whose possibility was anticipated by Hikida and Yamaguchi [13] from general supersymmetry arguments.

D_+-branes of type $(+, -, n, n)$ with $n = 1, 2, 3, 4$ preserve 4 dynamical supersymmetries by introducing a gauge field excitation, consistent with the result in [13]. Note that the dynamical supersymmetry in this case is preserved regardless of transverse locations of D-brane.

When $\Gamma^T = -\Gamma$, Eq. (25) shows that there is no chance for $q_\sigma^- |_{\partial \Sigma}$ to vanish and thus an open string on this D-brane does not preserve any dynamical supersymmetry at all.

We summarized our results on the kinematical and dynamical supersymmetry of D-branes in Table 2. The results on the dynamical supersymmetry $D_\pm$-branes preserving 16 supersymmetries without flux have been identified in [16] and are omitted in the Table 2.

SUPERSYMMETRY OF INTERSECTING D-BRANES

We now analyze the supersymmetry of intersecting D-branes. The supersymmetry of intersecting $D_\pm$-branes was completely identified in [17] using the Green-Schwarz worldsheet formulation which can also be applied to more general class of D-branes under consideration.

In general, the unbroken supersymmetry of intersecting D-branes is the 'intersection' of supersymmetries preserved by each brane. The intersection is characterized by the projection matrices $I_\pm$ in Eq. (20).

As for the dynamical supersymmetry of intersecting D-branes the (anti-)commutation relations among $\gamma^I = \{\gamma^r, \gamma^{r'}, \gamma^i, \gamma^{i'}\}$, Ω_0 and Ω_π are useful to find conserved dynamical supersymmetries:

$$\{\gamma^r, \Omega_0\} = \{\gamma^i, \Omega_0\} = [\gamma^{r'}, \Omega_0] = [\gamma^{i'}, \Omega_0] = 0, \tag{26}$$

TABLE 3. Supersymmetry of intersecting D-branes. $v = n_{D_-} + n_{D_+}$ is the number of unbroken kinematical supersymmetry where $n_{D_-} = \frac{1}{2}\text{Tr}(1+\gamma)P_-I_\pm$ (D_--type) and $n_{D_+} = \frac{1}{2}\text{Tr}(1+\gamma)P_+I_\pm$ (D_+-type). A D-brane with a gauge field condensate is denoted by the boldface.

D-brane type	Intersection	q^+	q^-
$\Gamma_0 = \Gamma_\pi = -1$	$D_-p - D_-q$	v	$\frac{1}{2}\text{Tr}(1-\gamma)I_\pm$
	$\mathbf{D}_-5 - \mathbf{D}_-5$	v	$\frac{1}{2}\text{Tr}(1-\gamma)I_\pm$
$\Gamma_0 = \Gamma_\pi = 1$	$D_+1 - \mathbf{D}_+p$	v	$\frac{1}{2}\text{Tr}(1-\gamma)P_\pm^{D+}I_\pm$
	$\mathbf{D}_+p - \mathbf{D}_+q$	v	$\frac{1}{2}\text{Tr}(1-\gamma)P_\pm^{D+}I_\pm$
$\Gamma_0 = \Gamma_\pi = -\gamma$	$OD_-5 - OD_-5$	v	$\frac{1}{2}\text{Tr}(1-\gamma)I_\pm$
$\Gamma_0 = \Gamma_\pi = \gamma$	$OD_+5 - OD_+5$	v	0
$\Gamma_0 = \Gamma_\pi = \pm\gamma^{1256}$	$\mathbf{OD}p - \mathbf{OD}q$	v	$\frac{1}{2}\text{Tr}(1-\gamma)P_\pm^{D+}P_\pm I_\pm$
$\Gamma_0 = -1, \Gamma_\pi = -\gamma$	$D_-p - OD_-5$	v	0
$\Gamma_0 = 1, \Gamma_\pi = \gamma$	$D_+p - OD_+5$	v	0
$\Gamma_0 = \pm\gamma^{1256}, \Gamma_\pi = \pm\gamma^{3478}$	$ODp - ODq$	v	0

$$\{\gamma^r, \Omega_\pi\} = \{\gamma^{r'}, \Omega_\pi\} = [\gamma^{r'}, \Omega_\pi] = [\gamma^i, \Omega_\pi] = 0. \tag{27}$$

We summarized the supersymmetry preserved by various configurations of intersecting D-branes in Table 3. The number of each type of kinematical supersymmetries depends on the total number of ND and DN directions in a way determined by the projection matrix $I_\pm P_-$ (D_--type) or $I_\pm P_+$ (D_+-type).

It was shown in [13, 14] that the plane wave background (1) admits supersymmetric curved D-branes as well as oblique D-branes. We also classified the supersymmetric curved D-branes. The detail analysis for the case with the background fields or with the boost can be referred in [23]

DISCUSSION

We presented the classification of supersymmetric D-branes in the type IIB plane wave background using the light-cone open string theory where only longitudinal D-branes are visible. We considered various D-branes, static, curved, and with backgound fields or boost. One can generate new symmetry related D-branes which are in general time-dependent [8], using the symmetries of the action (2) and the target spacetime (1) broken by D-branes, e.g., the translation and the boost generators along the transverse directions, $P^{r'}$ and $J^{+r'}$. A rotating D-brane and a giant graviton in Penrose limit can be described by these symmetry related boundary conditions which preserve the same amount of supersymmetry [8].

Here, we studied parallel and orthogonally intersecting D-branes only. It will be straightforward to extend our analysis to D-branes intersecting at general angles [17]. Since the rotational symmetry is reduced to $SO(4) \times SO(4)'$, there are only two kinds of supersymmetric intersection at general angles, resulting in less supersymmetric D-brane configurations. The classification is extended to the case with the D-branes with the background electric or magnetic fields. The case with the boost is similar to the case of the background fields. It remains to be further studied whether we can get interesting intersecting D-branes whose dual description in terms of gauge field theories give rise practical phenomenological models.

ACKNOWLEDGMENTS

We are supported by the grant from Korean Research Foundation Grant KRF 2003-015-C00111 and by the Science Research Center Program of the Korea Science and Engineering Foundation through the Center for Quantum Spacetime(CQUeST) of Sogang University with grant number R11 - 2005 - 021.

REFERENCES

1. Blau M, Figueroa-O'Farrill J, Hull C and Papadopoulos G 2002 *J. HIgh Energy Phys.* **01** 047 , [hepth/0110242] , 2002 *Class. and Quant. Grav.* **19** L87 , [hepth/0201081]
2. Metsaev R 2002 *Nucl. Phys.* B **625** 70 [hepth/0112044]
3. Metsaev R and Tseytlin A 2002 *Phys. Rev.* D **65** 126004 [hepth/0202109]
4. Berenstein D, Maldacena J and Nastase H 2002 *J. High Energy Phys.* **04** 013 [hepth/0202021]
5. Skenderis K and Taylor M 2002 *J. High Energy Phys.* **06** 025 [hepth/0205054]
6. Bain P, Meessen P and Zamaklar M 2003 *Class. and Quant. Grav.* **20** 913 [hepth/0205106]
7. Bergman O, Gaberdiel M and Green M 2003 *J. High Energy Phys.* **03** 002 [hepth/0205183/]
8. Skenderis K and Taylor M 2003 *Nucl. Phys.* B **665** 3 [hepth/0211011]
9. Billo M and Pesando I 2002 *Phys. Lett.* B **536** 121 [hepth/0203028]
10. Dabholkar A and Parvizi S 2002 *Nucl. Phys.* B **536** 121 [hepth/0203231]
11. Gaberdiel M and Green M 2003 *Ann. phys.* (NY) **307** 147 [hepth/0211122]
12. Skenderis K and Taylor M 2003 *J. High Energy Phys.* **07** 006 [hepth/0212184]
13. Hikida Y and Yamaguchi S 2002 *J. High Energy Phys.* **04** 013 [hepth/0210262]
14. Gaberdiel M, Green M, Schafer-Nameki S and Sinha A 2003 *J. High Energy Phys.* **10** 052 [hepth/0306056]
15. Sarkissian G and Zamaklar M 2004 *J. High Energy Phys.* **03** 005 [hepth/0308174]
16. Kim J , Lee B and Yang H 2003 *Phys. Rev.* D **68** 026004 [hepth/0302060]
17. Cha K, Lee B and Yang H 2003 *Phys. Rev.* D **68** 006 [hepth/0307146]
18. Cha K, Lee B and Yang H 2004 *J. High Energy Phys.* **03** 058 [hepth/0307146]
19. Michishita Y 2002 *J. High Energy Phys.* **10** 048 [hepth/0206131]
20. Lambert N and West P 1995 *Phys. Lett.* B **459** 515 [hepth/9905031]
21. M.B. Green and M. Gutperle, 1996 *Nucl. Phys.* B **476** 484, [hepth/9604091].
22. T. Mattik, *Branes in the plane wave background with gauge field condendsates*, [hepth/0501088].
23. B.-H. Lee, Jongwon Lee, Chanyong Park, and H.S. Yang, 2006 *J. High Energy Phys.* **01** 015 [hepth/0506091].

Twistor Strings and Supergravity

Mohab Abou-Zeid

*Theoretische Natuurkunde, Vrije Universiteit Brussel & The International Solvay Institutes,
Pleinlaan 2, 1050 Brussels, Belgium*

Abstract. Einstein gravity can be formulated in such a way that it leads to a perturbation theory about an asymmetric weak coupling limit that treats positive and negative helicities differently. The power counting rules for scattering amplitudes then suggest an interpretation in terms of a twistor string theory for gravity, with amplitudes supported on holomorphic curves in twistor space. After reviewing this formulation, I survey the recent construction of a family of new twistor string theories which are free from world-sheet anomalies and give the space-time spectra of Einstein supergravities, with second order field equations instead of the higher derivative conformal supergravities that arose from earlier twistor strings.

Keywords: Classical Theories of Gravity, Topological Strings, Superstrings and Heterotic Strings
PACS: 04.20.-q, 11.25.-w

INTRODUCTION

The string theories in twistor space proposed by Witten and by Berkovits [1, 2, 3] give a formulation of $N = 4$ supersymmetric Yang-Mills theory coupled to conformal supergravity. They provide an elegant derivation of a number of remarkable properties exhibited by the scattering amplitudes of these theories, giving important results for super-Yang-Mills tree amplitudes in particular. However, in these theories the conformal supergravity is inextricably mixed in with the gauge theory so that, in computations of gauge theory loop amplitudes, conformal supergravity modes propagate on internal lines [4]. There appears to be no decoupling limit giving pure super-Yang-Mills amplitudes, and although there has been considerable progress in studying the twistor-space Yang-Mills amplitudes at loops (see e. g. [5] and references therein), the results do not follow from the known twistor strings. A twistor string that gave Einstein supergravity coupled to super-Yang-Mills would be much more useful, and might be expected to have a limit in which the gravity could be decoupled to give pure gauge theory amplitudes. (By Einstein supergravity, we mean a supergravity with 2nd order field equations for the graviton, in contrast to conformal supergravity which has 4th order field equations.) Indeed, it is known that MHV amplitudes for Einstein (super)gravity [6] also have an elegant formulation in twistor space [1], suggesting that they could also arise from a twistor string theory.

Tree-level MHV amplitudes for (super)Yang-Mills theory [7, 8] have an elegant formulation in twistor space [9], and Witten considered the extension of this to general amplitudes in [1], where it was conjectured that amplitudes are non-zero only if all the external particles in a scattering process are represented by points in twistor space that lie on an algebraic curve of degree d given by

$$d = q - 1 + l, \tag{1}$$

where q is the number of negative helicity particles and l is the number of loops. An interesting way of understanding (1) in gauge theory [1] is that it follows naturally from the perturbation theory of the Chalmers-Siegel chiral formulation of Yang-Mills theory [10], in which positive and negative helicities are treated very differently. Moreover, the Chalmers-Siegel formulation

CP881, *Cairo International Conference on High Energy Physics (CICHEP II)*,
edited by S. Khalil

is precisely the form of the gauge theory that arises from the twistor string of [1]. There are analogous chiral formulations of gravity (which are reviewed in the next section), and it turns out that perturbation theory about them again leads to the relation (1), suggesting that such chiral formulations of gravity might arise from twistor string theories.

Recently, a family of new twistor string models which give Einstein (super)gravity coupled to Yang-Mills was constructed [11]; the results will be surveyed here. The new twistor string theories are constructed by gauging certain symmetries of the Berkovits twistor string, and they are free of world-sheet anomalies. Their structure is very similar to that of the Berkovits model, but the gauging adds new terms to the BRST operators so that the vertex operators have new constraints and gauge invariances.

The theories of [1, 2, 3] give target space theories that are anomalous in general, with the anomalies canceling only for 4-dimensional gauge groups. These anomalies presumably arise from inconsistencies in the corresponding twistor string model, but the mechanism for this is as yet unknown [4]. If there are such inconsistencies in the Berkovits twistor string that only cancel in special cases, there should be similar problems for the gauged models, and this may rule out some of the models, or restrict the choice of gauge group. This issue will not be discussed further here, but we hope to return to it elsewhere.

In [11] two classes of theories free from world-sheet anomalies were found. The first is formulated in $N = 4$ supertwistor space. Gauging a symmetry of the string theory generated by one bosonic and four fermionic currents gives a theory with the spectrum of $N = 4$ Einstein supergravity coupled to $N = 4$ super-Yang-Mills with arbitrary gauge group, while gauging a single bosonic current gives a theory with the spectrum of $N = 8$ Einstein supergravity, provided the number of $N = 4$ vector multiplets is six. In the Yang-Mills sector, the string theory is identical to that of Berkovits, so that it gives the same tree level Yang-Mills amplitudes. Both theories have the MHV 3-graviton interaction (with two positive helicity gravitons and one negative helicity one) of Einstein gravity. In the second class of string theories, gauging different numbers of bosonic and fermionic symmetries allows anomalies to be cancelled against ghost contributions for strings in twistor spaces with 3 complex bosonic dimensions and any number N of complex fermionic dimensions, corresponding to theories in four-dimensional space-time with N supersymmetries. The spectrum of states arising from ghost-independent vertex operators is as follows [11]. For $N = 0$, there is a theory with the bosonic spectrum of self-dual gravity together with self-dual Yang-Mills and a scalar, while for $N < 4$ there are supersymmetric versions of this self-dual theory. For $N = 4$, there is a theory whose spectrum is that of $N = 4$ Einstein supergravity coupled to $N = 4$ super-Yang-Mills with arbitrary gauge group.

A key difference between the models found in [11] and the twistor strings of refs. [1, 2, 3] is that space-time conformal invariance is broken. Recall that Penrose's non-linear graviton construction [12] provides an equivalence between 4-dimensional space-times $\mathcal{M}$ with self-dual Weyl curvature and certain complex 3-folds, the curved projective twistor spaces $P\mathcal{T}$ (for flat space-time, the corresponding twistor space $\mathbb{PT}$ is $\mathbb{CP}^3$). This gives an implicit construction of general conformally self-dual space-times. A special case of the conformally self-dual spaces are those that are Ricci-flat, so that the full Riemann tensor is self-dual. The corresponding twistor spaces $P\mathcal{T}$ then have extra structure. In particular, they have a fibration $P\mathcal{T} \rightarrow \mathbb{CP}^1$. The holomorphic one-form on $\mathbb{CP}^1$ pulls back to give a holomorphic one-form on $P\mathcal{T}$ which takes the form $I_{\alpha\beta}Z^\alpha dZ^\beta$ in homogeneous coordinates Z^α, for some $I_{\alpha\beta}(Z) = -I_{\beta\alpha}(Z)$ (which are the components of a closed 2-form on the non-projective twistor space $\mathcal{T}$). The dual bi-vector $I^{\alpha\beta} = \frac{1}{2}\varepsilon^{\alpha\beta\gamma\delta}I_{\gamma\delta}$ defines a Poisson structure and is called the *infinity twistor*. The models of [11] are modifications of the Berkovits twistor string which introduce explicit dependence on the infinity twistor, such that there are extra constraints and gauge invariances relating them to Einstein supergravities instead of conformal supergravities. Now the magnitude of the infinity twistor defines a length scale in space-time, and so determines the gravitational coupling κ. The twistor string theories of ref. [11] therefore have *two* independent coupling constants: the gravitational coupling κ, determined by the magnitude of the infinity twistor, and the Yang-Mills coupling g_{YM}, arising as in [4]. Then for the $N = 4$ theory there is a limit in which $\kappa \to 0$ and supergravity decouples from the super-Yang-Mills theory, so that, if the twistor string theory is consistent at loops, it will have a decoupling limit that gives $N = 4$ super-Yang-Mills loop amplitudes.

FORMULATIONS OF GRAVITY

Consider the formulation of Einstein gravity in terms of a vierbein $e^a{}_\mu$ and spin-connection $\omega_\mu{}^{bc}$, with corresponding one-forms e^a, ω^{bc}. In a second order formalism, one imposes the constraint that the torsion 2-form $T^a = 0$, and this determines the spin-connection in terms of the vierbein as $\omega_{\mu ab} = \Omega_{\mu ab}(e)$ where $\Omega_{\mu ab}(e)$ is the usual expression for the Lorentz connection in terms of the vierbein. The Einstein-Hilbert action is given in terms of the curvature 2-form R^{ab} by

$$\frac{1}{4\kappa^2} \int e^a \wedge e^b \wedge R^{cd}(\omega)\varepsilon_{abcd}. \tag{2}$$

The same action can be used in the first order formalism, in which the torsion is unconstrained and the vierbein $e_\mu{}^a$ and the connection $\omega_\mu{}^{ab}$ are treated as independent variables.

In Euclidean signature $(4,0)$, the spin group factorises as $Spin(4) = SU(2) \times SU(2)$ while in split signature it factorises as $Spin(2,2) = SU(1,1) \times SU(1,1)$. The spin-connection decomposes into the self-dual piece $\omega^{(+)ab}$ and the anti-self-dual piece $\omega^{(-)ab}$, which are the independent gauge fields for the two factors of the spin group. In 2-component spinor notation, where A, B transform under the first $SU(2)$ or $SU(1,1)$ factor and $\dot{A}, \dot{B}$ transform under the second, $\omega^{(+)ab}$ becomes ω^{AB} and $\omega^{(-)ab}$ becomes $\omega^{\dot{A}\dot{B}}$. The curvature 2-form can also be split into its self-dual and anti-self-dual pieces, and it is easily seen that $R^{(+)ab}$ depends only on $\omega^{(+)}$ while $R^{(-)ab}$ depends only on $\omega^{(-)}$. An equivalent form of the Einstein-Hilbert action (2) is given using $R^{(+)}$ instead of R by [16, 17]

$$\frac{1}{2\kappa^2} \int e^a \wedge e^b \wedge R^{(+)}_{ab}(\omega). \tag{3}$$

This gives the action (2) plus a topological term which can be written as

$$\frac{1}{2\kappa^2} \int d(T^a \wedge e_a). \tag{4}$$

Now (4) vanishes in the second order formalism in which one sets $T^a = 0$, and in the first order formalism is a total derivative that does not contribute to the field equations or Feynman diagrams. As $R^{(+)}$ depends only on $\omega^{(+)}$, the action (3) is independent of $\omega^{(-)}$ and depends only on the vierbein and the self-dual spin-connection. Moreover, the first order action is polynomial in these variables.

It is remarkable that one only needs the self-dual part of the spin-connection in order to formulate gravity. The torsion constructed from $e, \omega^{(+)}$ is

$$\tilde{T}^a = de^a + \omega^{(+)a}{}_b \wedge e^b. \tag{5}$$

If one imposes the constraint $\tilde{T}^a = 0$, one obtains

$$\omega^{(+)ab} = \Omega^{(+)ab}(e), \qquad \Omega^{(-)ab}(e) = 0. \tag{6}$$

This implies that

$$R^{(-)ab}(e) = 0, \tag{7}$$

where $R^{(-)ab}(e)$ is the anti-self-dual part of the curvature for the connection $\Omega(e)$. Then the Riemann curvature constructed from the vierbein is self-dual and hence Ricci-flat, so that the torsion constraint $\tilde{T}^a = 0$ imposes the field equations of self-dual gravity as well as solving for the spin-connection in terms of the vierbein [21].

Siegel [21] gave a remarkable asymmetric action for gravity that is analogous to the Chalmers-Siegel gauge theory action [10] by introducing a Lagrange multiplier field to impose the constraint $\tilde{T} = 0$. In the second order formalism, $\omega^{(+)ab}$ is given in terms of e by $\omega^{(+)ab} = \Omega^{(+)ab}(e)$ and the remaining part of $\tilde{T}^a = 0$ is imposed by a Lagrange multiplier $\sigma_\mu{}^{(-)ab}$ which is anti-self-dual, or in spinor notation $\sigma_\mu{}^{\dot{A}\dot{B}}$. This has the same index structure as the missing anti-self-dual spin-connection. Siegel's action can be written as

$$\int \sigma^{\dot{A}\dot{B}} \wedge \tilde{T}^A{}_{\dot{A}} \wedge e_{A\dot{B}}. \tag{8}$$

Varying σ imposes the self-dual gravity equation (6) so that e represents a graviton of helicity -2. Varying e gives

$$d\sigma^{\dot{A}\dot{B}} \wedge e_{A\dot{B}} = 0, \tag{9}$$

so that the Ricci tensor constructed from the linearised curvature $d\sigma$ for an anti-self-dual connection σ vanishes, and the Lagrange multiplier field represents a graviton of helicity $+2$. This action then describes particles of helicity ± 2, as in Einstein's theory, but the interactions are different for the two helicities, and in particular the theory is linear in σ. There is also a first-order form of this theory, in which $\omega^{(+)ab}$ is an independent field and a Lagrange multiplier is introduced to impose the full constraint $\tilde{T} = 0$ [21], which in turn implies the field equations [21]. Siegel also generalised (8) to give an asymmetric form of $N = 8$ supergravity, with Lagrange multipliers imposing torsion constraints of the supergravity theory [21].

Siegel's asymmetric theory of gravity can be put in a different form that arises as a weak-coupling limit of the Einstein theory, and gives a chiral perturbation theory of gravity similar to that arising from the Chalmers-Siegel action [15]. The gravity action (3) depends on the vierbein and $\omega^{(+)}$ only (and $\omega^{(-)}$ decouples completely); from now on we omit the superscript $(+)$ so that $\omega \equiv \omega^{(+)}$. Rescaling the connection by the gravitational coupling κ^2, we can write (3) in the form

$$\frac{1}{2} \int e^a \wedge e^b \wedge \left(d\omega_{ab} + \kappa^2 \omega_{ac} \wedge \omega^c{}_b \right). \tag{10}$$

Varying (10) independently with respect to $\omega^a{}_b$ and e^a_μ yields (6) and the Einstein equation

$$e^a \wedge \left(d\omega_{ab} + \kappa^2 \omega_{ac} \wedge \omega^c{}_b \right) = 0. \tag{11}$$

Taking the limit $\kappa \to 0$ in (10) yields a weak-coupling limit of gravity with action

$$\frac{1}{2} \int e^a \wedge e^b \wedge d\omega_{ab} = -\int e^a \wedge e^b \wedge \omega_{ac} \wedge \Omega^{(+)c}{}_b, \tag{12}$$

where $\Omega^{(+)} = \Omega^{(+)}(e)$ is the self-dual part of $\Omega(e)$. This is an action for two independent fields, the vierbein $e_\mu{}^a$ and the self-dual connection ω^{ac}; the latter now plays the role of a Lagrange multiplier field. Note that the self-duality of ω^{ac} implies that only the self-dual part $\Omega^{(+)}$ of $\Omega(e)$ occurs in the action. The field equation from varying the Lagrange multiplier field ω^{ac} sets the self-dual part of $\Omega(e)$ to zero,

$$\Omega^{(+)a}{}_b(e) = 0. \tag{13}$$

This implies that the self-dual part of the curvature constructed from the Levi-Civita connection $\Omega(e)$ vanishes:

$$R_{\mu\nu}{}^{ab}(\Omega^{(+)}) = 0, \tag{14}$$

so that the vierbein gives a metric with anti-self-dual Riemann curvature. The field equation for the vierbein gives

$$e_b \wedge d\omega^{ab} = 0. \tag{15}$$

Comparing with (11), this can be seen to be a version of the Einstein equation linearised around the anti-self-dual background spacetime described by the tetrad e^a, where the linearised graviton field is the self-dual connection ω^{ac}.

The fact that ω^{ab} and $\Omega^{(-)}(e)$ are respectively self-dual and anti self-dual means that they describe particles of opposite helicity: e describes a particle of helicity $+2$ and ω describes a particle of helicity -2. The linearized spectrum is the same for (12) and (2), but the interactions differ as (12) has no $--+$ vertex. The asymmetric theory (12) is equivalent to the Siegel theory.

The form of the action (10) has the weak-coupling limit (12), and one can consider perturbation theory in κ^2 about this weak-coupling limit, in complete analogy with the case of gauge theory [15]. As in that case [10, 1], it is useful to attribute to the independent fields e^a and ω^{ab} the weights $w[e] = 0$ and $w[\omega] = -1$ under a $U(1)$ transformation, corresponding to the 'anomalous' $U(1)$ R-symmetry S in the $N = 8$ supersymmetric extension of action (8) [21], with $S = 8w$. The chiral action (3) has weight $w = -1$ and the second term in (10) has weight $w = -2$. It is tempting to conjecture that the $S = -16$ interaction term is related to nonperturbative contributions in a new twistor string theory for Siegel's truncation of $N = 8$ supergravity. If one attributes weights $w = 1$ to κ^2 and $w = -1$ to the Planck constant $\hbar$, then the action rescaled by $1/\hbar$ has weight $w = 0$. The generating functional of scattering matrix elements at l-loops must be a sum of terms of the form

$$\hbar^{l-1} \tilde{f}(e) \kappa^{2d} \omega^q \tag{16}$$

for some function $\tilde{f}$, and for this to have total weight zero Witten's relation (1) must hold for each term in the effective action. The power q of ω is the number of negative helicity gravitons in an l-loop scattering process in the theory defined by (10). If the theory has a twistor string origin similar to that of [1], then the power d of κ^2 might arise as an instanton number, and the scattering would have support on curves in twistor space characterised by the integer d.

This formulation of gravity extends to one of $N = 8$ supergravity in which the Einstein term is written in the form (10) and the vector field kinetic terms take the Chalmers-Siegel form of the Yang-Mills action. In the weak coupling limit, it gives Siegel's chiral $N = 8$ supergravity [21].

Given the close analogy with the case of gauge theory [1], it is natural to conjecture that the gravity action (10), or its $N = 8$ supergravity generalisation, should also have an elegant twistor theory origin, and that the formula (1) has a similar twistor space interpretation.

THE NON-LINEAR GRAVITON AND THE INFINITY TWISTOR

Penrose's non-linear graviton construction provides a correspondence between curved twistor spaces and conformally anti-self-dual space-times, and so gives a general construction of such space-times. This arises from nontrivial deformations of the flat twistor correspondence in which, on the one hand, the space-time is deformed from flat space to one with a curved conformal structure with anti-self-dual Weyl curvature, and, on the other, the complex structure of a region in twistor space is deformed away from that of a region in projective space. One cannot deform the complex structure of the whole of flat twistor space as $P\mathcal{T}_0 = \mathbb{CP}^3$ is rigid and has no continuous deformations [18], but one can deform the complex structure of a suitable region of $P\mathcal{T}_0$ (such as a neighbourhood $\mathbb{PT}_0$ of a projective line in $\mathbb{CP}^3$). The complex structure of a space can be specified by a (1,1) tensor field J satisfying $J^2 = -1$ that is integrable, so that the Nijenhuis tensor $N(J)$ vanishes. Given the complex structure J_0 of $\mathbb{PT}_0$, one can construct a new complex structure

$$J = J_0 + \lambda J_1 + \lambda^2 J_2 + \dots \tag{17}$$

as a power series in a parameter λ, imposing the conditions $J^2 = -1$ and $N(J) = 0$. In holomorphic coordinates for J_0, $J^2 = 0$ implies that J_1 decomposes into a section j of $\Lambda^{(0,1)} \otimes T^{(1,0)}$ and its complex conjugate on $P\mathcal{T}_0$. The linearised condition $N(J) = 0$ is equivalent to $\bar{\partial}j = 0$. Furthermore, j represents an infinitesimal diffeomorphism if $j = \bar{\partial}\alpha$ for some section α of $T^{(1,0)}$. Thus a deformation corresponds to an element of the first Dolbeault cohomology group on twistor space with values in the holomorphic tangent bundle. Moreover, the linearised deformations J_1 are unobstructed to all orders and determine the tangent space to the moduli space of complex structures if certain second cohomology groups vanish, which they do when $\mathbb{PT}_0$ is a small enough neighbourhood of a line.

Witten's twistor string [1] is a topological string theory on (super)twistor space and has physical states corresponding to deformations of the complex structure of the target space $P\mathcal{T}_0$. The corresponding vertex operator constructed from J_1 is physical precisely when j represents an element of $H^1_{\bar{\partial}}(\mathbb{PT}_0)$. The twistor space string field theory action for Witten's theory has a term with a Lagrange multiplier imposing $N(J) = 0$ [4] and the corresponding term in the space-time action is

$$\int d^4x \sqrt{g} U^{ABCD} W_{ABCD}, \tag{18}$$

where W_{ABCD} is the anti-self-dual part of the Weyl tensor. If this were the complete gravity action, then U^{ABCD} would be a Lagrange multiplier imposing the vanishing of W_{ABCD}, so that the Weyl tensor would be self-dual. However, in addition there is a term $\int U^2$, which arises from D-instantons in Witten's topological B-model [4]. Integrating out U gives the conformal gravity action $\int W^2$.

In split $++--$ space-time signature, there is a three real dimensional submanifold $P\mathcal{T}_\mathbb{R}$ of complex twistor space $P\mathcal{T}$. In the flat case, $\mathbb{PT}_\mathbb{R} \subset \mathbb{PT}$ is the standard embedding of $\mathbb{RP}^3 \subset \mathbb{CP}^3$, and the information about deformations of the complex structure is encoded in an analytic vector field f on $P\mathcal{T}_\mathbb{R}$. Conformally self-dual space-times in split signature can also be constructed by deforming the embedding of $\mathbb{PT}_\mathbb{R}$ to some $P\mathcal{T}_\mathbb{R}$ in $\mathbb{PT}$ instead of deforming the complex structure of some region in $\mathbb{PT}$ to give $P\mathcal{T}$. The deformations of the anti-self-dual conformal structure correspond to deformations of the embedding of $P\mathcal{T}_\mathbb{R}$ in $\mathbb{CP}^3$ and are determined at first order by a vector field f on $P\mathcal{T}_\mathbb{R}$, or more precisely by a section of the normal bundle to $P\mathcal{T}_\mathbb{R} \subset \mathbb{CP}^3$. Berkovits' twistor string [2, 3] has open strings with boundaries on the real twistor space $\mathbb{PT}_\mathbb{R}$, and (conformal) supergravity physical states are created by an open string vertex operator constructed from a vector field f defined on $\mathbb{PT}_\mathbb{R}$, corresponding to deformations of the embedding of $\mathbb{PT}_\mathbb{R}$ in $\mathbb{PT}$.

There is an important variant of the Penrose construction that applies to the Ricci-flat case (in fact, this is the original non-linear graviton construction). A special case of the conformally self-dual spaces are those that are Ricci-flat, so that the full Riemann tensor is self-dual. The corresponding twistor spaces $P\mathcal{T}$ then have extra structure [20], as mentioned in the introduction; in particular, they have a fibration $P\mathcal{T} \to \mathbb{CP}^1$. Consider for example flat space-time $\mathcal{M} = \mathbb{R}^4$ in signature $+++++$, which has conformal compactification S^4 (this case is more straightforward to describe than that of split signature, which is explained in detail in [11]). The twistor space is $\mathbb{CP}^3$, which is a $\mathbb{CP}^1$ bundle over S^4: it is the projective primed spin bundle over the conformal compactification of $\mathcal{M}$. If conformal invariance is broken, then there is a distinguished point at infinity. Removing the point at infinity from S^4 to leave $\mathbb{R}^4$ amounts to removing the fibre over this point in the twistor space, leaving $\mathbb{PT}' = \mathbb{CP}^3 - \mathbb{CP}^1$, the projective primed spin bundle over $\mathbb{R}^4$. However, $\mathbb{PT}'$ is also a bundle over $\mathbb{CP}^1$ with fibres $\mathbb{C}^2$, the planes through the missing $\mathbb{CP}^1$. A projective line joining two points X^α and Y^β in twistor space can be represented by a bivector $X^{[\alpha}Y^{\beta]}$, and the infinity twistor is the bivector corresponding to the projective line over the point at infinity in S^4. Choosing a point at infinity, or an infinity twistor, breaks the conformal group down to the Poincaré group. For Minkowski space, the infinity twistor determines the light-cone at infinity in the conformal compactification. A similar situation obtains more generally: the infinity twistor breaks conformal invariance.

The self-dual Ricci-flat space-times are obtained by seeking deformations of the complex structure of twistor space as in the generic case, but now the Ricci-flatness condition places further restrictions on the deformations allowed. In the split signature setup [11], the vector field f on $\mathbb{RP}^3$ is required to be a Hamiltonian vector field with respect to the infinity twistor, so that in homogeneous coordinates we can write

$$f^\alpha = I^{\alpha\beta} \frac{\partial h}{\partial Z^\beta} \tag{19}$$

for some function h of homogeneity degree 2 on $\mathbb{RP}^3$. In the linearised theory, such a function h corresponds to a positive-helicity graviton in space-time via the Penrose transform, and the non-linear graviton construction gives the generalisation of this to the non-linear theory [19]. In the Dolbeault picture, the tensor J_1 is given by a $(0,1)$-form j^α of the form

$$j^\alpha = I^{\alpha\beta} \frac{\partial h}{\partial Z^\beta} \tag{20}$$

where h is a $(0,1)$-form representing an element of $H^1(\mathbb{PT}', \mathcal{O}(2))$.

These facts suggest seeking a twistor string that is a modification of either the Berkovits or the Witten string theories which introduces explicit dependence on the infinity twistor, such that there are extra constraints on the vertex operators imposing that the deformation of the complex structure be of the form (19). Then the leading term in the corresponding space-time action should have a term with a Lagrange multiplier imposing self-duality, not just conformal self-duality, and further terms quadratic in the multiplier (from instantons in Witten's approach) could then give Einstein gravity. The formulation of Einstein gravity reviewed in the previous section is of just this form. The twistor string theories of [11] are modifications of the Berkovits twistor string with the desired properties. We expect that similar refinements of Witten's twistor string should are also possible; this will be discussed elsewhere.

GAUGED BERKOVITS TWISTOR STRING THEORIES

The key ingredient in the construction of the new twistor string models [11] is that the 1-form on twistor space

$$k = I_{\alpha\beta} Z^\alpha dZ^\beta \tag{21}$$

corresponding to the infinity twistor is used to construct a current, and the corresponding symmetry is gauged. In the supersymmetric setup, supertwistor space $P\mathscr{T}$ is fibred over $\mathbb{CP}^{1|0}$ or $\mathbb{CP}^{1|N}$, and in the latter case a local basis of N fermionic 1-forms on $\mathbb{CP}^{1|N}$ pull back to N locally defined fermionic 1-forms k^i on super-twistor space. The resulting gauge-fixed theory is given by the Berkovits twistor string theory plus some extra ghosts, and there are extra terms in the BRST operator involving these ghosts. The dynamics and vertex operators are of the same form as for the Berkovits twistor string, but the extra terms in the BRST charges give extra constraints and gauge invariances for the vertex operators, including the constraint which reduces conformal gravity to Einstein gravity. Variants of the theory are obtained by also gauging some fermionic currents constructed from the k^i.

Gauging different numbers of bosonic and fermionic symmetries allows anomalies to be cancelled against ghost contributions for strings in twistor spaces with 3 complex bosonic dimensions and any number N of complex fermionic dimensions, corresponding to theories in four-dimensional space-time with N supersymmetries. The spectrum of states arising from ghost-independent vertex operators is as follows. For $N = 0$, there is a theory with the bosonic spectrum of self-dual gravity together with self-dual Yang-Mills and a scalar, and for $N \leq 4$ there are supersymmetric versions of this self-dual theory. As twistor theory has been particularly successful in formulating self-dual gravity [12] and self-dual Yang-Mills [13], it seems fitting that these theories should emerge from twistor string theory. It is intriguing that some of the theories found in [11] have similar structure to $\mathscr{N} = 2$ string theories [14]. The case of $N = 4$ is particularly interesting as in that case the spectrum is parity invariant and is that of $N = 4$ Einstein supergravity coupled to $N = 4$ super-Yang-Mills with arbitrary gauge group. The $N = 4$ theory has a decoupling limit giving pure Yang-Mills, opening the prospect of a twistor string formulation of super-Yang-Mills loop amplitudes.

The form (21) is of weight $w = 2$, but a 1-form of general weight w can be made by multiplying by a function $v(Z)$ of weight $w - 2$ (so that w is a section of $\mathscr{O}(w-2)$) to give

$$\hat{k} = v(Z) I_{\alpha\beta} Z^\alpha dZ^\beta. \tag{22}$$

Similarly, fermionic 1-forms $\hat{k}^i$ of general weigths w_i can be obtained by multiplying the k^i by suitable functions. (In general the multiplier functions needed in this construction can have singularities when $w < 2$, leading to potential problems, but there are non-singular functions of negative weigths in e. g. the case of real twistor spaces [11]). It turns out that gauging the symmetries associated with such 1-forms of general weigths gives many new twistor string models which formally are free from world-sheet anomalies, and for which the specrum and interactions are independent of the multiplier functions provided these are chosen to have no zeroes or poles [11]. An important special case of the construction consists in choosing the multipliers in such a way that all 1-forms $\hat{k}, \hat{k}^i$ have weights 0. In particular, there are two models of special interest that are formulated in $N = 4$ supertwistor space. Gauging a symmetry of the string theory generated by the bosonic current constructed from the $N = 4$ generalisation of (22) and by four fermionic currents constructed from the 1-forms $\hat{k}^i$ gives a theory with the spectrum of $N = 4$ Einstein supergravity coupled to $N = 4$ super-Yang-Mills with arbitrary gauge group, while gauging only that same bosonic current gives a theory with the spectrum of $N = 8$ Einstein supergravity, provided the number of $N = 4$ vector multiplets is six. In the Yang-Mills sector, the string theory is identical to that of Berkovits, so that it gives the same tree level Yang-Mills amplitudes. Both theories have the MHV 3-graviton interaction (with two positive helicity gravitons and one negative helicity one) of Einstein gravity.

Thus for $N = 4$, there are two twistor theories, both of which have the spectrum of $N = 4$ supergravity coupled to $N = 4$ super-Yang-Mills. One is the theory obtained by gauging 1-forms with nonzero weigths and the other is the theory obtained by gauging weightless forms. The supergravity sector in both cases has the spectrum of $N = 4$ Einstein supergravity, and in the latter case there is at least one non-trivial interaction [11]. However, there are a number of different supersymmetric theories with this spectrum, and a key question is whether the interactions which arise in the twistor string theories are those of the standard non-chiral $N = 4$ supergravity theory, or those of an $N = 4$ theory with chiral interactions generalising the chiral formulation of Einstein gravity presented in the second section of this paper (there is also the possibility of a free theory). Answering this question requires further analysis of the scattering amplitudes, and we will return to this elsewhere. However, the theory based on weigthless 1-forms has the usual non-chiral Yang-Mills interactions and has a non-trivial cubic gravitational coupling, so it is presumably the full Yang-Mills theory coupled to either chiral or non-chiral $N = 4$ supergravity. The usual non-chiral interacting $N = 4$ supergravity coupled to Yang-Mills theory has no anomalies, but it is expected to have ultra-violet divergences. Nonetheless, it has a limit in which gravity decouples to leave $N = 4$ super-Yang-Mills, and this is believed to be a consistent ultra-violet finite field theory. The theory of chiral $N = 4$ supergravity coupled to $N = 4$ super-Yang-Mills is likely to have better ultra-violet behaviour than the full supergravity (and might conceivably be finite) and it has a similar decoupling limit so that, whichever supergravity theory arises, there should be a decoupling limit giving pure $N = 4$ super-Yang-Mills amplitudes. This limit in the twistor theory is given by scaling the infinity twistor so that $I^{\alpha\beta} \to 0$. Then from (19), for any supergravity wave-function h, the corresponding f^α will vanish and so any amplitude involving h will vanish. It will be interesting to check that this leads to a full decoupling of gravity at all orders in perturbation theory. There is then the intriguing possibility that this twistor string can give $N = 4$ super-Yang-Mills in this limit.

The theory based on gauging the current constructed from the $N = 4$ extension of the 1-form (22) gives the spectrum of $N = 4$ supergravity plus four $N = 4$ gravitino multiplets, together with super-Yang-Mills for any gauge group. There are then 8 gravitini of helicity $+3/2$ and 8 gravitini of helicity $-3/2$, so that the theory should be an $N = 8$ supergravity theory. Again, there is the possibility of either a theory with chiral interactions, or a non-chiral one. If it is a standard non-chiral $N = 8$ supergravity, the total number of vector fields should be 28 and this requires the number of Yang-Mills multiplets to be six. This suggests that, if the twistor string gives a consistent non-chiral theory, there must be a constraint fixing the number of vector multiplets to be 6. The Berkovits string is expected to have a constraint fixing the number of vector multiplets to be 4, to cancel the anomalies of conformal supergravity [4], and both constraints could have the same (as yet unknown) origin. Alternatively, the theory arising could be Siegel's chiral $N = 8$ supergravity [21], in which the negative helicity fields appear linearly. In [21], Siegel argued that the $\mathcal{N} = 2$ string gives $N = 4$ chiral Yang-Mills from the open string sector and $N = 8$ chiral supergravity from the closed string sector, and that the chirality of the interactions implied that the supergravity and super-Yang-Mills fields do not couple, so that one can consistently have $N = 8$ chiral supergravity and an arbitrary number of $N = 4$ chiral Yang-Mills multiplets. It will be interesting to see whether either of these interacting $N = 8$ supergravity theories arise here. If the space-time theories arising from the perturbative string theory are chiral supergravities, then it is possible that non-perturbative effects could give rise to the non-chiral interactions, as they do for Yang-Mills in Witten's topological twistor string [1].

ACKNOWLEDGEMENTS

The author thanks Christopher Hull and Lionel Mason for very enjoyable collaborations, and the organizers of CICHEP II for hospitality in Cairo and financial support. This work was supported in part by a PPARC Postdoctoral Research Fellowship with grant reference PPA/P/S/2000/00402, by the 'FWO-Vlaandere' through projects G.0034.02 and G.0428.06, by the Belgian Federal Science Policy Office through the Interuniversity Attraction Pole P5/27, and by the European Union FP6 RTN programme MRTN-CT-2004-005104.

REFERENCES

1. E. Witten, *Perturbative gauge theory as a string theory in twistor space*, Commun. Math. Phys. **252** (2004), [arXiv:hep-th/0312171].
2. N. Berkovits, *An alternative string theory in twistor space for N = 4 super-Yang-Mills*, Phys. Rev. Lett. **B93** 011601 (2004), [arXiv:hep-th/0402045].
3. N. Berkovits and L. Motl, *Cubic twistorial string field theory*, JHEP **0404** (2004) 056, [arXiv:hep-th/0403187].
4. N. Berkovits and E. Witten, *Conformal supergravity in twistor-string theory*, JHEP **0408** 009 (2004), [arXiv:hep-th/0406051].
5. F. Cachazo and P. Svrček, *Lectures on twistor strings and perturbative Yang-Mills theory*, Proc. Sci. **RTN2005** (2005) 005, [arXiv:hep-th/0504194].
6. F. A. Berends, W. T. Giele and H. Kuijf, *On relations between multi-gluon and multigraviton scattering*, Phys. Lett. B **211** (1988) 91.
7. S. Parke and T. Taylor, *An amplitude for N gluon scattering*, Phys. Rev. Lett. **56** (1986) 2459.
8. F. A. Berends and W. T. Giele, *Recursive calculations for processes with N gluons*, Nucl. Phys. **B306** (1988) 759.
9. V. P. Nair, *A current algebra for some gauge theory amplitudes*, Phys. Lett. **B214** (1988) 215.
10. G. Chalmers and W. Siegel, *The self-dual sector of QCD amplitudes*, Phys. Rev. D **54** (1996) 7628 [arXiv:hep-th/9606061].
11. M. Abou-Zeid, C. M. Hull and L. J. Mason, *Einstein supergravity and new twistor string theories*, [arXiv:hep-th/0606272].
12. R. Penrose, *Nonlinear gravitons and curved twistor theory*, Gen. Rel. Grav. **7** (1976) 31.
13. R. S. Ward, *On selfdual gauge fields*, Phys. Lett. A **61** (1977) 81.
14. H. Ooguri and C. Vafa, *Geometry of N=2 strings*, Nucl. Phys. B **361**, 469 (1991).
15. M. Abou-Zeid and C. M. Hull, *A chiral perturbation expansion for gravity*, JHEP **0602**, 057 (2006), [arXiv:hep-th/0511189].
16. N. Berkovits and W. Siegel, *Covariant field theory for self-dual strings*, Nucl. Phys. B **505** (1997) 139 [arXiv:hep-th/9703154].
17. W. Siegel, *Fields*, [arXiv:hep-th/9912205].
18. K. Kodaira, *Complex manifolds and deformations of complex structures*, Springer (1986) reprinted in Classics in Mathematics (2005).
19. L. J. Mason, *The relationship between spin-2 fields, linearized gravity and linearized conformal gravity*, Twistor Newsletter **23** (1987) 67 reprinted as §I.2.18 *Further advances in twistor theory, Volume I: The Penrose transform and its applications*, L. J. Mason and L. P. Hughston eds., Pitman research notes in Maths **231**, Longman (1990).
20. S. Huggett and K. Tod, *An introduction to twistor theory*, Second Edition, Cambridge University Press, Cambridge (1994).
21. W. Siegel, *Selfdual N=8 supergravity as closed N=2 (N=4) strings*, Phys. Rev. D **47** (1993) 2504, [arXiv:hep-th/9207043].

4D Effective Theory and Geometrical Approach

A. Salvio

SISSA and INFN, Via Beirut 2-4, 34014 Trieste, Italy.

Abstract. We consider the 4D effective theory for the light Kaluza-Klein (KK) modes. The heavy KK mode contribution is generally needed to reproduce the correct physical predictions: an equivalence, between the effective theory and the D-dimensional (or *geometrical*) approach to spontaneous symmetry breaking (SSB), emerges only if the heavy mode contribution is taken into account. This happens even if the heavy mode masses are at the Planck scale. In particular, we analyze a 6D Einstein-Maxwell model coupled to a charged scalar and fermions. Moreover, we briefly review non-Abelian and supersymmetric extensions of this theory.

Keywords: Extra dimensions, Kaluza-Klein modes
PACS: 11.25.Mj, 04.50.+h.

INTRODUCTION

The low energy limit of a higher dimensional theory is usually studied by taking into account only the light mode contribution. The masses are derived from the bilinear part of the effective action and the role of the heavy modes in the actual values of the masses and the couplings of the effective theory for the light modes are seldom taken into account. However, we know that the process of "integrating out" the heavy modes [1] has the effect of modifying the couplings of the light modes or introducing additional terms that are suppressed by inverse powers of the heavy masses [2].

In a first part of this contribution we will summarize the study of the heavy mode contribution to the low energy dynamics of higher dimensional models, performed in Ref. [3]. There two methods have been used. The first one, which is called the *4D effective theory approach*, starts from a solution of a higher dimensional theory and develops an action functional for the light modes. This effective action generally has a local symmetry that should be broken by Higgs mechanism. Our interest is in the broken phase of the effective theory. The procedure is essentially what is adopted in the effective description of higher dimensional theories including superstring and M-theory compactifications. In this construction the heavy KK modes are generally ignored simply by reasoning that their masses are of the order of the compactification mass and this can be as heavy as the Planck mass.

In the second approach, which we shall call the *geometrical approach*, we shall find a solution of the higher dimensional equations with the same symmetry group as the one of the broken phase of the 4D effective theory. We shall then study the physics of the light modes around this solution. The result for the low energy physics will turn out to be different from the first approach. The difference is precisely due to the fact that in constructing the effective theory along the lines of the first approach the contribution of the heavy KK modes has been ignored.

This statement has been explicitly proved, at the classical level, for a quite general

CP881, *Cairo International Conference on High Energy Physics (CICHEP II),*
edited by S. Khalil
© 2007 American Institute of Physics 978-0-7354-0382-6/07/$23.00

higher dimensional scalar model with lagrangian $\mathscr{L} = -\frac{1}{2}\partial_M\Phi\partial^M\Phi + V(\Phi)$, where Φ is a set of scalar fields, and then extended to a more interesting (Abelian) gauge and gravitational theory [3]. Here we will briefly report the latter case, which, in the low energy limit, reduces to a framework that is similar to the electroweak part of the Standard Model. In this framework, the heavy KK mode contribution can be geometrically interpreted as the deformation of the internal space.

Moreover, in a final section, we shall review possible extensions of these results to non-Abelian theories without fundamental scalars or to supersymmetric versions of 6D gauge and gravitational theories. In the former framework the Higgs field is identified with the internal components of the non-Abelian gauge field [4] and the complete 6D gauge symmetry relaxes the dependence of the Higgs mass on the ultraviolet cutoff. The latter class of theories can be used as toy models for string theory compactifications [5] and has shown some promise in addressing the cosmological constant problem [6].

6D EINSTEIN-MAXWELL-SCALAR MODEL

Here we analyze a 6D model, which includes the Einstein-Hilbert gravity, a Maxwell field A and a complex charged scalar ϕ. The bosonic action reads[1]

$$S_B = \int d^6X\sqrt{-G}\left[\frac{1}{\kappa^2}R - \frac{1}{4}F^2 - |\nabla\phi|^2 - V(\phi)\right].$$

where $F = dA$, $\nabla\phi = (d + ieA)\phi$. We choose $V(\phi) = m^2|\phi|^2 + \xi|\phi|^4 + \lambda$, where m^2 is a real mass squared, ξ is a real and positive parameter and λ represents a 6D cosmological constant. This system is a simple generalization of the 6D Einstein-Maxwell model of Ref. [7], where it was proved that the space-time $(Minkowski)_4 \times S^2$ is a solution of the equations of motion (EOMs), in the presence of a monopole background. This solution is

$$
\begin{aligned}
ds^2 &= \eta_{\mu\nu}dx^\mu dx^\nu + a^2\left(d\theta^2 + \sin^2\theta d\varphi^2\right), \\
A &= \frac{n}{2e}(\cos\theta \mp 1)d\varphi, \\
\phi &= 0,
\end{aligned}
\tag{1}
$$

where a is the radius of S^2 and n is the monopole number ($n = 0, \pm 1, \ldots$). Besides 4D Poincaré invariance, this background preserves an $SU(2) \times U(1)$ symmetry, which turns out to be the gauge symmetry of the low energy 4D effective theory [7]. The group factor $SU(2)$ has a geometrical origin as the isometry group of the internal space, whereas $U(1)$ represents the bulk gauge symmetry.

Moreover it is possible to introduce a couple of fermions $\psi_\pm$, with 6D chirality ± 1, and standard Yukawa couplings: $\mathscr{L}_{Yuk} = g_Y\overline{\psi_+}\psi_-\phi^\dagger + g_Y\phi\overline{\psi_-}\psi_+$, where g_Y is assumed to be real for simplicity. Furthermore we assume that the corresponding $U(1)$ fermion

---

[1] Our conventions are $\eta_{MN} = (-1, +1, +1, +1, \ldots)$ and $R_{MN} = \partial_P\Gamma^P_{MN} - \partial_M\Gamma^P_{PN} + \ldots$.

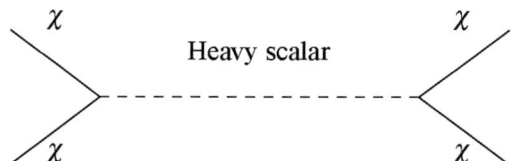

FIGURE 1. Heavy mode contribution to the quartic couplings. Reprinted from [3].

charges are $e_+ = e/2$ and $e_- = 3e/2$, because this corresponds to a simple fermion harmonic expansion over S^2.

A complete study of the fluctuations around Solution (1) shows that the low energy 4D bosonic spectrum presents the following states: the graviton, the $SU(2) \times U(1)$ gauge fields and a complex scalar field χ, coming from ϕ, in the $(|n|+1)$-dimensional representation of $SU(2)$. For example, for $n = 2$ we obtain a triplet and, henceforth, we will assume $n = 2$, as the geometrical approach turns out to be simple in this case. Moreover, in the fermion sector, we have a right-handed singlet and a left-handed triplet.

4D Higgs Mechanism. If the gauge symmetry is unbroken, the states that we have mentioned above are exactly massless, apart from χ. Indeed we can introduce a small[2] mass μ for χ by choosing a suitable value of m. In order to create a small mass for fermions and gauge fields, usually one computes an action functional for the light modes, including bilinear terms and interactions, and then studies the Higgs mechanism in the corresponding 4D theory. In our case, this can be achieved by generalizing the zero-mode ansatz method of Ref. [7], to include the light scalar χ. In particular, the lagrangian for χ turns out to be of the form $\mathscr{L}_\chi = -(D_\mu\chi)^\dagger D^\mu\chi - \mathscr{U}(\chi)$, where D_μ is the $SU(2) \times U(1)$ covariant derivative and $\mathscr{U}$ is the scalar potential for χ. In our model the $U(1)$ and $SU(2)$ gauge constants, which appear in D_μ, are respectively given by

$$g_1 = \frac{e}{\sqrt{4\pi a}}, \quad g_2 = \sqrt{\frac{3}{16\pi}\frac{\kappa}{a^2}}, \tag{2}$$

whereas the potential is $\mathscr{U}(\chi) = \mu^2\chi^\dagger\chi + \lambda_1(\chi^\dagger\chi)^2 + \lambda_2|\chi^T\chi|^2 + \ldots$, where the dots represent higher order operators and the λ_i are the quartic coupling constants allowed by the $SU(2) \times U(1)$ gauge symmetry. We have $\lambda_1 = \lambda_H + c_1\lambda_G$, and $\lambda_2 = -(\lambda_H + c_2\lambda_G)/3$, where

$$\lambda_H = \frac{9}{20\pi a^2}\xi, \quad \lambda_G = \frac{9\kappa^2}{80\pi a^4}.$$

---

[2] It is small in the sense that $\mu \ll 1/a$.

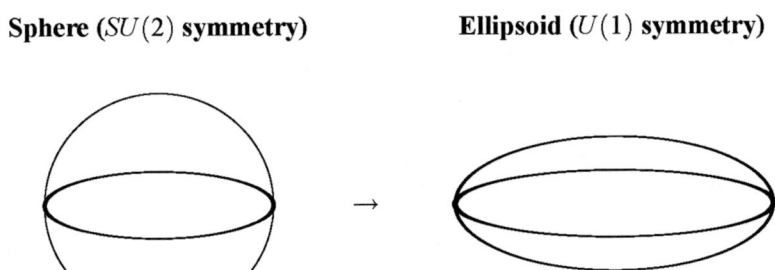

Sphere (*SU*(2) symmetry)   **Ellipsoid (*U*(1) symmetry)**

FIGURE 2. Electroweak symmetry breaking in the geometrical approach.

We observe that λ_H and λ_G represent respectively the light mode and the heavy mode contribution to λ_i. The constants c_i parametrize the latter contribution, and can be explicitly computed by evaluating diagrams of the form given in Fig. 1.

Now we focus on the SSB of $SU(2) \times U(1)$ down to $U(1)$, and we assume $c_i = 0$. In this phase ($\mu^2 < 0$), χ acquires a non-vanishing vacuum expectation value (VEV), which, for $c_i = 0$, is given by

$$|<\chi>|^2 = \frac{3}{4}\frac{-\mu^2}{\lambda_H}. \tag{3}$$

The corresponding vector, fermion and scalar spectrum, at the leading non trivial order in $\sqrt{-a^2\mu^2}$, is shown in the second column of Table 1, apart from the massless gauge field associated to the residual $U(1)$.

Geometrical Approach.   Now we want to compare the 4D effective theory with the geometrical approach to SSB. By definition the latter involves a solution of the higher dimensional EOMs that has the same symmetry as the effective theory in the broken phase. At the leading non trivial order in the small parameter $\eta^{1/2} \equiv \sqrt{-a^2\mu^2}$, we find[3]

$$
\begin{aligned}
ds^2 &= \eta_{\mu\nu}dx^\mu dx^\nu + a^2\left[(1+|\eta|\beta\sin^2\theta)d\theta^2 + \sin^2\theta d\varphi^2\right], \\
A &= \frac{1}{e}(\cos\theta \mp 1)d\varphi, \\
\phi &= \eta^{1/2}\alpha\exp(i\varphi)\sin\theta,
\end{aligned}
\tag{4}
$$

where $\beta = \kappa^2|\alpha|^2$ and

$$|\alpha|^2 = \frac{9}{32\pi a^4}\frac{1}{|\lambda_H - \lambda_G|}.$$

---

[3] This solution was discussed in Ref. [8], but incorrectly.

TABLE 1. The vector (V), fermion (F) and scalar (S) spectra.

Mass squared	4D Effective Theory	Geometrical Approach
M_V^2	$\dfrac{3e^2}{8\pi a^2}\dfrac{-\mu^2}{\lambda_H}$	$\dfrac{3e^2}{8\pi a^2}\dfrac{-\mu^2}{\lambda_H-\lambda_G}$
$M_{V\pm}^2$	$\dfrac{9e^2}{16\pi a^2}\dfrac{-\mu^2}{\lambda_H}$	$\dfrac{9e^2}{16\pi a^2}\dfrac{-\mu^2}{\lambda_H-\lambda_G}$
M_F^2	$\dfrac{3g_Y^2}{16\pi a^2}\dfrac{-\mu^2}{\lambda_H}$	$\dfrac{3g_Y^2}{16\pi a^2}\dfrac{-\mu^2}{\lambda_H-\lambda_G}$
$M_{F\pm}^2$	0	0
M_S^2	$-2\mu^2$	$-2\mu^2$
$M_{S\pm}^2$	$-\mu^2$	$-\mu^2\dfrac{\lambda_H+\lambda_G}{\lambda_H-\lambda_G}$

Consistency requires that if $\mu^2 > 0$ then $\lambda_H < \lambda_G$, whereas if $\mu^2 < 0$ then $\lambda_H > \lambda_G$. We are interested in $\mu^2 < 0$, as it corresponds to the gauge symmetry breaking in the 4D effective theory approach. Therefore we assume $\lambda_H > \lambda_G$. The VEV of χ, which corresponds to Solution (4), is

$$|<\chi>|^2 = \frac{3}{4}\frac{-\mu^2}{\lambda_H - \lambda_G}$$

and we observe that it is equal to (3), apart from the shift $\lambda_H \to \lambda_H - \lambda_G$. Moreover, the metric that appears in Configuration (1) describes an S^2, whereas in (4) we have the metric of an ellipsoid. This distortion corresponds to the electroweak symmetry breaking in the geometrical approach, as it is shown in Fig. 2.

The low energy vector, fermion and scalar spectrum[4], which corresponds to Solution (4) is presented in the third column of Table 1, apart from the massless gauge field. We observe that, for vectors and fermions, the only difference between the 4D effective theory and the geometrical approach is the shift $\lambda_H \to \lambda_H - \lambda_G$, as for the VEV of χ. However, concerning the scalar spectrum, we have $M_S^2/M_{S\pm}^2 = 2$, in the 4D effective theory approach, whereas $M_S^2/M_{S\pm}^2 = 2(1-\delta)/(1+\delta)$, where $\delta \equiv \lambda_G/\lambda_H$, in the geometrical approach. Since a ratio of masses is a measurable quantity, there is a physical disagreement between the two approaches. The error is measured by λ_G/λ_H and we can roughly estimate its magnitude: if we require g_1 and g_2 in (2) to be of the order of 1, and we also consider the relation between κ and the 4D Planck length, we obtain that λ_G is of order of 1. Therefore the condition $\lambda_G/\lambda_H \ll 1$ becomes $\lambda_H \gg 1$, which is a strong coupling regime. Probably this range is not allowed if one requires to study the 4D effective theory by using perturbation theory. We conclude that the heavy mode contribution to the low energy dynamics is in general non negligible even

---

[4] This spectrum has been computed by using the formalism presented in Ref. [9].

in standard KK theories, where the heavy mode masses are naturally at the Planck scale. Finally, we observe that this contribution can be interpreted in a geometrical way, as the internal space deformation of the 6D solution: indeed, if we put $\beta = 0$ but we keep $\alpha \neq 0$ in (4), which corresponds to neglecting the S^2 deformation, the spectra in Table 1 turn out to be equal.

NON-ABELIAN AND SUPERSYMMETRIC EXTENSIONS

Gauge-Higgs Unification. This scenario consists of models without fundamental scalars, which, in some sense, geometrize the Higgs mechanism. Explicit realizations, which include dynamical gravity, are presented in Ref. [4]. In particular, the authors analyzed a 6D Einstein-Yang-Mills model, which is a non-Abelian extension, without bulk scalars, of our theory. In a simple set up the bulk gauge group is chosen to be $SU(3)$ and a non-Abelian generalization of Solution (1) can break $SU(3)$ down to $SU(2) \times U(1)$. The internal components of the bulk gauge fields contain a doublet of $SU(2)$, which can be naturally interpreted as a Higgs field. In this way the Higgs mass is protected from dangerous power-law radiative corrections by the bulk gauge symmetry.

In the 4D effective theory approach the Higgs doublet triggers the SSB of $SU(2) \times U(1)$ down to the electromagnetic $U(1)$. The results presented in the present paper and in [3] suggest that this method provides the correct 6D predictions for the observable quantities. This is because, in our model, the solution of the EOMs of the 4D effective theory can be lifted back to a solution of the complete 6D theory, if the heavy modes are properly taken into account.

6D Supergravities. Other extensions of our work can be done in the context of supersymmetric versions of 6D gauge and gravitational theories. In particular, 6D gauged[5] supergravities have attracted much interest for several reasons. One of them is that the flat 6D space-time *is not* a solution of the corresponding EOMs and the most symmetric solution is $(Minkowski)_4 \times S^2$, which has been shown recently to be the *unique* maximally symmetric solution of such models [10]. Therefore, these theories provide a theoretical explanation for the background that we have considered in the previous section.

Moreover, 6D gauged supergravity compactifications share some properties with 10D supergravity compactifications, whilst remaining relatively simple, and so it can be used as a toy model for 10D string theory compactifications [5], in particular they can give rise to chiral fermions in 4D. Furthermore, like in string theory, the requirement of anomaly freedom is a strong guiding principle to construct consistent models. Indeed the minimal version of such gauged supergravity, *the Salam-Sezgin model* [11], suffers from the breakdown of local symmetries due to the presence of gravitational, gauge and mixed anomalies, which render this model inconsistent at the quantum level [12]; but it can be transformed in an anomaly free model by choosing the gauge group and the supermultiplet in a suitable way [13]. Therefore, the extension of our analysis to

---

[5] "gauged" means that a subgroup of the R-symmetry group is promoted to a local symmetry.

this context could be a first step towards the study of the heavy modes in string theory compactifications.

Moreover, such 6D supergravities have been recently investigated in connection with attempts to find a solution to the cosmological dark energy problem [6]. Some 3-branes solutions and their perturbations, which can be relevant for this scenario, have been studied in Refs. [10, 14]. These backgrounds are deformations of Background (1), like our ellipsoid solution in the geometrical approach, but involving a warp factor and conical defects. This similarity suggests that our computation can be extended to 6D gauged supergravities expanded around these 3-brane solutions. If the heavy mode contribution is physically relevant, the underlying 6D physics should manifest itself in the low energy dynamics.

ACKNOWLEDGMENTS

This work was supported in part by the Swiss Science Foundation. We are appreciative of the hospitality at the IPT of Lausanne and the support by INFN. We would also like to thank S. Randjbar-Daemi, M. Shaposhnikov and K. Zuleta for illuminating and stimulating discussions.

REFERENCES

1.   K. G. Wilson, Rev. Mod. Phys. **47**, 773 (1975); S. Weinberg, Phys. Lett. B **91** (1980) 51.
2.   T. Appelquist and J. Carazzone, Phys. Rev. D **11** (1975) 2856.
3.   S. Randjbar-Daemi, A. Salvio and M. Shaposhnikov, Nucl. Phys. B **741** (2006) 236 [arXiv:hep-th/0601066].
4.   G. R. Dvali, S. Randjbar-Daemi and R. Tabbash, Phys. Rev. D **65** (2002) 064021 [arXiv:hep-ph/0102307].
5.   Y. Aghababaie, C. P. Burgess, S. L. Parameswaran and F. Quevedo, JHEP **0303** (2003) 032 [arXiv:hep-th/0212091].
6.   Y. Aghababaie, C. P. Burgess, S. L. Parameswaran and F. Quevedo, Nucl. Phys. B **680** (2004) 389 [arXiv:hep-th/0304256]; C. P. Burgess, arXiv:hep-th/0510123.
7.   S. Randjbar-Daemi, A. Salam and J. A. Strathdee, Nucl. Phys. B **214** (1983) 491.
8.   J. Sobczyk, Class. Quant. Grav. **4** (1987) 37.
9.   S. Randjbar-Daemi and M. Shaposhnikov, Nucl. Phys. B **645** (2002) 188 [arXiv:hep-th/0206016].
10. G. W. Gibbons, R. Guven and C. N. Pope, Phys. Lett. B **595** (2004) 498 [arXiv:hep-th/0307238];
11. A. Salam and E. Sezgin, Phys. Lett. B **147** (1984) 47.
12. L. Alvarez-Gaume and E. Witten, Nucl. Phys. B **234** (1984) 269.
13. S. Randjbar-Daemi, A. Salam, E. Sezgin and J. A. Strathdee, Phys. Lett. B **151** (1985) 351; S. D. Avramis, A. Kehagias and S. Randjbar-Daemi, JHEP **0505** (2005) 057 [arXiv:hep-th/0504033]; S. D. Avramis and A. Kehagias, JHEP **0510** (2005) 052 [arXiv:hep-th/0508172]; R. Suzuki and Y. Tachikawa, J. Math. Phys. **47** (2006) 062302 [arXiv:hep-th/0512019].
14. Y. Aghababaie *et al.*, JHEP **0309** (2003) 037 [arXiv:hep-th/0308064]; C. P. Burgess, F. Quevedo, G. Tasinato and I. Zavala, JHEP **0411** (2004) 069 [arXiv:hep-th/0408109]; A. J. Tolley, C. P. Burgess, D. Hoover and Y. Aghababaie, JHEP **0603** (2006) 091 [arXiv:hep-th/0512218]; A. J. Tolley, C. P. Burgess, C. de Rham and D. Hoover, arXiv:hep-th/0608083; H. M. Lee and A. Papazoglou, Nucl. Phys. B **747**, 294 (2006) [arXiv:hep-th/0602208]; S. L. Parameswaran, S. Randjbar-Daemi and A. Salvio, arXiv:hep-th/0608074.

Baryogenesis by Heavy Quarks: Q-genesis

Jihn E. Kim

Department of Physics and Astronomy, Seoul National University, Seoul 151-747, Korea

Abstract. In this talk, I present a new mechanism for baryogenesis the Q-genesis in which the heavy quarks are the source of baryon number [1]. There exists a narrow allowed region for the Q-genesis. [1]

Keywords: Baryogenesis, Heavy Quark
PACS: 12.60.-i, 98.80.Ft

INTRODUCTION

From the observed facts in the heaven, astrophysics and cosmology deal with cosmic microwave background radiation (CMBR), abundant light elements, galaxies and inter-galactic molecules, and dark matter (DM) and dark energy (DE) in the universe. In this talk, I present a recent work [1] regarding the light elements in this list whose source can be baryon number (B) from heavy quark decay.

Sakharov's three conditions for generating $\Delta B \neq 0$ from a baryon symmetric universe are

- Existence of $\Delta B \neq 0$ interaction,
- C and CP violation, and
- Evolution in a nonequilibrium state.

GUTs seemed to provide the basic theoretical framework for baryogenesis, because in most GUTs $\Delta B \neq 0$ interaction is present. Introduction of C and CP violation is always possible if not forbidden by some symmetry. The third condition on the non-equilibrium state evolution can be possible in the evolving universe but it has to be checked with specific interactions.

Thus, a cosmological evolution with $\Delta B \neq 0$ and C and CP violating particle physics model can produce a nonvanishing ΔB. The problem is "How big is the generated ΔB". Here, for nucleosynthesis, we need

$$\Delta B \simeq 0.6 \times 10^9 n_\gamma. \tag{1}$$

For example, the SU(5) GUT with X and Y gauge boson interactions are not generating the needed magnitude when applied in the evolving universe. In the SU(5) GUT, two quintet Higgs are needed for the required magnitude. With this scenario, GUTs with colored scalars seemed to be the theory for baryogenesis for some time.

---

[1] Talk presented at CICHEP-II, Cairo, Egypt, Jan. 15, 2006.

CP881, *Cairo International Conference on High Energy Physics (CICHEP II)*,
edited by S. Khalil

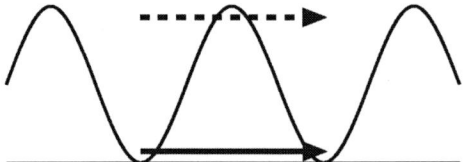

FIGURE 1. Vacuum tunneling at high (dashed arrow) and low (solid arrow) temperatures.

But high temperature QFT aspects changed this view completely. The spontaneously broken electroweak sector of the standard model (SM) does not allow instanton solutions. When the $SU(2)_W$ is not broken, there are electroweak $SU(2)$ instanton solutions. Tunneling via these electroweak instantons is extremely suppressed, $\sim \exp(-2\pi/\alpha_w)$. This tunneling amplitude is the zero temperature estimate. At high temperature where the electroweak phase transition occurs, the transition rate can be huge, and in cosmology this effect must be considered [2]. The tunneling amplitude due to sphaleron effect is large at high and small at low temperatures as shown in Fig. 1. This sphaleron effect transforms $SU(2)$ doublets. The 't Hooft vertex for this process must be a SM singlet, which is shown in Fig. 2.

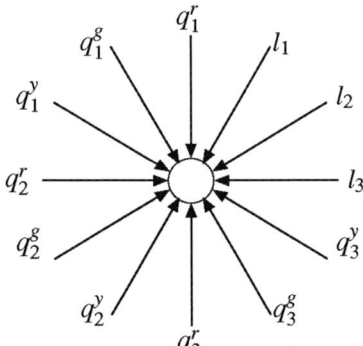

FIGURE 2. The sphaleron process via 't Hooft interaction.

The baryon number violating interaction washes out the baryon asymmetry produced during the GUT era. Since the above sphaleron interaction violates $B+L$ but conserves $B-L$, if there were a net $B-L$, there results a baryon asymmetry below the weak scale. The partition of $(B-L)$ into B and L below the electroweak scale is the following if the complete washout of $B+L$ is achieved,

$$\Delta B = \tfrac{1}{2}(B-L)_{\text{orig.}}, \quad \Delta L = \tfrac{1}{2}(B-L)_{\text{orig.}} \tag{2}$$

$B-L$ must be present in the beginning. If so, the leptogenesis uses this transformation of the $(B-L)$ number(obtained from heavy N decays) to baryon number. The ν-genesis also uses this transformation.

Thus, in some L to B transformation models, we need to generate a net $(B-L)$ number at the GUT scale (Q-genesis does not need this). The $SU(5)$ GUT conserves $B-L$ and hence cannot generate a net $B-L$.

So, for the baryon number generation one has to go beyond the SU(5) GUT. There exist three examples of baryon number generation from fermion sources: (1) Leptogenesis [3], (2) Neutrino genesis [4], and (3) Q-genesis [1].

INTRODUCTION OF HEAVY QUARKS

We note that SU(2) singlets avoid the sphaleron process and hence singlet quarks survive the electroweak era. With this idea, we must pursue along the line of heavy quarks that achieves

- Heavy quarks must mix with light quarks so that after the electroweak phase transition they can generate the quark number $(B/3)$.
- They must be sufficiently long lived.
- Of course, a correct order of $\Delta B \neq 0$ should be generated.

The SU(2) singlet quarks were considered before in connection with (i) flavor changing neutral currents (FCNC) [5], and recently for the BELLE data [6].

For the absence of FCNC at tree level, the electroweak isospin T_3 eigenvalues must be the same. Thus, introducing L-hand quark singlets will potentially introduce the FCNC problem. But in most discussions, the smallness of mixing angles with singlet quarks has been overlooked. Since the quark singlets can be superheavy compared to 100 GeV, the small mixing angles are natural, rather than being unnatural.

For definiteness, let us consider $Q_{em} = -\frac{1}{3}$ heavy quark D for which we must satisfy

1. ΔD generation mechanism is possible,
2. $10^{-10}\text{s} < \tau_D < 1$ s,
3. Sphaleron should not wash out all ΔD, and
4. The FCNC bound is satisfied.

Theoretically, is it natural to introduce such a heavy quark(s)? It is so. For example, in E6 GUT there exist Ds in $\mathbf{27}_F$. Trinification GUT also has particle D, viz.

$$\mathbf{27} \rightarrow \mathbf{16} + \mathbf{10} + \mathbf{1} \rightarrow \mathbf{10} + \mathbf{5}^* + \mathbf{1} + \mathbf{5} + \mathbf{5}^* + \mathbf{1}$$
$$\rightarrow (q + u^c + e^c) + N_5 + (l + d^c) + (D + L_2) + (D^c + L_1) + N_{10} \tag{3}$$

When we consider this kind of vector-like quarks $(D + D^c)$, there are three immediate related physical problems to deal with: heavy quark axion, the Nelson-Barr type, and FCNC. But here, we will consider the FCNC problem only.

We will consider one family first for d-type, $(t, b)_L^T$, b_R. It gives all the needed features. The mass matrix is

$$M_{-1/3} = \begin{pmatrix} m & J \\ 0 & M \end{pmatrix} \tag{4}$$

The entry 0 in the above is a natural choice, since it can be achieved by redefinition of R-handed singlets. The above mass matrix can be diagonalized by considering $MM^\dagger$, to give

$$\begin{pmatrix} |m_b|^2 \\ |m_D|^2 \end{pmatrix} = \tfrac{1}{2}(|M|^2 + |m|^2 + |J|^2) \mp \sqrt{[(|M|+|m|)^2 + |J|^2] \cdot [(|M|-|m|)^2 + |J|^2]} \tag{5}$$

which tends to $m_b \to m$ and $m_D \to M$ in the limit of $M^2 \gg |m|^2, |mJ|$. So, with vanishing phases, we have the following eigenstates,

$$|b\rangle \simeq \begin{pmatrix} 1 \\ -J/M \end{pmatrix}, \quad |D\rangle \simeq \begin{pmatrix} J/M \\ 1 \end{pmatrix}. \tag{6}$$

Since J is the doublet VEV and M is a parameter or a singlet VEV, the mixing angle can be sufficiently small. This is the well-known decoupling of vectorlike quarks. It can be generalized to three ordinary quarks and n heavy quarks. The $(3+n) \times (3+n)$ matrix

$$M_{-1/3} = \begin{pmatrix} M_d & J \\ J' & M_D \end{pmatrix}$$

can take the following form by redefining R-handed b and D fields,

$$M_{-1/3} = \begin{pmatrix} M_d & J \\ 0 & M_D \end{pmatrix} \tag{7}$$

For an easy estimate, below we express J as

$$J = f m_b.$$

Q-GENESIS BY HEAVY QUARKS

Here, we generate the D number as usual in GUTs cosmology through the Sakharov mechanism, and in the end we will identify the D number as the $3B$ number. The relevant interaction we introduce is

$$g_{Di} X_i u^c D^c + g_{ei} X_i^* u^c e^c + \text{h.c.} \tag{8}$$

which can introduce ΔD through the interference term from Fig. 3 plus self energy diagrams.

The interference is needed to generate a nonvanishing D number. The cross term contributes as $g_{e1}^* g_{e2} g_{D1}^* g_{D2}$. If we allow arbitrary phases in the Yukawa couplings, the relative phases of g_{D1} and g_{D2} can be cancelled only by the relative phase redefinition of X_2 and X_1. The same applies to g_{e1} and g_{e2}. Thus, the phase η appearing in $g_{e1}^* g_{e2} g_{D1}^* g_{D2}$ is physical. The D number generated in this way is

$$\frac{n_D}{n_\gamma} \simeq 0.5 \times 10^{-2} \varepsilon \tag{9}$$

$$\varepsilon \sim 10^{-2} \frac{\eta}{8\pi} [F(x) - F(1/x)], \quad x = \frac{M_{X_1}}{M_{X_2}} \tag{10}$$

where $F(x) = 1 - x \ln(1 + (1/x))$.

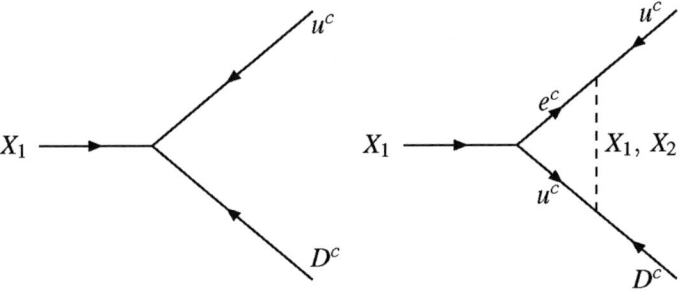

FIGURE 3. Diagrams contributing to $X \to u^c + D^c$.

LIFETIME ESTIMATE

The dominant decay of D proceeds via

$$D \to tW, bZ, bH^0 \tag{11}$$

from which we obtain

$$\Gamma_D = \frac{\sqrt{2}G_F}{8\pi}|J|^2 m_D, \quad \varepsilon = \frac{fm_b}{m_D} \tag{12}$$

The lifetime of D must be made longer than $2x10^{-11}$ s (beginning of electroweak phase transition), and it should be made shorter than 1 s (beginning of nucleosynthesis), i.e. 2×10^{-11} s $< \tau_D < 1$ s, which gives

$$\frac{1}{(10^6 m_{D,GeV})^{3/2}} < |\varepsilon| < \frac{1}{(2.7 \times 10^2 m_{D,GeV})^{3/2}}, \quad m_{D,GeV} = \frac{m_D}{\text{GeV}}. \tag{13}$$

The mixing of D with b is of order ε. For one period of oscillation, we expect that a fraction $|\varepsilon|^2$ of D is expected to transform to b. Since the period keeping the electroweak phase in cosmology is of order,

$$\frac{1}{H} = \sqrt{\frac{3M_P^2}{\rho}} \simeq \sqrt{\frac{3}{g_*} \frac{M_W}{M_P^2}}.$$

Thus, the following fraction is expected to be washed out via D oscillation into b,

$$\sim \frac{M_P m_b^2}{M_W^2 m_D} f^2 \simeq 10^{16} \frac{f^2}{m_{D,GeV}}.$$

For m_D of order $> 10^6$ GeV, we need $f < 10^{-5}$. In this range, some heavy quarks are left and the sphaleron does not erase the remaining D number.

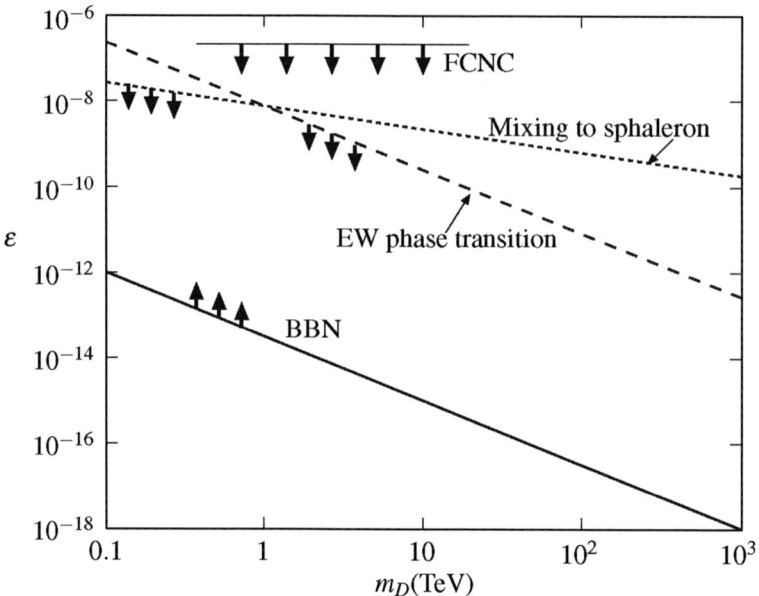

FIGURE 4.   Allowed parameter space in terms of mixing angle and D quark mass.

FCNC

For the FCNC, we consider the following processes which are compared with the existing experimental bounds

$$Z \to b\bar{b}, \quad |z_{bb}| = 0.996 \pm 0.005 \quad \to \quad |\varepsilon| \le 0.009$$

$$B \to X_s l^+ l^-, \quad \text{from tree} \quad |z_{sb}| = \frac{J_b J_s}{m_D^2} < 1.4 \times 10^{-3}$$

$$K^+ \to \pi^+ \nu\bar{\nu}, \quad |z_{sd}| \le 7.3 \times 10^{-6} \tag{14}$$

where the last bound comes from

$$\frac{\text{Br.}(K^+ \to \pi^+ \nu\bar{\nu})_{FCNC}}{\text{Br.}(K^+ \to \pi^0 e^+ \nu)} = \frac{3|z_{sd}|^2}{2\lambda^2} \le 2 \times 10^{-9}.$$

Therefore, from the FCNC processes, we obtain the bound on the coupling f,

$$\frac{1}{(4.8 \times 10^9 \sqrt{m_{D,GeV}})} < |f| < \frac{1}{(2.1 \times 10^4 \sqrt{m_{D,GeV}})}. \tag{15}$$

Fig. 4 summarizes the allowed parameter space in terms of mixing angle and D quark mass. The remarkable fact is that there exist a band allowed by the current data.

Z_2 SYMMETRY

To implement a small f naturally, we can impose a discrete symmetry such as a parity symmetry Z_2,

$$Z_2 : \begin{cases} b_{L,R} \to b_{L,R}, \ D_{L,R} \to D_{L,R} \\ \phi \to \phi, \ \varphi \to -\varphi \end{cases} \tag{16}$$

where ϕ and φ are Higgs doublets giving mass and mixing, respectively. The discrete symmetry forbids a mixing between doublets. We introduce a soft term, violating the discrete symmetry, which can mix them. Consistently with the discrete symmetry except the soft term, the potential is given by

$$\begin{aligned} V &= (m_\delta^2 \varphi^* \phi + \text{h.c.}) - \mu^2 \phi^* \phi + M_\varphi^2 \varphi^* \varphi \\ &\quad + \lambda_1 (\varphi^* \varphi)^2 + \lambda_2 (\phi^* \phi)^2 + \cdots \end{aligned} \tag{17}$$

The soft term violates the Z_2 symmetry. If the Z_2 is exact, there is no mixing between D and b and then D is absolutely stable: $\langle \phi \rangle = v \neq 0$, and $\langle \varphi \rangle = 0$. But the existence of the soft term violates the Z_2 and a tiny VEV, $\langle \varphi \rangle$, is generated. The estimate of mixing is

$$\langle \varphi \rangle = \frac{m_\delta^2 v}{M_\varphi^2}, \ f_{\text{off}} q_L \varphi D_R \quad \to \quad J = f_{\text{off}} \frac{m_\delta^2 v}{M_\varphi^2}$$

from which we estimate

$$f = \frac{\sqrt{2} f_{\text{off}} m_\delta^2}{f_b M_\varphi^2} \tag{18}$$

which can lead to naturally small values of $|J| = f m_b$ and $\varepsilon = J/M_D$. Typical value for m_D is given by constraints. The mass M_φ^2 can be superheavy.

Another point to be stressed, but unrelated to Q-genesis, is that we can construct a DM model of D with an exact Z_2 symmetry.

A related idea in the effective theory, but not the same at the renormalizable level, can be found in [7], where R-parity violating SUSY is used with B carrying singlet scalars S,

$$W_{NR} \sim \frac{1}{M_T} u^c d^c d^c S + \text{h.c.}$$

where T is our heavy quark D. But here T is much heavier than S. It has the similar idea of evading the electroweak phase transition era.

CONCLUSION

We presented a new mechanism for baryogenesis: the Q-genesis. We obtained the constraints on the parameter space. With a Z_2 discrete symmetry, the smallness of the parameter is naturally implemented.

TABLE 1. Comparison of a few baryogenesis mechanisms.

	$B - L = 0$	$B - L \neq 0$	Sphaleron
leptogenesis [3]		Yes	$(v, e)_L$ converts to B
v-genesis [4]	Possible	Possible	$(v, e)_L$ converts to B
Q-genesis [1]	Possible	Possible	Q decay produces B

Finally, in Table 1, this Q-gensis mechanism is compared with other baryogenesis mechanisms. Leptogenesis and v-genesis seem to be the plausible ones since the observed neutrino masses need R-handed neutrinos. Survival hypothesis sides with leptogenesis. Q-genesis depends on the unobserved Q, but this may be needed in solutions of the strong CP problem. Also, it appears in E_6 and trinification GUTs. One should consider all kinds of SM singlet particles and vector-like representations for the baryon asymmetry in the universe:[2]

$$N, v_R \quad : \quad \text{fermion singlets}$$
$$Q_L + Q_R \quad : \quad \text{vectorlike fermion}$$
$$N_L + N_R \quad : \quad \text{vectorlike fermion with Dirac mass}$$
$$S \quad : \quad \text{singlet scalar carrying } B \text{ number}$$

In Table 1, the $B - L$ number is that of light fermions. So, in the neutrino-genesis, one counts the R-handed light neutrino(v_R) number also in the $B - L$ number. In the Q-genesis, whatever sphaleron does on the light lepton number, still there exists baryon number generation from the Q decay, which occurs independently from the light fermion number.

ACKNOWLEDGMENTS

This work is supported in part by the KRF grants, No. R14-2003-012-01001-0, No. R02-2004-000-10149-0, and No. KRF-2005-084-C00001.

REFERENCES

1. H. D. Kim, J. E. Kim, and T. Morozumi, Phys. Lett. **B616** (2005) 108.
2. V. A. Kuzmin, V. A. Rubakov, and M. E. Shaposhnikov, Phys. Lett. **B155** (1985) 36.
3. M. Fukugita and T. Yanagida, Phys. Lett. **B174** (1986) 45.
4. K. Dick, M. Lindner, M. Ratz and D. Wright, Phys. Rev. Lett. **84**, 4039 (2000).
5. K. Kang and J. E. Kim, Phys. Lett. **B64** (1976) 93.
6. L. T. Handoko and T. Morozumi, Mod. Phys. Lett. **A10** (1995) 309 and Mod. Phys. Lett. **A10** (1995) 1733(E).
7. S. Davidson and R. Hempfling, Phys. Lett. **B391** (1997) 287.
8. I. Affleck and M. Dine, Nucl. Phys. **B249** (1985) 361.

---

[2] Note colored scalars can appear in the Affleck-Dine mechanism [8].

Electroweak symmetry breaking in TeV-scale string models

Noriaki Kitazawa

Department of Physics, Tokyo Metropolitan University,
Hachioji, Tokyo 192-0397, Japan
e-mail: kitazawa@phys.metro-u.ac.jp

Abstract. We propose a scenario of the electroweak symmetry breaking by one-loop radiative corrections in a class of string models with D3-branes at non-supersymmetric orbifold singularities with the string scale in TeV region. As a test example, we consider a simple model based on a D3-brane at locally $\mathbf{C}^3/\mathbf{Z}_6$ orbifold singularity, and the electroweak Higgs doublet fields are identified with the massless bosonic modes of the open string on that D3-brane. They have Yukawa couplings with three generations of left-handed quarks and right-handed up-type quarks which are identified with the massless fermionic modes of the open string on the D3-brane. We calculate the one-loop correction to the Higgs mass due to the non-supersymmetric string spectrum and interactions, and qualitatively suggest that the negative mass squared can be generated. The problems which must be solved to proceed quantitative calculations are pointed out.

Keywords: string phenomenology, electroweak symmetry breaking
PACS: 11.25.Wx, 12.60.Fr

INTRODUCTION

The dynamics of the electroweak symmetry breaking is still unknown. The standard model has the naturalness or fine-tuning problem, and its minimal supersymmetric extension also requires a certain level of fine-tuning[1]. Although the technicolor dynamics (without supersymmetry) [2, 3] is a candidate of the "inevitable" electroweak symmetry breaking, like the chiral symmetry breaking in QCD, it has some problems to be overcome (for recent works, see refs.[4, 5], for example). The dynamical electroweak symmetry breaking using controllable dynamics of supersymmetric gauge theories is proposed in refs.[6, 7].

In this article we propose an alternative dynamics of "inevitable" electroweak symmetry breaking in string theory without supersymmetry with the string scale in TeV region. This is the idea which has already proposed in ref.[8]. The authors calculated one-loop effective potential for certain scalar fields in a non-supersymmetric brane-anti-brane system ($D9$-$\overline{D5}$ system), and found that these scalar fields can have non-zero vacuum expectation values. These scalar fields correspond to Wilson lines and brane moduli, and the one-loop effective potential can be obtained by a modification of the open string vacuum amplitude at one loop. The vacuum expectation values of these scalar fields break original gauge symmetry $USp(16) \times USp(16)$. In this letter we consider the similar phenomena in different systems of D-branes at singularities[9], in which the Higgs doublet fields do not correspond to Wilson lines and brane moduli. The two point function of the Higgs doublet fields, namely the correction to the Higgs masses, have to be directly

CP881, *Cairo International Conference on High Energy Physics (CICHEP II)*,
edited by S. Khalil

73

calculated to see whether they can have vacuum expectation values or not.

In the next section, we briefly review the system of D3-branes at orbifold singularities. In the third section, we calculate the one-loop two point function of the gauge boson from the open string in ten dimensions. This calculation gives a good guidance to the calculation of the Higgs mass correction. In the last section, we calculate the one-loop correction to the mass of the Higgs doublet field which is realized on the D3-brane at non-supersymmetric locally $\mathbf{C}^3/\mathbf{Z}_6$ orbifold singularity. We suggest that the negative mass squared can be generated. Some concluding comments are also included in this section.

D-BRANES AT SINGULARITIES

We introduce one specific semi-realistic system based on a D3-brane at locally $\mathbf{C}^3/\mathbf{Z}_6$ orbifold singularity. A complete and self-contained introduction to the system of D3- and D7-branes located at singularities, is given in ref.[9].

Consider type IIB theories with six dimensions compactified to orbifolds or orientifolds, and assume that there is a locally $\mathbf{C}^3/\mathbf{Z}_6$ orbifold singularity in the compact six dimensional space. The open string states on the D3-brane at that singularity are modified by the $\mathbf{Z}_6$ projection. If we simply have N coincident D3-brane, the massless states in its four-dimensional world-volume belong to a $\mathcal{N}=4$ supersymmetric $U(N)$ gauge multiplet. The $\mathbf{Z}_6$ projection breaks the supersymmetry and the gauge symmetry (to the group with the same rank, $U(N_1) \times U(n_2) \times U(n_3)$ with $n_1 + n_2 + n_3 = N$, for example), and some states of the original gauge fields remain the states of gauge fields, some other states correspond to massless matter fields in bi-fundamental representations under the new gauge symmetry, and there are some states which are completely projected out. If the projection keeps four-dimensional $\mathcal{N}=1$ supersymmetry, the resultant numbers of bosonic and fermionic degrees of freedom are equal, and the structure of $\mathcal{N}=1$ supermultiplet remains. There are some consistent ways of projection in which there is no correspondence between bosonic and fermionic states and supersymmetry is completely broken by the spectrum.

In the language of $\mathcal{N}=1$ supermultiplets the original massless field contents on N D3-brane is a $U(N)$ gauge vector multiplet and three chiral multiplets in the adjoint representation under $U(N)$. The scalar components of three chiral multiplets can be understood as the position moduli of D3-brane, and they can be understood as the local complexified coordinates of the compact six dimensional space. There is $SU(4)_R$ global symmetry under which these six bosonic degrees of freedom belong sextet and the four fermionic degrees of freedom, three chiral fermions in three chiral multiplets and gaugino, belong quartet. If the $\mathbf{Z}_6$ transformation is the subgroup of $SU(4)_R$, the projection is consistent under the time evolution. The $\mathbf{Z}_6$ transformation on the three local complexified coordinates is described by three integers $b_1, b_2, b_3 = 0, 1, \cdots, 5$ with the transformation matrix $\mathrm{diag}(e^{2\pi i b_1/6}, e^{2\pi i b_2/6}, e^{2\pi i b_3/6})$. This is the transformation matrix on a sextet for any possible values of b_i. The $\mathbf{Z}_6$ transformation on four fermionic degrees of freedom is described by four integers $a_1, a_2, a_3, a_4 = 0, 1, \cdots, 5$ with $a_1 + a_2 + a_3 + a_4 = 0$ mod 6 with the transformation matrix $\mathrm{diag}(e^{2\pi i a_1/6}, e^{2\pi i a_2/6}, e^{2\pi i a_3/6}, e^{2\pi i a_3/6})$. Two sets of integers have the relations of $b_1 = a_2 + a_3$, $b_2 = a_3 + a_1$ and $b_3 = a_1 + a_2$. The condition to have $\mathcal{N} = 1$ supersymmetry is $a_4 = 0$, and in this case $b_r = -a_r$ with $r = 1, 2, 3$. Both three complexified bosonic and three complexified fermionic open string world-sheet fields, which correspond to three local complexified coordinates in compact six dimensional space, transform in the same way as three local complexified coordinates. The transformation of the space-time fermion is realized by the various spin combinations of the vacuum states of Ramond sector.

The open string Chan-Paton factor may transform under $\mathbf{Z}_6$. The transformation is described as

$$|ij\rangle \longrightarrow (\gamma_3)_{ii'}|i'j'\rangle(\gamma_3^{-1})_{j'j}, \tag{1}$$

where $\gamma_3 = \mathrm{diag}(I_{n_0}, e^{2\pi i/6}I_{n_1}, \cdots e^{2\pi i \cdot 5/6}I_{n_5})$ with $n_0 + n_1 + \cdots + n_5 = N$ and I_n is $n \times n$ unit matrix.

The massless open string states, which are singlet under $\mathbf{Z}_6$ transformation, correspond to the massless fields. In addition to the gauge bosons of the gauge symmetry of $U(n_0) \times U(n_1) \times \cdots \times U(n_5)$, we have massless matter fields in bi-fundamental representation:

$$\text{complex scalars} \quad \sum_{r=1}^{3}\sum_{i=0}^{5}(n_i, \bar{n}_{i-b_r}), \tag{2}$$

$$\text{Weyl fermions} \quad \sum_{\alpha=1}^{3}\sum_{i=0}^{5}(n_i, \bar{n}_{i+a_\alpha}), \tag{3}$$

where n_i and $\bar{n}_i$ mean fundamental and anti-fundamental representation of $U(n_i)$, respectively.

We proceed to much more concrete model. We take $N = 6$ and $n_0 = 1$, $n_1 = 3$, $n_2 = 2$ and $n_3 = n_4 = n_5 = 0$. For $\mathbf{Z}_6$ projection, we take $b_1 = b_2 = b_3 = 2$, $a_1 = a_2 = a_3 = 1$ and $a_4 = -3$. This set up gives non-supersymmetric spectrum. We have gauge symmetry of $U(3) \times U(2) \times U(1)$ and massless matter fields

$$\begin{aligned}
\text{Higgs doublet fields:} &\quad H_r &\quad 3 \times (1, 2, -1), &\tag{4}\\
\text{left-handed quarks:} &\quad q_{Lr} &\quad 3 \times (3, 2^*, 0), &\tag{5}\\
\text{right-handed quarks:} &\quad u^c{}_{Lr} &\quad 3 \times (3^*, 1, +1), &\tag{6}
\end{aligned}$$

where we omit to describe the charges of $U(1)$ factors of $U(3)$ and $U(2)$. There are Yukawa couplings among these fields which are obtained by the $\mathbf{Z}_6$ projection of the interactions due to the superpotential in the original $\mathcal{N} = 4$ supersymmetric theory. The Yukawa coupling constants are equal to the gauge coupling constant. Note that all the gauge coupling constants of $U(3) \times U(2) \times U(1)$ are equal at tree level.

There are four point Higgs self-couplings which are remnants of the D-term scalar potential in the original supersymmetric theory.

$$V = \frac{g^2}{4}\sum_{r,s=1,2,3}\left((H_r^\dagger H_s)(H_s^\dagger H_r) + (H_r^\dagger H_r)(H_s^\dagger H_s)\right), \tag{7}$$

where g is the gauge coupling constant of D3-brane. There are no flat directions on the vacuum expectation values of Higgs doublet fields. This means that Higgs doublet fields are not D-brane moduli.

This is not the complete system. We have to consider the R-R (Ramond-Ramond) tadpole cancellation to make a consistent string theory. R-R tadpole cancellation conditions contain chiral anomaly cancellation conditions. We have to consider both twisted and untwisted tadpoles. Twisted tadpoles can be cancelled out by introducing appropriate D7-branes. Inclusion of D7-branes means introduction of additional gauge symmetry and massless matter fields which ensure the chiral anomaly cancellation about the gauge symmetry on D3-brane. To consider the untwisted tadpole cancellation we have to specify a concrete compactification space. Since the construction of the concrete realistic models is not the issue of this letter, and the existence of the untwisted tadpole does not affect our forthcoming discussions, we do not discuss the untwisted tadpole cancellation.

ONE-LOOP TWO POINT FUNCTION IN TEN DIMENSIONS

We consider two point functions of gauge bosons in the zero momentum limit, namely the mass corrections to gauge bosons. Although we know the result: it should vanish because of the gauge invariance, it is instructive to calculate the mass correction to the Higgs doublet fields.

Before going to the calculation in string theory, it is worth to mention the calculation in four dimensional $\mathcal{N} = 1$ supersymmetric U(1) gauge field theory. The quadratically divergent contributions come from two boson loop diagrams and one fermion loop diagram:

$$\Pi^{\mu\nu}_{\text{boson}} = (1-2) \times \eta^{\mu\nu} \int \frac{d^4 p}{(2\pi)^4 i} \frac{1}{p^2}, \tag{8}$$

$$\Pi^{\mu\nu}_{\text{fermion}} = \eta^{\mu\nu} \int \frac{d^4 p}{(2\pi)^4 i} \frac{1}{p^2}, \tag{9}$$

where $\eta^{\mu\nu} = \text{diag}(-1,1,1,1)$. These are the contributions of one chiral multiplet. We see the cancellation of these two corrections due to supersymmetry. If we take Euclidean momentum cutoff regularization, boson and fermion give positive and negative correction to the mass squared of the gauge boson, respectively. If we take the gauge invariant regularization, dimensional regularization, for example, the corrections vanish individually.

We use the techniques for one-loop calculation in open string theory, which is described in the text book of ref.[10], because it has good correspondence with the one-loop calculation in field theory. The one-loop contributions of the bosonic states and fermionic states to the two point function of the massless vector mode of the open string (gauge field) are respectively described by

$$A^{\text{NS}} = \int \frac{d^{10} p}{(2\pi)^{10} i} \text{tr}(\Delta V(1)\Delta V(1) P_{\text{GSO}}), \tag{10}$$

$$A^{\text{R}} = -\int \frac{d^{10} p}{(2\pi)^{10} i} \text{tr}(SW(1)SW(1) P_{\text{GSO}}), \tag{11}$$

where NS and R mean Neveu-Schwarz and Ramond sector, respectively, P_{GSO} is the Gliozzi-Scherk-Olive projection operator, propagators are defined as

$$\Delta \equiv \int_0^1 x^{L_0-1} dx, \tag{12}$$

$$S \equiv iG_0\Delta, \tag{13}$$

and vertex operators are defined as

$$W(1) = g_O e_\mu \psi^\mu e^{ik\cdot X}, \tag{14}$$

$$V(1) \equiv \{G_0, W(1)\}$$

$$= \frac{g_O}{\sqrt{2\alpha'}} e_\mu \left(i\dot{X}^\mu + 2\alpha'(k\cdot\psi)\psi^\mu \right) e^{ik\cdot X} \tag{15}$$

with open string coupling constant g_O and polarization vector e_μ. The argument 1 of vertex operators means the complex valuable $z = \exp(-i(\sigma_1 + i\sigma_2))$ with the value of the Euclidean world-sheet coordinates $\sigma_1 = \sigma_2 = 0$. Taking the value of $\sigma_1 = 0$ means that we are considering planner diagrams: both two vertex operators are attached to one of two boundaries of annulus (or cylinder). We do not discuss the non-planner diagram, because it is not important to the calculation of the mass correction to the Higgs doublet fields in the next section. The dot operation in eq.(15) means the differentiation by σ_2.

We evaluate only the leading terms in the internal momentum integration of eqs.(10) and (11). These are dominant contributions because of the following reasons. The open string one-loop calculation is essentially to count the number of possible states in the loop with weight $\exp(-2\pi t L_0)$ and to integrate over the cylinder modulus $0 \le t < \infty$ ($2\pi t$ is the circumference of the cylinder). The integrant of the modulus integration mainly has value in the regions of small t because of the exponential weight. The leading internal momentum integration gives an enhancement factor for the small t region, which is absent in the sub-leading momentum integration. Therefore the leading term in the internal momentum integration can be considered as the dominant contribution. In four dimensional field theory, this corresponds to evaluate only the quadratically divergent terms like in eqs.(8) and (9).

The results of the calculation is the following.

$$A^{\mathrm{NS}} \simeq \frac{1}{2}\frac{g_O^2}{\alpha'^5} e^\mu e_\mu \int_0^1 \frac{d\rho}{\rho} \left(\frac{\pi}{-\ln\rho}\right)^5 \frac{1}{\eta(it)^8} \left(\left(\frac{\theta_3(it)}{\eta(it)}\right)^4 - \left(\frac{\theta_4(it)}{\eta(it)}\right)^4 \right), \tag{16}$$

$$A^{\mathrm{R}} \simeq -\frac{1}{2}\frac{g_O^2}{\alpha'^5} e^\mu e_\mu \int_0^1 \frac{d\rho}{\rho} \left(\frac{\pi}{-\ln\rho}\right)^5 \frac{1}{\eta(it)^8} \left(\frac{\theta_2(it)}{\eta(it)}\right)^4, \tag{17}$$

where $\rho = \exp(-2\pi t)$, η is Dedekind eta function and θ_2, θ_3 and θ_4 are Jacobi theta functions. Here we have taken the limit of zero external momentum. The total correction vanishes because of the relation $(\theta_3)^4 - (\theta_4)^4 - (\theta_2)^4 = 0$, which means the balance of the number of bosonic and fermionic states, namely supersymmetry. More rigorous calculation requires to include the contribution from non-planner and nonorientable diagrams in type I theory with the care of Chan-Paton factor as well as the sub-leading

77

contribution of the internal momentum integration. However, this level of calculation is enough to give a guide to the calculation of the one-loop mass correction of the Higgs doublet fields. In case of no supersymmetry in the spectrum, boson and fermion contributions are not necessary balanced. If the boson contribution is dominant, the correction to the mass squared is positive, and if the fermion contribution is dominant, the correction to the mass squared is negative.

ONE-LOOP CORRECTION TO THE HIGGS MASS

The calculation is very similar to that in the previous section. The essential difference is the boundary condition of the world-sheet fields as well as Z_6 projection. Among the ten world-sheet boson and ten world-sheet fermion fields, the fields corresponding to the parallel direction to D3-brane follow Neumann boundary condition, and others, namely the fields corresponding to the transverse direction to D3-brane follow Dirichlet boundary condition for both edges of the open string. One important fact is that there is no string center of mass momentum of the transverse direction to D3-brane.

The amplitudes to be calculated are as follows.

$$A_{\text{Higgs}}^{\text{NS}} = \int \frac{d^4p}{(2\pi)^4 i} \text{tr} \left(\Delta V(1)^{(-)} \Delta V(1)^{(+)} P_{\text{GSO}} P_{Z_6} \right), \tag{18}$$

$$A_{\text{Higgs}}^{\text{R}} = -\int \frac{d^4p}{(2\pi)^4 i} \text{tr} \left(SW(1)^{(-)} SW(1)^{(+)} P_{\text{GSO}} P_{Z_6} \right), \tag{19}$$

where P_{Z_6} is the Z_6 projection operator, vertex operators are defined as

$$W(1)^{(+)} = g_O u^{i_1}{}_{i_2} \psi^{(+)} e^{ik \cdot X}, \tag{20}$$

$$V(1)^{(+)} \equiv \{G_0, W(1)^{(+)}\}$$

$$= \frac{g_O}{\sqrt{2\alpha'}} u^{i_1}{}_{i_2} \left(i\dot{X}^{(+)} + 2\alpha'(k \cdot \psi) \psi^{(+)} \right) e^{ik \cdot X} \tag{21}$$

with

$$X^{(\pm)} = \frac{1}{\sqrt{2}} \left(X^4 \pm iX^5 \right), \qquad \psi^{(\pm)} = \frac{1}{\sqrt{2}} \left(\psi^4 \pm i\psi^5 \right) \tag{22}$$

for one of three Higgs doublet states, and the momentum k_μ can have non-zero value only for $\mu = 0, 1, 2, 3$. The vertex operators $W(1)^{(-)}$ and $V(1)^{(-)}$ are Hermite conjugates of $W(1)^{(+)}$ and $V(1)^{(+)}$, respectively. The indexes i_1 and i_2 of the factor $u^{i_1}{}_{i_2}$ denote the Chan-Paton factor of U(1) and U(2), respectively. By considering the flow of the Chan-Paton charge, it is easily understood that only the planner diagram contributes.

Now, we calculate the leading terms of internal momentum integration, which are dominant contributions. We find that there is no leading term in Neveu-Schwarz amplitude (boson loop), because $X(1)^{(\pm)}$ has no zero modes due to Dirichlet boundary condition. Therefore, we may only calculate Ramond amplitude (fermion loop), and make sure whether it vanishes or not. If it is not zero, we have a negative correction to the Higgs mass squared, and the electroweak symmetry breaking by the Higgs vacuum expectation value through the one-loop quantum effect is expected. With a special care of $\mathbf{Z}_6$ projection, we obtain the following result.

$$A^R_{\text{Higgs}} \simeq -3 \cdot \frac{1}{2} \frac{g_O^2}{\alpha'^2} u^{\dagger i_2}{}_{i_1} u^{i_1}{}_{i_2} \int_0^1 \frac{d\rho}{\rho} \left(\frac{\pi}{-\ln\rho} \right)^2 \tag{23}$$

$$\times \frac{1}{6} \sum_{\gamma=0}^5 16 \prod_{m=1}^\infty \left(\frac{1+\rho^m}{1-\rho^m} \right)^2 \left(\frac{1+(e^{2\pi i/3})^\gamma \rho^m}{1-(e^{2\pi i/3})^\gamma \rho^m} \right)^3 \left(\frac{1+(e^{-2\pi i/3})^\gamma \rho^m}{1-(e^{-2\pi i/3})^\gamma \rho^m} \right)^3,$$

where the first factor 3 is the Chan-Paton factor. The only $U(3)$ charged stats, including massless left-handed quarks and right-handed up-type quarks, contribute the loop. Although certainly this amplitude is not zero, we have to consider more model dependent contributions due to the twisted R-R tadpole cancellations. That is a required contribution which completely cancels the one-loop correction in case with supersymmetry.

Here, to cancel the twisted R-R tadpoles, we introduce 36 D7-brane whose world-volume is our four dimensional space-time and second and third complex planes in six dimensional compact space. We take the $\mathbf{Z}_6$ action to the Chan-Paton factor of this D7-brane as $\gamma_7 = \text{diag}(I_{u_0}, e^{2\pi i/6} I_{u_1}, \cdots e^{2\pi i \cdot 5/6} I_{u_5})$ with $u_0 = 6$, $u_1 = 0$, $u_2 = 3$ and $u_3 = u_4 = u_5 = 9$. This D7-brane gives new gauge symmetries $U(6) \times U(3) \times U(9)_1 \times U(9)_2 \times U(9)_3$ with very small gauge coupling constants, since we take the string scale in TeV range. This symmetries emerge as global symmetries at low energies. We have new massless and massive states form the open string with one edge on D3-brane and another edge on D7-brane. Although we do not explain detailed spectrum here, since the concrete model building is not the aim of this letter, we would like to stress that there are no massless fermion states which have Yukawa couplings with Higgs doublet fields. This is an important difference from the case with supersymmetry. The leading one-loop correction to the Higgs mass squared from this open string is obtained as follows.

$$A'^R_{\text{Higgs}} \simeq -9 \cdot \frac{1}{2} \frac{g_O^2}{\alpha'^2} u^{\dagger i_2}{}_{i_1} u^{i_1}{}_{i_2} \int_0^1 \frac{d\rho}{\rho} \left(\frac{\pi}{-\ln\rho} \right)^2 \cdot \frac{1}{6} \sum_{\gamma=0}^5 \left((e^{2\pi i/3})^\gamma + (e^{-2\pi i/3})^\gamma \right)$$

$$\times 16 \prod_{m=1}^\infty \left(\frac{1+\rho^m}{1-\rho^m} \right)^2 \left(\frac{1+(e^{2\pi i/3})^\gamma \rho^m}{1-(e^{2\pi i/3})^\gamma \rho^m} \right) \left(\frac{1+(e^{-2\pi i/3})^\gamma \rho^m}{1-(e^{-2\pi i/3})^\gamma \rho^m} \right)$$

$$\times \left(\frac{1+(e^{2\pi i/3})^\gamma \rho^{m-1/2}}{1-(e^{2\pi i/3})^\gamma \rho^{m-1/2}} \right)^2 \left(\frac{1+(e^{-2\pi i/3})^\gamma \rho^{m-1/2}}{1-(e^{-2\pi i/3})^\gamma \rho^{m-1/2}} \right)^2, \tag{24}$$

where the first factor 9 is the Chan-Paton factor, and the factor $((e^{2\pi i/3})^\gamma + (e^{-2\pi i/3})^\gamma)$ means the non-trivial transformation of Ramond sector vacuum under the $\mathbf{Z}_6$ action, which is related with the fact that no massless fermion states have Yukawa couplings with the Higgs doublet. Only the massive states with $U(9)_1$ or $U(9)_3$ charges contribute. The last two factors in the infinite product are the realizations of the Dirichlet-Neumann boundary condition of the open string in second and third complex planes in compact six dimensional space. The contribution of eq.(24) certainly does not cancel out the contribution of eq.(23), though the divergent contributions from twisted R-R tadpoles are cancelled out. The finite contribution of eq.(24) should be negative, since it is also due to the fermion one loop. We suggest that the negative mass squared of the Higgs doublet field can be generated, and the electroweak symmetry breaking can be expected, in this class of non-supersymmetric models with D-branes at singularities.

There are some concluding comments in order.

There may still exist a divergence in the total correction of eq.(23) plus eq.(24) due to the uncanceled twisted NS-NS (Neveu-Schwarz-Neveu-Schwarz) tadpoles. Since there is no supersymmetry, R-R tadpole cancellation does not necessary result NS-NS tadpole cancellation. The existence of NS-NS tadpole means that some redefinitions of backgrounds are required [11, 12, 13, 14]. It is probable that twisted moduli obtain vacuum expectation values as a result of the background redefinition by the uncanceled twisted NS-NS tadpoles. The vacuum expectation values of twisted moduli give some modification of Higgs potential of eq.(7) due to the emergence of the Fayet-Iliopoulos terms for anomalous $U(1)$ gauge symmetries in the original supersymmetric theory [15, 16]. This effect by itself can give vacuum expectation value to Higgs doublet fields even without negative mass squared at one loop. The vacuum expectation values of twisted moduli may also result blowing-up the orbifold singularity (see for example [17]), and the actual geometrical D3-brane reconfiguration by the vacuum expectation value of Higgs doublet fields may be understood at this blown-up orbifold singularity.

There is another possibility that the present non-supersymmetric $\mathbf{C}^3/\mathbf{Z}_6$ orbifold singularity is unstable and decays to some non-singular point. In other words, there might be tachyon modes in twisted NS-NS closed string sector localized at the non-supersymmetric singularity, which means the instability of the singularity[18]. The non-supersymmetric orbifold singularities suffer from this phenomenon in general. We need to find stable non-supersymmetric singularities without tachyons or non-supersymmetric singularities with very long life time.

There are three Higgs doublet fields, H_1, H_2 and H_3, in our $\mathbf{Z}_6$ model, and our calculations concern one of them, H_1. The contribution of eq.(23) is applicable for all three Higgs doublet fields, but the contribution, which is related with the R-R tadpole cancellations, may be different depending on models. In our model with one D7-brane, the result of eq.(24) is only for H_1, and H_2 and H_3 will obtain a different result. This kind of asymmetric configuration of D7-branes for twisted R-R tadpole cancellation gives asymmetric corrections to Higgs doublet fields. It may be possible that only one Higgs doublet field, which is a linear combination of Higgs doublet fields, has vacuum expectation value, and others are heavy. This is a phenomenologically preferable situation, since the existence of many Higgs doublet fields with vacuum expectation values causes a problem, flavor-changing neutral current problem, in general. Such a Higgs doublet field may have non-trivial Yukawa couplings for the hierarchical masses and flavor mixings, though the original Higgs doublet fields usually have trivial Yukawa couplings.

Inclusion of sub-leading term is important for more quantitative and detailed analysis. Modern techniques, like path integral formalism, might be better for this aim.

ACKNOWLEDGMENT

I would like to thank O. Yasuda for useful comments.

REFERENCES

1. G. L. Kane, J. Lykken, B. D. Nelson and L. T. Wang, Phys. Lett. B **551** (2003) 146 [arXiv:hep-ph/0207168].
2. S. Weinberg, Phys. Rev. D **19** (1979) 1277.
3. L. Susskind, Phys. Rev. D **20** (1979) 2619.
4. D. K. Hong, S. D. H. Hsu and F. Sannino, Phys. Lett. B **597** (2004) 89 [arXiv:hep-ph/0406200].
5. N. D. Christensen and R. Shrock, arXiv:hep-ph/0509109.
6. R. Harnik, G. D. Kribs, D. T. Larson and H. Murayama, D **70** (2004) 015002 [arXiv:hep-ph/0311349].
7. N. Haba and N. Okada, arXiv:hep-ph/0409113.
8. I. Antoniadis, K. Benakli and M. Quiros, Nucl. Phys. B **583** (2000) 35 [arXiv:hep-ph/0004091].
9. G. Aldazabal, L. E. Ibanez, F. Quevedo and A. M. Uranga, JHEP **0008** (2000) 002 [arXiv:hep-th/0005067].
10. M. B. Green, J. H. Schwarz and E. Witten, "Superstring Theory," Vol. 1 and Vol. 2, Cambridge Univ. Pr. (1987).
11. W. Fischler and L. Susskind, Phys. Lett. B **171** (1986) 383.
12. W. Fischler and L. Susskind, Phys. Lett. B **173** (1986) 262.
13. S. R. Das and S. J. Rey, Phys. Lett. B **186** (1987) 328.
14. E. Dudas, G. Pradisi, M. Nicolosi and A. Sagnotti, Nucl. Phys. B **708** (2005) 3 [arXiv:hep-th/0410101].
15. M. R. Douglas and G. W. Moore, arXiv:hep-th/9603167.
16. M. R. Douglas, B. R. Greene and D. R. Morrison, Nucl. Phys. B **506** (1997) 84 [arXiv:hep-th/9704151].
17. M. Cvetic, L. L. Everett, P. Langacker and J. Wang, JHEP **9904** (1999) 020 [arXiv:hep-th/9903051].
18. A. Adams, J. Polchinski and E. Silverstein, JHEP **0110** (2001) 029 [arXiv:hep-th/0108075].

Standard Model and SU(5) GUT with Local Scale Invariance and the Weylon

Hitoshi Nishino and Subhash Rajpoot[1]

Department of Physics & Astronomy, California State University,
1250 Bellflower Blvd., Long Beach, CA 90840

Abstract. Weyl's scale invariance is introduced as an additional local symmetry in the standard model of electroweak interactions. An inevitable consequence is the introduction of general relativity coupled to scalar fields *à la* Dirac and an additional vector particle we call the Weylon. Once Weyl's scale invariance is broken, the phenomenon (a) generates Newton's gravitational constant G_N and (b) triggers the conventional spontaneous symmetry breaking mechanism that results in masses for all the fermions and bosons. The scale at which Weyl's scale symmetry breaks is of order Planck mass. If right-handed neutrinos are also introduced, their absence at present energy scales is attributed to their mass which is tied to the scale where scale invariance breaks. Some implications of these ideas are noted in grand unification based on the gauge symmetry SU(5).

Keywords: Weyl, Scale Invariance, Standard Model, General Relativity, Extra Gauge Bosons
PACS: 12.60.-i, 12.60.Cn, 12.60.Fr, 04.20.-q

INTRODUCTION

This work falls under the category of *unconventional persuits*. Nevertheless the research is respectable and as I will show, leads to some very interesting and profound results.

The notion that the standard model [1] is the underlying theory of elementary particle interactions, excluding gravity, is the prevailing consensus supported by all experiments of the present time. The only missing ingredient is the elusive Higgs particle [2]. It is conceivable that the symmetry breaking mechanism is indeed spontaneous and the Higgs particle will be discovered. However, there are reasons, both aesthetic and otherwise, that necessitate the extensions of the standard model. Seeking unity of all particle interactions (grand unification) and explaining the ultimate instability of matter (proton decay) [3] are examples that fall in the former category while neutrino oscillations [4, 5] is an example that falls in the latter category.

At a much deeper level, the very notion of the origin of scales in physics is yet another fundamental issue yearning explanation. The problem reduces to comprehending the origin of just one fundamental scale, all other scales being different manifestations of this fundamental scale. To this end, either Weyl's scale invariance symmetry [6, 7] or the much larger symmetry, the fifteen parameter group of conformal invariance [8, 9, 10], are thought to play a significant role as fundamental symmetries of Nature. A glance at the elementary particle mass spectrum attests to the fact that scale invariance and conformal invariance are badly broken symmetries of Nature. In the past, these

---

[1] Conference speaker

CP881, *Cairo International Conference on High Energy Physics (CICHEP II)*,
edited by S. Khalil
© 2007 American Institute of Physics 978-0-7354-0382-6/07/$23.00

symmetries were employed to gain insight on the origin of Newton's gravitational constant G_N, a dimensionful quantity, as a symmetry breaking effect, induced either spontaneously or due to quantum corrections [11, 12, 13].

In this work we attempt at combining gauge and scale symmetries in an extension of the standard model in which not only gravity but also the entire particle mass spectrum of the standard model are generated in terms of just one fundamental scale associated with scale symmetry breaking. The approach is modest in that we exercise economy and consider extending the standard model with only Weyl's local scale invariance [6, 7], the doomed symmetry that gave birth to the gauge principle and ultimately paved the way for implementing gauge invariance as we know it and practise it today. As will be shown, in the absence of fine-tuning, the scale at which the scale invariance symmetry breaks turns out to be of order Planck mass $M_P \approx 1.2 \times 10^{19}$ GeV. The extended model predicts the existence of an additional vector particle we will call the Weylon. Its mass is tied to the scale at which Weyl's symmetry breaks and is also of order M_P.

Implementing scale invariance in the standard model has been previously considered [14]. The main result there was the elimination of the Higgs boson from the standard model particle spectrum. Here we present a different philosophy of the same work which has been recently considered [15]. In the present model, the standard model Higgs particle is not eliminated, and is the sought-after particle. In other words, *after* scale breaking our model at low energy describes *the* standard model of elementary particles *supplemented* with the Einstein-Hilbert action for gravitational interactions.

SCALE INVARIANCE

Under scale invariance the parallel transport of a vector around a closed loop in four dimensional space-time not only changes its direction but also its length. In such a manifold the line element ds has no absolute meaning because a comparison of lengths at two different points involves the scale factor $\Lambda(x)$ where $\Lambda(x)$ is the parameter of scale transformations. The fundamental metric tensor $g_{\mu\nu}$ transforms as

$$g_{\mu\nu}(x) \to \widetilde{g}_{\mu\nu}(x) = e^{2\Lambda(x)} g_{\mu\nu}(x) \ . \tag{1}$$

However, the ratio of two infinitesimal lengths is well defined when both lengths refer to the same point. This implies that the angle θ between two infinitesimal vectors dx and δx remains unchanged since

$$\cos\theta \ = \ \frac{g_{\mu\nu}dx^\mu \delta x^\nu}{\sqrt{g_{\alpha\beta}dx^\alpha dx^\beta}\sqrt{g_{\lambda\sigma}\delta x^\lambda \delta x^\sigma}} \ . \tag{2}$$

Thus, in reality, scale transformations lead to the larger fifteen parameter conformal transformations under which the coordinates x^μ undergo the following transformations

Translations ;

$$x^\mu \to x^\mu + a^\mu \quad \text{(4 parameters)}, \tag{3}$$

Lorentz Transformations ;

$$x^\mu \to L^\mu{}_\nu x^\nu \quad \text{(6 parameters)}, \tag{4}$$

Accelerations;

$$x^\mu \to x^\mu + \frac{a^\mu x^2}{1 - 2a^\alpha x_\alpha + x^2 a^2} \quad \text{(4 parameters)}, \tag{5}$$

and Dilatations;

$$x^\mu \to e^\Lambda x^\mu \quad \text{(1 parameter)} . \tag{6}$$

The generators of these transformations are

$$
\begin{aligned}
M_{\mu\nu} &= x_\mu P_\nu - x_\nu P_\mu \quad \text{(Lorentz Rotations)}, \\
P_\mu &= -i\partial_\mu \quad \text{(Translations)}, \\
K_\mu &= 2x_\mu x^\nu P_\nu - x^2 P_\mu \quad \text{(Accelerations)}, \\
D &= x^\mu P_\mu \quad \text{(Dilatation)} .
\end{aligned}
\tag{7}
$$

These satisfy the broader algebra

$$
\begin{aligned}
[M_{\mu\nu}, M_{\rho\sigma}] &= g_{\mu\rho} M_{\nu\sigma} - g_{\nu\sigma} M_{\mu\rho} - g_{\nu\rho} M_{\mu\sigma} - g_{\mu\sigma} M_{\nu\rho} , \\
[M_{\mu\nu}, P_\sigma] &= g_{\nu\sigma} P_\mu - g_{\mu\rho} P_\nu , \\
[M_{\mu\nu}, K_\lambda] &= g_{\nu\lambda} K_\mu - g_{\mu\lambda} K_\nu , \\
[M_{\mu\nu}, D] &= 0 , \\
[P_\mu, P_\nu] &= 0 , \\
[P_\mu, K_\nu] &= 2(g_{m\nu} D - M_{\mu\nu}) , \\
[P_\mu, D] &= P_\mu , \\
[K_\mu, K_\nu] &= 0 , \\
[K_\mu, D] &= -K_\mu , \\
[D, D] &= 0 .
\end{aligned}
\tag{8}
$$

In what follows we will only deal with the restricted symmetry associated with the generators $M_{\mu\nu}$, P_μ (the Poincaré group) and the one parameter group associated with Weyl's scale transformations.

SCALE INVARIANT ACTION

Under Weyl's scale invariance as a local symmetry the electroweak symmetry $SU(2) \times U(1)$ is extended to

$$G = SU(2) \times U(1) \times \widetilde{U}(1) , \tag{9}$$

where $\widetilde{U}(1)$ represents the local non-compact Abelian symmetry associated with Weyl's scale invariance. The additional particles introduced are the vector boson S_μ associated with $\widetilde{U}(1)$ and a real scalar field σ [16, 17, 18, 19, 20] that transforms as a singlet under G. The distinct feature of the new symmetry is that under it fields transform with a real phase whereas under the $SU(2) \times U(1)$ symmetries fields transform with complex phases.

Under $\tilde{U}(1)$ a generic field in the action is taken to transform as $e^{w\Lambda(x)}$ with a scale dimension w. Thus under $G = SU(2) \times U(1) \times \tilde{U}(1)$ the transformation properties of the entire particle content of the extended model are the following: The e-family $(g = 1)$,

$$\Psi_L^{1q} = \begin{pmatrix} u \\ d \end{pmatrix} \sim (2, \tfrac{1}{3}, -\tfrac{3}{2}) \; ; \quad \Psi_L^{1l} = \begin{pmatrix} v_e \\ e \end{pmatrix} \sim (2, -1, -\tfrac{3}{2}) \; ;$$

$$\Psi_{1R}^{1q} = u_R \sim (1, \tfrac{4}{3}, -\tfrac{3}{2}) \; ; \quad \Psi_{2R}^{1q} = d_R \sim (1, -\tfrac{2}{3}, -\tfrac{3}{2}) \; ;$$

$$\Psi_{2R}^{1l} = e_R \sim (1, -2, -\tfrac{3}{2}) \; , \tag{10}$$

and similarly for the μ-family $(g = 2)$ and the τ-family $(g = 3)$. All of these fermions have the same scale dimension $w = -3/2$. The scalar boson sector comprises the Higgs doublet Φ and the real scalar σ,

$$\Phi \sim (2, -1, -1) \; ; \quad \sigma \sim (1, 0, -1) \; , \tag{11}$$

with the common scale dimension $w = -1$. We introduce W_μ, B_μ and S_μ as the gauge potentials respectively associated with the $SU(2)$, $U(1)$, $\tilde{U}(1)$ symmetries. We suppress the $SU(3)$ of strong interactions as neglecting it will not affect our results and conclusions. The four dimensional volume element transforms as

$$d^4x \sqrt{-g} \to e^{4\Lambda(x)} \, d^4x \sqrt{-g} \; . \tag{12}$$

Since the vierbein $e_\mu{}^m$ and its inverse $e_m{}^\mu$ satisfy $e_\mu{}^m e_{vm} = g_{\mu v}$ and $e_m{}^\mu e_{n\mu} = \eta_{mn}$ where $(\eta_{mn}) = \mathrm{diag.}\,(1, -1, -1, -1)$ is the tangent space metric, it follows that the transformation properties of $e_\mu{}^m$ and its inverse $e_m{}^\mu$ under Weyl's symmetry are

$$e_\mu{}^m \to e^{\Lambda(x)} \, e_\mu{}^m \; , \quad e_m{}^\mu \to e^{-\Lambda(x)} \, e_m{}^\mu \; . \tag{13}$$

The action I of the model is [15]

$$
\begin{aligned}
I = \int d^4x \sqrt{-g} \Bigg[&-\tfrac{1}{4} g^{\mu\rho} g^{v\sigma} (W_{\mu v} W_{\rho\sigma} + B_{\mu v} B_{\rho\sigma} + U_{\mu v} U_{\rho\sigma}) \\
&+ \sum_{\substack{f=q,l \\ g=1,2,3 \\ i=1,2}} \left(\overline{\Psi}_L^{\,gf} e_m{}^\mu \gamma^m D_\mu \Psi_L^{gf} + \overline{\Psi}_{iR}^{\,gf} e_m{}^\mu \gamma^m D_\mu \Psi_{iR}^{gf} \right) + g^{\mu v} (D_\mu \Phi)(D_v \Phi^\dagger) + \tfrac{1}{2}(D_\mu \sigma)^2 \\
&+ \sum_{\substack{f=q,l \\ g,g'=1,2,3 \\ i=1,2}} \left(\mathbf{Y}_{gg'}^{f} \overline{\Psi}_L^{\,gf} \Phi \Psi_{iR}^{g'f} + \mathbf{Y}_{gg'}^{'f} \overline{\Psi}_L^{\,gf} \tilde{\Phi} \Psi_{iR}^{g'f} \right) + \mathrm{h.c.} \\
&- \tfrac{1}{2}(\beta \phi^\dagger \Phi + \zeta \sigma^2) \tilde{R} + V(\Phi, \sigma) \Bigg] \; ,
\end{aligned}
\tag{14}
$$

where $\widetilde{\Phi} \equiv i\sigma_2\phi^*$, the indices (g, g′) are for generations, the indices f = (q, l) refer to (quark, lepton) fields, $\mathbf{Y}^f_{gg'}$ or $\mathbf{Y}'^f_{gg'}$ are quark, lepton Yukawa couplings that define the mass matrices after symmetry breaking, the index $i = 1, 2$ is needed for right-handed fermions, while β and ζ are dimensionless couplings. The various D's acting on the fields represent the covariant derivatives constructed in the usual manner using the principle of minimal substitution. Explicitly,

$$D_\mu \Psi^{gf}_L = \left(\partial_\mu + ig\tau \cdot W_\mu + \tfrac{i}{2}g' Y^{gf}_L B_\mu - \tfrac{3}{2}fS_\mu - \tfrac{1}{2}\widetilde{\omega}_\mu{}^{mn}\sigma_{mn}\right)\Psi^{gf}_L ,$$

$$D_\mu \Psi^{gf}_{iR} = \left(\partial_\mu + \tfrac{i}{2}g' Y^{gf}_{iR} B_\mu - \tfrac{3}{2}fS_\mu - \tfrac{1}{2}\widetilde{\omega}_\mu{}^{mn}\sigma_{mn}\right)\Psi^{gf}_{iR} ,$$

$$D_\mu \Phi = \left(\partial_\mu + ig\tau \cdot W_\mu - \tfrac{1}{2}g' B_\mu - fS_\mu\right)\Phi ,$$

$$D_\mu \sigma = \left(\partial_\mu - fS_\mu\right)\sigma . \tag{15}$$

The Y^{gf}_L's , Y^{gf}_{iR}'s represent the hypercharge quantum numbers (e.g., f = q, g = 1, i = 1, $Y^{1q}_L = 1/3$, $Y^{1q}_{1R} = 4/3$, etc.), g, g′, f are the respective gauge couplings of $SU(2)$, $U(1)$, $\widetilde{U}(1)$. The $W_{\mu\nu}$ and $B_{\mu\nu}$ are the filed strengths associated with the gauge fields W_μ, B_μ of $SU(2)$, $U(1)$ while

$$U_{\mu\nu} \equiv \partial_\mu S_\nu - \partial_\nu S_\mu \tag{16}$$

is the field strength associated with Weyl's $\widetilde{U}(1)$. It is gauge invariant, since S_μ transforms as

$$S_\mu \to S_\mu - \tfrac{1}{f}\partial_\mu \Lambda . \tag{17}$$

The gauge fields and the field strengths carry scale dimension $w = 0$. The spin connection $\widetilde{\omega}_\mu{}^{mn}$ [21] is defined in terms of the vierbein $e_\mu{}^m$

$$\widetilde{\omega}_{mrs} \equiv \tfrac{1}{2}(\widetilde{C}_{mrs} - \widetilde{C}_{msr} + \widetilde{C}_{srm}) ,$$

$$\widetilde{C}_{\mu\nu}{}^r \equiv (\partial_\mu e_\nu{}^r + fS_\mu e_\nu{}^r) - (\partial_\nu e_\mu{}^r + fS_\nu e_\mu{}^r) , \tag{18}$$

while the affine connection $\widetilde{\Gamma}^\rho{}_{\mu\nu}$ is defined by

$$\widetilde{\Gamma}^\rho{}_{\mu\nu} = \tfrac{1}{2}g^{\rho\sigma}\left[(\partial_\mu + 2fS_\mu)g_{\nu\sigma} + (\partial_\nu + 2fS_\nu)g_{\mu\sigma} - (\partial_\sigma + 2fS_\sigma)g_{\mu\nu}\right] . \tag{19}$$

The Riemann curvature tensor $\widetilde{R}^\rho{}_{\sigma\mu\nu}$ is

$$\widetilde{R}^\rho{}_{\sigma\mu\nu} = \partial_\mu\widetilde{\Gamma}^\rho{}_{\nu\sigma} - \partial_\nu\widetilde{\Gamma}^\rho{}_{\mu\sigma} - \widetilde{\Gamma}^\lambda{}_{\mu\sigma}\widetilde{\Gamma}^\rho{}_{\nu\lambda} + \widetilde{\Gamma}^\lambda{}_{\nu\sigma}\widetilde{\Gamma}^\rho{}_{\mu\lambda} , \tag{20}$$

where $\widetilde{\Gamma}^\rho{}_{\mu\nu}$, $\widetilde{R}^\rho{}_{\sigma\mu\nu}$ and the Ricci tensor $\widetilde{R}^\rho{}_{\mu\rho\nu} = \widetilde{R}_{\mu\nu}$ have scale dimension $w = 0$, while the scalar curvature $\widetilde{R} = g^{\mu\nu}\widetilde{R}_{\mu\nu}$ has the form

$$\widetilde{R} = R - 6fD_\mu S^\mu + 6f^2 S_\mu S^\mu ,$$

$$D_\kappa S^\mu = \partial_\kappa S^\mu + \widetilde{\Gamma}^\mu{}_{\kappa\nu}S^\nu , \tag{21}$$

and transforms with scale dimension $w = -2$. The potential $V(\phi, \sigma)$ is given by

$$V(\Phi, \sigma) = \lambda\,(\Phi^\dagger\Phi)^2 - \mu\,(\Phi^\dagger\Phi)\,\sigma^2 + \xi\,\sigma^4 \,, \tag{22}$$

where λ, μ, ξ are dimensionless couplings.

BREAKING OF SCALE INVARIANCE AND IMPLICATIONS

The scalar potential in this model consists of quartic terms only as required by Weyl's scale invariance. Yet the desired descent, a two stage process, of G to $U(1)_{\mathrm{em}}$

$$G = SU(2) \times U(1) \times \widetilde{U}(1) \;\rightarrow\; SU(2) \times U(1) \;\rightarrow\; U(1)_{\mathrm{em}} \tag{23}$$

is possible. In the primary stage of symmetry breaking, scale invariance symmetry is broken. This occurs spontaneously and is achieved by setting

$$\sigma(x) = \tfrac{1}{\sqrt{2}}\Delta \,, \tag{24}$$

where Δ is a constant for the symmetry breaking scale associated with Weyl's $\widetilde{U}(1)$. It is to be noted that this phenomenon of spontaneous scale breaking is conceptually no different from conventional spontaneous symmetry breaking. In conventional spontaneous symmetry breaking, the term quadratic in the Higgs field changes sign suddenly from positive to negative while in spontaneous scale breaking under discussion here the scalar field σ freezes suddenly. The primary stage of symmetry breaking also determines Newton's gravitational constant G_N,

$$\zeta\,\Delta^2 = \tfrac{1}{4\pi G_N} \,. \tag{25}$$

Thus $\Delta \approx 0.3 \times M_P / \sqrt{\zeta}$ and barring any fine-tuning $\Delta \approx \mathcal{O}(M_P)$, if we take $\zeta \approx \mathcal{O}(1)$. At this stage the scalar field σ becomes the goldstone boson [22, 23]. The vector particle associated with $\widetilde{U}(1)$ breaking, the Weylon, absorbs the goldstone field and becomes massive with mass M_S given by

$$M_S = \sqrt{\tfrac{3f^2}{4\pi G_N}} \approx 0.5 \times f M_P \,. \tag{26}$$

Thus $M_S \approx \mathcal{O}(M_P)$ in the absence of fine-tuning $f \approx \mathcal{O}(1)$. Weyl's $\widetilde{U}(1)$ symmetry decouples completely and the scalar potential after the primary stage of symmetry breaking takes the form

$$V(\Phi) = -\mu\,\Delta^2(\Phi^\dagger\Phi) + \lambda\,(\Phi^\dagger\Phi)^2 + \tfrac{\xi}{4}\Delta^4 \,. \tag{27}$$

It is to be noted that this form of the potential, apart from the vacuum energy density term contributing to the cosmological constant, is of the same form as the standard Higgs potential in the standard model. All the conventional particles are still massless at this stage. With G_N defined, it is appropriate to work in the weak field approximation.

Henceforth we set $\sqrt{g}g_{\mu\nu} \approx \eta_{\mu\nu} + \mathcal{O}(\kappa)$ where $\kappa^2 = 16\pi G_N$. The secondary stage of symmetry breaking is spontaneous in the conventional sense. This takes place when $\Phi \to \langle\Phi\rangle$ where

$$\langle\Phi\rangle = \frac{1}{\sqrt{2}} \begin{pmatrix} \eta \\ 0 \end{pmatrix} \ , \tag{28}$$

$$\eta = \sqrt{\frac{\mu\Delta^2}{\lambda}} \ , \tag{29}$$

and η is the electroweak symmetry breaking scale of order 250 GeV. In the standard model, μ and λ are unrelated while in this model they are related,

$$\frac{\mu}{\lambda} = \left(\frac{\eta}{\Delta}\right)^2 \approx 2.4 \times \zeta \, G_F^{-1} M_P^{-2} \approx 10^{-33} \times \zeta \ . \tag{30}$$

After spontaneous symmetry breaking (SSB), the conventional particles acquire masses as in the standard model,

$$M_W = \tfrac{1}{2}g\eta \ , \quad M_Z = \frac{M_W}{\cos\theta_W} \ ,$$
$$\mathbf{M}^{\mathrm{f}}_{gg'} = \tfrac{1}{\sqrt{2}}\mathbf{Y}^{\mathrm{f}}_{gg'}\eta \ , \quad \mathbf{M}'^{\mathrm{f}}_{gg'} = \tfrac{1}{\sqrt{2}}\mathbf{Y}'^{\mathrm{f}}_{gg'}\eta \ , \tag{31}$$

where θ_W is the weak angle and $\mathbf{M}^{\mathrm{f}}_{gg'}$, $\mathbf{M}'^{\mathrm{f}}_{gg'}$ are the quark $(\mathrm{f} = q)$ and the charged lepton $(\mathrm{f} = l)$ mass matrices in terms of the Yukawa couplings $\mathbf{Y}^{\mathrm{f}}_{gg'}$ and $\mathbf{Y}'^{\mathrm{f}}_{gg'}$. At this stage neutrinos are still massless. In this model there is still left over the conventional Higgs particle h_0 with mass given by

$$M_{h_0} = \sqrt{\mu}\Delta \approx 0.3 \times \sqrt{\tfrac{\mu}{\zeta}} \, M_P \ , \tag{32}$$

which is undetermined as μ and ζ are still free parameters. It is interesting to note that in this model the mass of the Higgs particle is tied to the scale associated with the breaking of Weyl's $\tilde{U}(1)$ symmetry which is of order Planck mass. In principle, M_{h_0} can be as large as M_P posing problems with unitarity. However, although the standard model is a renormalizable theory [24, 25], the present model is not. This puts into doubt the validity of the unitarity constraint derived in the renormalizable standard model and extrapolated to the non-renormalizable extended model considered here. After SSB, the mass of the Weylon gets shifted,

$$M_S \to \sqrt{\frac{3f^2}{4\pi G_N}\left(1 + \frac{\beta\eta^2}{\zeta\Delta^2}\right)} \ . \tag{33}$$

However, the additional contribution is negligibly small as $\eta^2/\Delta^2 \approx 10^{-33}$. Apart from being superheavy, another distinct property of the Weylon is that it completely decouples from the fermions of the standard model.

NEUTRINO MASSES

At the present time, one fundamental issue is that of neutrino masses and their lightness as compared to the masses of other particles. In the standard model and the model under consideration, neutrinos are strictly massless as neither right-handed neutral lepton fields nor unconventional scalar fields are present. A popular extension of the standard model that addresses the issue of neutrino masses and mixings in an aesthetically appealing way introduces right-handed neutrinos $\Psi_{1R}^{1l} = \nu_{eR}$, $\Psi_{1R}^{2l} = \nu_{\mu R}$, $\Psi_{1R}^{3l} = \nu_{\tau R}$ that lead to seesaw masses [26] for the the conventional neutrinos. This scenario is usually entertained in the $SO(10)$ grand unified theory, where the right-handed neutrinos acquire super heavy masses. The super heavy scale is determined by the stage at which the internal symmetry $SO(10)$ breaks, and has nothing to do with gravitational interactions. If right-handed neutrino fields are also introduced in the present model, the seesaw mechanism can naturally be accommodated due to the presence of the singlet field σ. The relevant interaction Lagrangian is

$$L_\nu = \sum_{\substack{g,g'=1,2,3 \\ i=1}} \left(\mathbf{Y}_{gg'}^l \overline{\Psi}_L^{gl} \Phi \Psi_{iR}^{g'l} + \text{h.c.} + \tfrac{1}{2} \mathbf{Y}_{gg'}^{RR} \sigma_{1R}^{gl\,T} C \sigma \Psi_{1R}^{g'l} \right) . \tag{34}$$

Lepton number is explicitly broken by the last term. Scale breaking gives superheavy Majorana masses to the right-handed neutrinos and SSB subsequently gives Dirac masses that connects the left- and right-handed neutrinos leading to the following familiar 6×6 mass matrix

$$\mathbf{M}_\nu = \frac{1}{\sqrt{2}} \begin{pmatrix} \mathbf{0} & \mathbf{Y}_{gg'}^l \eta \\ \mathbf{Y}_{g'g}^{l*} \eta & \mathbf{Y}_{gg'}^{RR} \Delta \end{pmatrix} , \tag{35}$$

the eigenvalues of which are three seesaw masses for the light neutrinos and three heavy neutrinos with enough parameters to fit the observed solar and atmospheric neutrino oscillation phenomena. In the present model, the scale of right-handed neutrino masses is tied to the scale Δ associated with Weyl's $\widetilde{U}(1)$ breaking which in turn is tied to Newton's constant G_N. This is unlike the see-saw GUT scenario where right-handed neutrino masses are tied to the GUT scale at which the grand unification internal symmetry breaks. Thus in our scale invariant model the absence of right-handed neutrinos from the low energy scales is attributed to their superheavy masses which are naturally of $\mathcal{O}(M_P)$. Perhaps this is an indication that right-handed neutrinos (and also gauge-mediated right-handed currents) and gravitational interactions are ultimately related.

We stress that our model needs only quartic potential for the scalar fields Φ and σ only with dimensionless couplings as its foundation. The scale-breaking parameter Δ then induces the quadratic terms in the resulting potential (20). Whereas in the standard model μ and λ are not related, our model relates them in terms of Δ via (30).

We note that the symmetry breaking scheme depicted in the model under consideration would apply universally to theories that accommodate local scale invariance and generate Newton's constant G_N as a symmetry breaking effect. In the conventional

SSB mechanism the scalar potential contains terms that are quadratic in scalar fields. Such terms are either added explicitly by hand or generated via quantum corrections.

Our contention is that the present model presents a viable scheme in which gravity is unified, albeit in a semi-satisfactory way, with the other interactions. In the standard model physical fields and the couplings like electric charge $e = 1/\sqrt{g^{-2} + g'^{-2}}$ and Fermi constant $G_F = g^2/(8M_W^2)$ get defined *after* SSB. Similarly, in the present model, not only e and G_F, but also G_N gets defined *after* symmetry breaking, thus conforming to the main theme in physics that all phenomena observed in Nature are symmetry breaking effects. In the complete theory of all interactions, the model described here will emerge as an effective theory representing the four fundamental interactions in the low energy limit.

SCALE INVARIANT SU(5) GUT

In theories unifying all the elementary particle interactions and possessing both local scale invariance and internal symmetry invariance, it is a scale invariance breaking that would precede spontaneous symmetry breaking. This is because since all such theories would contain the scalar curvature R, Newton's constant G_N would be generated as the primary symmetry breaking effect. After scale breaking, the resulting potential would contain the necessary terms quadratic in scalar fields to effect SSB, similar to the discussion in the text, resulting in the GUT scale $M_G \approx M_P$, intermediate scale(s) M_I ($M_I, M_{II}, M_{III}, \cdots$) and the electroweak scale $M_W \approx \sqrt{1/G_F}$ with the hierarchy $M_G > M_I > M_{II} > M_{III} > \cdots > M_W$.

As a concrete example we illustrate this scenario in a scale invariant $SU(5)$ model. The $SU(5)$ GUT consists of the usual gauge bosons in the **24**, the fermions in the **5** and the **$\overline{10}$**, and the scalar fields in the **5** ($\equiv H$) and the **24** ($\equiv \Phi$) representations of $SU(5)$. To make scale invariant $SU(5)$ GUT, we extend the gauge symmetry from $SU(5)$ to

$$G = SU(5) \times \tilde{U}(1) \tag{36}$$

and add a real scalar σ that is a singlet of $SU(5)$. The scale invariant Lagrangian is straightforward to write down along the lines discussed in the text. The most important term is the scalar potential $V(H, \Phi, \sigma)$ where

$$
\begin{aligned}
V(H, \Phi, \sigma) =\ & \lambda_H (H^\dagger H)^2 + \lambda_\Phi (\operatorname{Tr}\Phi^2)^2 + \lambda'_\Phi \operatorname{Tr}(\Phi^4) + \lambda_\sigma \sigma^4 \\
& + \lambda_{H\Phi} H^\dagger H \operatorname{Tr}\Phi^2 + \lambda_{H\sigma} H^\dagger H \sigma^2 + \lambda_{\Phi\sigma} (\operatorname{Tr}\Phi^2)\sigma^2 \\
& + \lambda'_{H\Phi} (H^\dagger \Phi^2 H) + \lambda_{\sigma H\Phi}\, \sigma H^\dagger \Phi H + \lambda'_{\sigma\Phi}\, \sigma \operatorname{Tr}\Phi^3 \ .
\end{aligned}
\tag{37}
$$

This is the most general potential consistent with the symmetries of the theory. Notice the important fact that all terms are quartic in the scalar fields. The primary descent occurs when the singlet σ acquires a VEV *i.e.*, $\langle \sigma(x) \rangle = \Delta/\sqrt{2}$. In this stage scale invariance is spontaneously broken and

$$G = SU(5) \times \tilde{U}(1) \xrightarrow{\langle\sigma\rangle \equiv M_P} SU(5) \tag{38}$$

After this stage of symmetry breaking the potential is the usual one of the $SU(5)$ GUT and consists of the usual fields H and Φ. Dimensionful couplings linear and quadratic in the mass dimension appear. The potential, after rescaling, now contains terms quadratic, cubic and quartic in H and Φ and has the required rich structure to trigger spontaneous symmetry breaking in the conventional sense with the secondary stage and the ternary stage characterized by the vacuum expectation values of Φ and H,

$$SU(5) \xrightarrow{\langle \Phi \rangle \equiv M_I} SU(3) \times SU(2) \times U(1) \xrightarrow{\langle H \rangle \equiv M_W} SU(3) \times U(1)_{em} \quad (39)$$

This model is now the usual $SU(5)$ model with an additional gauge boson, the Weylon, Conceptually, there are marked differences. The standard $SU(5)$ theory fell out of repute because it predicted low weak angle $\sin^2 \theta_W$ and rapid Proton decay, predictions that turned out to be contrary to empirical observations. The present model may not suffer from such defects. The main reason is that the scale invariant $SU(5)$ model described here is semi-renormalizable. It is an effective theory that will eventually emerge from a unified scheme of all interactions that successfully incorporates quantum gravity. Thus the renormalization effects that sent the standard $SU(5)$ theory to disrepute do not apply to the scale invariant $SU(5)$ model discussed here. The additional renormalization effects due to gravitational interactions may easily provide the patch necessary to restore the standard $SU(5)$ model back to its full glory. Donohue [27] has argued that treating conventional field theory models with quantum gravity included (such as the one described here) leads to viable effective theories with quantum corrections due to gravitation interactions as legitimate contributions to the part of the theory that has conventional renormalizable interactions. Consider the one loop renormalized gauge couplings in the $SU(5)$ model with additional contributions $\delta_x = \delta_x(M_P, M_I, M_W), x = 1, 2, 3$ resulting from the complete theory,

$$\frac{1}{g_1^2(M_W)} = \frac{1}{g^2} + b_1 \ln \frac{M_I}{M_W} + \delta_1 \ ,$$
$$\frac{1}{g_2^2(M_W)} = \frac{1}{g^2} + b_2 \ln \frac{M_I}{M_W} + \delta_2 \ ,$$
$$\frac{1}{g_3^2(M_W)} = \frac{1}{g^2} + b_3 \ln \frac{M_I}{M_W} + \delta_3 \ , \quad (40)$$

where the b_i's are the usual one loop β-function coefficients, g_i's are the renormalized gauge couplings of $SU(3), SU(2), U(1)$ at the weak scale M_W and the $\delta_x, x = 1, 2, 3$ are the additionl contributions satisfying the constraint $\delta_1 = \delta_2 = \delta_3 = \delta$ at the renormalization point $\mu = M_P$. Also,

$$\frac{1}{e^2(M_W)} = \frac{1}{g_2^2(M_W)} + \frac{5}{3g_1^2(M_W)} \quad \text{and} \quad \sin^2 \theta_W(M_W) = \frac{e^2(M_W)}{g_2^2(M_W)} \ . \quad (41)$$

Since gravitational interactions do not contribute to electric charge, the definition of e remains defined in terms of g_1 and g_2. With these modifications the predictions for the weak angle and the intermediate GUT scale are

$$\sin^2 \theta_W(M_W) = \sin^2 \theta_W(M_W)\,|_{SU(5)} + \kappa_1 \left[(b_2 - b_3)\delta_1 + (b_3 - b_1)d_2 + (b_1 - b_2)d_3\right] \ ,$$

$$\ln\frac{M_I}{M_W} = \ln\frac{M_G}{M_W}\Big|_{SU(5)} + \kappa_2(5\delta_1 + 3d_2 - 8d_3) , \tag{42}$$

where $\big|_{SU(5)}$ are the expressions as in the conventional $SU(5)$ GUT, $\kappa_1 = 20\pi\alpha_{em}/(8b_3 - 3b_2 - 5b_1)$, $\kappa_2 = 8\pi^2/3(8b_3 - 3b_2 - 5b_1)$ and the δ_i's are the additional contributions. As input we take the weak mixing angle to be equal to the experimental value, $\sin^2\theta_W(M_W) = 0.23$, $\alpha_{em} \approx 1/128$, $\alpha_s \approx 0.11$ and the intermediate scale to be the value $M_G/M_W = 10^{15}$ that meets the present limit on the lifetime of the Proton. With this, the constraints on the various δ_i's are

$$0.16\delta_1 - 0.40\delta_2 + 0.26\delta_3 = 1 ,$$
$$0.40\delta_1 + 0.24\delta_2 - 0.63\delta_3 = 1 . \tag{43}$$

Tiny effects due to gravitational interactions can easily amplify the various δ_i's at the renormalization point $\mu = M_W$ to provide the required patch such that the scale invariant $SU(5)$ model fares better that the conventional $SU(5)$ GUT. That this is indeed the case has been recently demonstrated by Robinson and Wilczek [28] who, working in the philisophy advocated by Donoghue [27], compute the one loop contributions due to graviton exchange to the renormalization of the gauge couplings and show that the graviton contributions work in the right direction as implied in this work.

To conclude, we have accommodated Weyl's scale invariance as a local symmetry in the standard electroweak model. This inevitably leads to the introduction of general relativity. The additional particles are one vector particle we call the Weylon and a real scalar singlet that couples to the scalar curvature $\widetilde{R}$ *à la* Dirac [16]. The scale at which Weyl's scale invariance breaks defines Newton's gravitational constant G_N. Weyl's vector particle, *i.e.*, the Weylon absorbs the scalar singlet σ and acquires mass $\mathcal{O}(M_P)$ in the absence of fine tuning. The scalar potential is unique in the sense that it consists of terms only quartic in the scalar fields and dimensionless couplings. Yet, as we have demonstrated, symmetry breaking is possible such that the left-over symmetry is $U(1)_{em}$ and all particle masses are consistent with present day phenomenology. If right-handed neutrinos are also introduced, the light neutrinos acquire seesaw masses and the suppression factor in the neutrino masses is of $\mathcal{O}(M_P)$. As a concrete example, $SU(5)$ GUT with local scale invariance is presented and the implications noted.

I don't know about you, but

"Herman Weyl would have been very happy"

to see his work revived in the light of our present understanding of elementary particle interactions. After all, his gauge idea may turn be out not as futile as once perceived.

ACKNOWLEDGMENTS

We would like to thank Jyoti and Ravi for reading the manuscript. This work is supported in part by NSF Grant # 0308246.

REFERENCES

1. S.L. Glashow, Nucl. Phys. **22** (1961) 519; S. Weinberg, Phys. Rev. Lett. **18** (1967) 507; A. Salam, in *'Elementary Particle Physics'*, N. Svartholm, ed. (Nobel Symposium No. 8, Almqvist & Wiksell, Stockholm, 1968), p. 367.
2. P.W. Higgs, Phys. Lett. **12** (1964) 132; Phys. Lett. **13** (1964) 508; Phys. Rev. **145** (1966) 1156.
3. J.C. Pati and A. Salam; Phys. Rev. **8** (1973) 1240; J.C. Pati and A. Salam, Phys. Rev. Lett. **31** (1973) 661; Phys. Rev. **10** (1974) 275; H. Georgi and S.L. Glashow, Phys. Rev. Lett. **32** (1974) 438.
4. B. Pontecorvo, J. Exptl. Theor. Phys. **33** (1957) 549; Sov. Phys. JETP **6** (1958) 429.
5. B. Pontecorvo, J. Exptl. Theor. Phys. **34** (1958) 247; Sov. Phys. JETP **7** (1958) 172.
6. H. Weyl, S.-B. Preuss. Akad. Wiss. **465** (1918); Math. Z. **2** (1918) 384; Ann. Phys. **59** (1919) 101; *Raum, Zeit, Materie'*, vierte erweiterte Auflage: Julius Springer (1921).
7. A.S. Eddington, *The Mathematical Theory of Relativity*, Cambridge University Press, (1922).
8. E. Cunningham, Proc. London Math. Soc. **8** (1909) 77.
9. H. Bateman, Proc. London Math. Soc. **8** (1910) 223.
10. For more references, also see T. Fulton, F. Rohrlich and l. Witten, Rev. Mod. Phys. **34** (1962) 442.
11. Ya. B. Zel'dovich, Zh. Eksp. Teor. Fiz. Pis'ma Red., **6** (1967) 316 (JETP Lett. **6** (1967) 316.
12. A. Sakharov, Doc. Akad. Nauk. SSSR **177** (1968) 70 (Sov. Phys. Dokl. **12** (1968) 1040.
13. A. Sakharov, Teor. Mat. Fiz. **23** (1975) 435.
14. C. Pilot and S. Rajpoot, 'Gauge and Gravitational Interactions with Local Scale Invariance', in *Proc. 7th Mexican School of Particles and Fields and 1st. Latin American Symposium on High-Energy Physics* (VII-EMPC and I-SILAFAE - Dedicated to Memory of Juan Jose Giambiagi), Merida, Yucatan, Mexico, 1996, ed. J.C. D'Oliva, M. Klein-Kreisler, H. Mendez (AIP Conference proceedings: 400, 1997), p. 578.
15. H. Nishino and S. Rajpoot, *'Broken Scale Invariance in the Standard Model'*, hep-th/0403039.
16. P.A.M. Dirac, Proc. Roy. Soc. (London) **A333** (1973) 403.
17. R. Utiyama, Prog. Theor. Phys. **50** (1973) 2080.
18. T. W. B. Kibble, J. Math. Phys. **2** (1961) 212.
19. D.K. Sen and K.A. Dunn, J. Math. Phys. **12** (1971) 578.
20. Y.M. Cho, Phys. Rev. Lett. **68** (1992) 3133.
21. V. Fock, Zeit. für Phys. **57** (1929) 261.
22. J. Goldstone, Nuovo Cimento **19** (1961) 154.
23. J. Goldstone, A. Salam, and S. Weinberg, Phys. Rev. **127** (1962) 965.
24. G. 't Hooft and M. Veltman, Nucl. Phys. **44** (1972) 189, *i*bid. **B50** (1972) 318.
25. B.W. Lee and J. Zinn-Justin, Phys. Rev. **D5** (1972) 3121, 3137, 3155.
26. M. Gell-Mann, P. Ramond and R. Slansky, in *S*upergravity, Proceedings of the Workshop, Stony Brook, New York, 1979, ed. P. van Nieuwenhuizen and D.Z. Freedman (North-Holland, Amsterdam, 1979), p.315; T. Yanagida, in 'Proc. Workshop on Unified Theory and the Baryon Number of the Universe', Tsukuba, Japan, 1979, *e*d. O. Sawada and A. Sugamoto (KEK Report No. 79-18, Tsukuba, 1979), p. 95.
27. J. F. Donoghue, Phys. Rev. D **50**, 3874 (1994).
28. S. P. Robinson and F. Wilczek, Phys. Rev. Lett. **96** (2005) 231601.

Neutrino Mass and Mixing: Leptons versus Quarks

Alexei Yu. Smirnov

International Centre for Theoretical Physics, Strada Costiera 11, 34014 Trieste, Italy
Institute for Nuclear Research, RAS, Moscow, Russia

Abstract.
Understanding similarities and differences of properties of quark and leptons is one of the milestones on the way to underlying physics. Several observations, if not accidental, can strongly affect implications of data: (i) nearly tri-bimaximal character of lepton mixing, (ii) special neutrino symmetries, (iii) the QLC-relations. We consider possible connections between quarks and leptons which include the quark-lepton symmetry and unification, approximate universality, and quark-lepton complementarity. Presence of new neutrino states and their mixing with the left or/and right handed neutrinos can be the origin of additional differences of quarks and leptons.

Keywords: Masses, mixing, flavor symmetries, unification
PACS: `14.60.-z,14.65.-q`

INTRODUCTION

One of the key issues on the way to underlying physics is a comparison of properties of quarks and leptons and understanding their similarities and differences. This comparison has two aspects of the fundamental importance:

- understanding the fermion masses and mixings;
- uncovering the path of further unification - unification of quarks and leptons, particles and forces.

Are quarks and leptons similar or fundamentally different? Still whole spectrum of possibilities exists from the weakly broken quark-lepton universality to existence of different structures and symmetries in these two sectors.

In this paper we confront properties of quarks and leptons. We then discuss their possible connections: (1) symmetry and unification; (2) universality; (3) complementarity; (4) diversity, that is, existence of new structures which can produce difference in two sectors.

CP881, *Cairo International Conference on High Energy Physics (CICHEP II)*,
edited by S. Khalil

TABLE 1. The best fit values of mixing angles in the quark and lepton sectors at m_Z scale in degrees. Shown are also the sums of the angles with 1σ error bars.

angles	quarks	leptons	sum
θ_{12}	$12.8°$	$33.9°$	$46.7° \pm 2.4°$
θ_{23}	$2.3°$	$41.6°$	$43.9°{\,}^{+5.1°}_{-3.6°}$
θ_{13}	$0.5°$	$< 8.0°$	$< 8.5°$

LEPTONS VERSUS QUARKS

Confronting mixing and masses

To compare mixings in the quark and lepton sector we use the standard parametrization of mixing matrices:

$$V_f = V_{23}(\theta_{23})I_\delta V_{13}(\theta_{13})V_{12}(\theta_{12}), \quad f = CKM, \ PMNS, \tag{1}$$

where V_{ij} is the matrix of rotations in the ij- plane, and I_δ is the diagonal matrix of the CP-violating phases. (Notice that V_{PMNS} corresponds to $V^\dagger_{CKM}$).

The Table I presents the mixing angles in the quark and lepton sectors from the analysis of ref. [1]. Similar results have been obtained by other groups [2, 3]. Shown are also the sums of the corresponding angles. Apparently, the mixing patterns in these two sectors are strongly different. The only common feature is that the 1-3 mixings (between the "remote" generations) are small in both cases.

Several comments are in order. The b.f. value of the 1-2 mixing angle, $\theta_{12} = 33.9°$, deviates from the maximal mixing by more than 6σ [2].

The 2-3 mixing is consistent with maximal one. The deviation of the b.f. value from maximal mixing is characterized by $D_{23} \equiv 0.5 - \sin^2\theta_{23} = 0.03 - 0.06$. Still large deviation is allowed: $-0.17 < D_{23} < 0.21$ and relative shift can be as large as

$$D_{23}/\sin^2\theta_{23} \sim 0.4 \quad (2\sigma). \tag{2}$$

The 1-3 leptonic mixing is consistent with zero. The most conservative 3σ bound is $\sin^2\theta_{13} < 0.048$ [1], and at 1σ we have $\sin\theta_{13} < 0.13$. So, apparently the quark feature $\theta_{13} \sim \theta_{12} \times \theta_{23}$ does not work here. An interesting benchmark for θ_{13} is the ratio of the solar and atmospheric neutrino mass scales, $\sin\theta_{13} = \sqrt{r} \equiv \sqrt{\Delta m^2_{21}/\Delta m^2_{31}} = 0.17$,

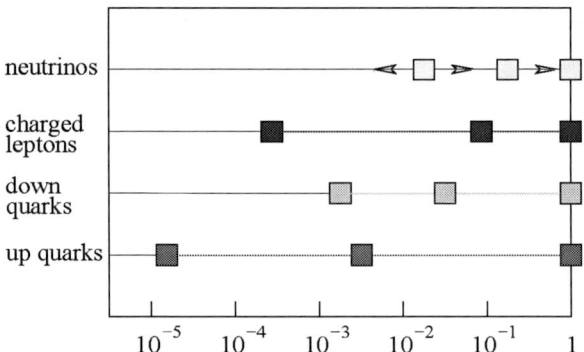

FIGURE 1. Mass hierarchies of quarks and leptons. The mass of heaviest fermion of a given type is taken to be 1.

which is allowed at about 2σ level. An additional (model dependent) factor of the order 0.3 - 2 may appear in this relation. Much smaller values of $\sin\theta_{13}$ would imply most probably certain symmetry of the mass matrix.

Let us consider now the masses. The latest analysis of the cosmological data (including the WMAP 3 years result) gives the upper bound on the sum of masses of active neutrinos [4] $\sum_i m_i < 0.14$ eV, 95%C.L. which already starts to disfavor the degenerate spectrum of neutrinos. On the other hand, if the Heidelberg-Moscow result [5] is confirmed and if it is due to exchange of the light Majorana neutrinos, the neutrino mass spectrum should be strongly degenerate with a common mass $m_0 \sim (0.2-0.6)$ eV. This would be in conflict with the cosmological bound.

The solar and the atmospheric mass differences squared give the lower bound on ratio of the second and third neutrino masses:

$$\frac{m_2}{m_3} \geq \sqrt{r} = 0.15 - 0.20. \tag{3}$$

This should be compared with ratios for charged leptons and quarks (at m_Z scale): $m_\mu/m_\tau = 0.06$, $m_s/m_b = 0.02 - 0.03$, $m_c/m_t = 0.005$. Apparently, the neutrino hierarchy (3) is the weakest one. This is consistent with possible mass-mixing relation: large mixings are associated to weak mass hierarchy. In fig. 1 we show the mass ratios for three generations. The strongest hierarchy and geometric relation $m_u \times m_t \sim m_c^2$ exist for the upper quarks. It seems the observed pattern of masses is an interplay of some regularities (flavor alignment) and randomness ("anarchy"). That may indicate the perturbative picture when the lowest order masses and mixing are universal, whereas corrections have more complicated ("random") flavor structure.

In what follows we will discuss certain observed features which can strongly affect interpretation of the results.

Tri-bimaximal mixing

In the lepton sector experimental results are in a very good agreement with the so called tri-bimaximal mixing [6]. The corresponding mixing matrix is

$$U_{tbm} = U_{23}^m U_{12}(\theta_{12}) = \frac{1}{\sqrt{6}} \begin{pmatrix} 2 & \sqrt{2} & 0 \\ -1 & \sqrt{2} & \sqrt{3} \\ 1 & -\sqrt{2} & \sqrt{3} \end{pmatrix}, \tag{4}$$

where $\sin^2 \theta_{12} = 1/3$ is about 1σ larger than the best experimental fit value. Here ν_2 is tri-maximally mixed. Mixing parameters turn out to be some simple numbers $0, 1/\sqrt{3}, 1/\sqrt{2}$ and can appear as Clebsh-Gordon coefficients.

In the case of normal mass hierarchy ($m_1 \approx 0$) the mass matrix which leads to the tri-bimaximal mixing has the following form

$$m_\nu \approx \frac{m_3}{2} \begin{pmatrix} 0 & 0 & 0 \\ 0 & 1 & -1 \\ 0 & -1 & 1 \end{pmatrix} + \frac{m_2}{3} \begin{pmatrix} 1 & 1 & 1 \\ 1 & 1 & 1 \\ 1 & 1 & 1 \end{pmatrix}, \tag{5}$$

where $m_2 \approx \sqrt{\Delta m_{21}^2}$ and $m_3 \approx \sqrt{\Delta m_{31}^2}$. It is the sum of two singular matrices with certain symmetries which gives some hint of its origin.

Matrix (4) was motivated by certain geometric consideration. If description of the data by (4) is not accidental and certain principle/symmetry is behind, we should conclude on substantial differences in the quark and lepton sectors. Though some models have been proposed which reproduce the tri-bimaximal mixing and include also quark [7].

Complementarity

According to the Table I, the sums of the mixing angles of quarks and leptons for the 1-2 and 2-3 generations agree with $45°$. The quark and lepton mixings sum up to maximal mixing [8, 9]. Possible implications of these equalities called the quark-lepton complementarity relation (QLC) will be considered in sect. 3.3. Notice that the QLC relations written for angles are are essentially parametrization independent. Indeed, due to smallness of 1-3 mixings in the quark and lepton sectors the relations can be written as $\arcsin(V_{us}) + \arcsin(V_{e2}) = \pi/4$. The mixing matrix elements V_{us} and V_{e2} are physical parameters.

Neutrino symmetry

Several observations may testify for special symmetry(ies) associated to neutrinos. In particular, (i) maximal or nearly maximal 2-3 mixing, (ii) zero 1-3 mixing, both indicate the same underlying symmetry: Both features can be consequences of the $\nu_\mu - \nu_\tau$ permutation symmetry of the neutrino mass matrix [10] in the flavor basis. The permutation symmetry can be a part of, e.g., discrete S_3, A_4 or D_4 groups which in turn, are the subgroups of continuous SO(3).

Important fact is that the symmetry is realized for neutrinos only, and only in the flavor basis where the charge lepton mass matrix is diagonal. The symmetry is broken in the charged lepton sector by inequality of masses of muon and tau lepton. Model should be constructed in such a way that symmetry is weakly broken in the neutrino sector but strongly broken for the charged leptons. This implies different transformation properties of the right handed components of neutrinos and charged leptons, since the left components form the SU(2) doublets. This, in turn, contradicts the L-R symmetry, and consequently, the $SO(10)$ type of unification. Still such symmetry transformations can be consistent with the $SU(5)$ unification. Alternatively, one can consider more sophisticated fermionic or/and Higgs sectors. It is also non-trivial to extend the symmetry to the quark sector which prevents from any simple Grand Unification.

A modification of the $\nu_\mu \leftrightarrow \nu_\tau$ symmetry has been proposed recently that can be the universal symmetry of quarks and leptons [11]. The symmetry is formulated in the basis which differs from the flavor basis and therefore should be considered as the $2-3$ family symmetry. It is argued that beside maximal (large) 2-3 leptonic mixing, smallnes of the V_{cb} element of the CKM-mixing matrix testifies for this symmetry as well.

Generic feature of the models is that introduction of symmetry is motivated by maximal or nearly maximal lepton mixing. However realizations of the symmetry in a majority of gauge models show that large mixing appears eventually as a result of tuning of parameters and *not as consequence of symmetry*. This clearly makes whole context to be inconsistent.

Two remarks are in order.

(i) Symmetry is realized in terms of the mass (Yukawa coupling) matrices. It turns out that structure of the mass matrix is very sensitive to even small deviations of the 2-3 mixing from maximal and 1-3 mixing from zero. Taking the best fit values of parameters from [1] $\sin^2 \theta_{13} = 0.01$, $\sin^2 \theta_{23} = 0.44$, we obtain the matrix of the absolute values of masses in meV [12]:

$$M = \begin{pmatrix} 3.2 & 6.0 & 0.6 \\ \dots & 24.8 & 21.4 \\ \dots & \dots & 30.7 \end{pmatrix} \tag{6}$$

which should be compared with the symmetry matrix: here the 12 and 13 elements are strongly different and 33- element is greater than 22- element by $20-25\%$.

(ii) The present measurements allow substantial deviations of θ_{23} from maximal and θ_{13} from zero. That, in turn, allows even stronger deviation of the matrix from the symmetric form. So, it is not excluded that neutrino symmetry approach is simply misleading.

Additional structure?

The features discussed above: tri-bimaximal mixing, neutrino symmetry, quark-lepton complementarity may indicate that quarks and leptons are fundamentally different and some additional structures exist that lead to this difference. The main question here is whether these features/relations are real or accidental? "Real" in a sense that simple and direct symmetry or principle exist which lead to the relations. "Accidental" in a sense that relations are an interplay (sum) of several independents effects or contributions.

Quarks and leptons have similar gauge structure, which establishes clear correspondence of the leptons and quarks. On the other hand, the quarks and leptons have strongly different mass and mixing patterns. The hope is that all particular features of neutrino mass spectrum and lepton mixing can be reduced eventually to the neutrality of neutrinos: zero electric and color charges. This neutrality opens unique possibility for neutrinos to

- have the Majorana mass terms, and
- mix with singlets of the SM symmetry group.

Both features are realized in the seesaw mechanism [13]. The second one may have two different effects: (i) modify the mass matrix of active neutrinos, (ii) produce certain dynamical effects in the neutrino conversion (if new states are light).

Is this enough to explain all salient properties of neutrinos? Do the data really indicate existence of new physics structure (new particles, interactions, symmetries)? Is this additional structure the seesaw, or something beyond seesaw is involved? In this connection a general context could be that beyond the SM apart from the RH neutrinos some other fermions (singlets of the SM symmetry group) exist. These fermions can have various origins in physics beyond the SM, being related to Grand Unification, supersymmetry, existence of extra dimension, *etc.*. Existence of large number of singlets is a generic consequence of string theory. Masses of these singlets can be essentially at any scale, from zero to the Planck mass. They can mix in general with both LH and RH neutrino components.

The singlets and their mixing with SM neutrinos may be a missed structure which explains the difference of quark and lepton properties on the top of strong interactions.

QUARK-LEPTON CONNECTIONS

Quark-lepton symmetry

There is an apparent correspondence between quarks and leptons. Each quark has its own counterpartner in the leptonic sector. Leptons can be treated as the 4th color following the Pati-Salam $SU(4)_C$ unification symmetry [14].

Further unification is possible, when quarks and leptons form multiplets of larger gauge group. The most appealing possibility is SO(10) [15], where all known components of quarks and leptons as well as the RH neutrinos form unique 16-plet. It is difficult to believe that these features are accidental. Though, it is not excluded that the quark-lepton connection has rather complicated form.

The quark-lepton symmetry is not equivalent to the quark-lepton unification. Indeed, in the $SU(5)$ GU models the quark-lepton correspondence ($v \leftrightarrow u$, $d \leftrightarrow l$) is explicitly broken by different $SU(5)$-gauge transformation properties: $u, u^c \sim \mathbf{10}$, whereas $v \sim \mathbf{5}$, $v^c \sim \mathbf{1}$, then $d \sim \mathbf{10}$, $d^c \sim \mathbf{5}$ but $l \sim \bar{\mathbf{5}}$, $l^c \sim \mathbf{10}$. This unification leads to diversity which is not seen in the low energy effective theory.

The difference of the gauge properties can lead to (i) different mass hierarchies of upper and down quarks, and also charge leptons and neutrinos [16]; (ii) different mixings of quarks and leptons. In fact, the loopsided mechanism of large mixing realizes this possibility [17].

Generically, GUT's provide with all ingredients that are necessary for the seesaw mechanism: (i) RH neutrino components; (ii) large mass scale; (iii) lepton number violation.

Besides this, generically GUT's give relations between masses and mixings of leptons and quarks. They lead to equalities of masses if a single Higgs multiplet is involved in the Yukawa couplings, with well known example of the $b - \tau$ unification at the GUT scale. In general, when several different Higgs multiplets are involved, one gets "sum rules" between masses and mixings of quarks and leptons [18].

However, GUT's do not explain the flavor structures. Apart from some exceptional cases (*e.g.*, antisymmetric representations) no flavor structure is produced by GUT's. Existing attempts to combine GUT's and various horizontal or family symmetries (especially neutrino symmetries) have not produced yet substantial results.

Quark-lepton universality

Can we speak on the quark-lepton universality in a complete theory, in spite of big differences of mass and mixing patterns? Is it possible that not only the gauge but also Yukawa interactions of quark and leptons are very similar?

The idea behind is that the matrix of Yukawa couplings, Y, has the following universal form

$$Y = Y_0 + \delta Y_f, \quad f = u, d, D, l, N, \tag{7}$$

where $\delta Y_f \ll Y_0$ and Y_0 is the universal matrix for all fermions, Y_D is the Dirac type neutrino Yukawa matrix, Y_M is the Majorana type matrix for the RH neutrinos. Similarity (universality) of quarks and leptons is realized in terms of the matrices of Yukawa couplings and not of observables - mass ratios and mixing angles. The key point is that similar mass matrices can lead to substantially different mixing angles and masses (eigenvalues) if the matrices are nearly singular (rank-1) [19, 20]. It is the universal matrix Y_0 that is the singular one. The singular matrices are "unstable" in a sense that small perturbations can lead to strong variations of mass ratios and mixing angles (the latter - in the context of seesaw).

As an important example we take

$$Y_0 = \begin{pmatrix} \lambda^4 & \lambda^3 & \lambda^2 \\ \lambda^3 & \lambda^2 & \lambda \\ \lambda^2 & \lambda & 1 \end{pmatrix}, \quad \lambda \sim 0.2 - 0.3. \tag{8}$$

This matrix has only one non-zero eigenvalue and no physical mixing appears at this stage.

Let us introduce perturbations, ε, in the following form

$$Y_{ij}^f = Y_{ij}^0(1 + \varepsilon_{ij}^f), \quad f = u, d, e, v, N, \tag{9}$$

where Y_{ij}^0 is the element of the original singular matrix. This form can be justified, e.g., in context of the Froggatt-Nielsen mechanism [21]. (The key element is the form of perturbations (9) which distinguishes the ansatz (8) from other possible schemes with singular matrices.) It has been shown that small perturbations $\varepsilon \le 0.25$ are enough to explain large difference in mass hierarchies and mixings of quarks and leptons [20].

The seesaw plays crucial role here: It generates not only small neutrino masses but also large lepton mixing. Indeed, according to the seesaw $m \propto M_R^{-1}$, and nearly singular matrix of the RH neutrinos leads to enhancement of the lepton mixing [22]. In this approach maximal lepton mixing is accidental.

The quark-lepton universality can be introduced differently as universality of the *mixing matrices* [23]. (In the lowest order that should be equivalent to certain universality of the mass matrices.)

Quark-lepton complementarity (QLC)

As it was mentioned in sec. 2.3, within 1σ the data are in agreement with the quark-lepton complementary relations

$$\theta_{12} + \theta_C = \frac{\pi}{4}, \quad \theta_{23} + \arcsin V_{cb} = \frac{\pi}{4}. \tag{10}$$

For various reasons it is difficult to expect exact equalities (10). However certain correlation clearly shows up:

- the 2-3 leptonic mixing is close to maximal one because the 2-3 quark mixing is very small;
- the 1-2 leptonic mixing deviates from maximal one substantally because the 1-2 quark mixing (*i.e.*, Cabibbo angle) is relatively large.

Can it be accidental? A general scheme for the QLC relations is that

$$\text{``lepton mixing} = \text{bi} - \text{maximal mixing} - \text{CKM''}, \tag{11}$$

where the bi-maximal mixing matrix is [24]:

$$U_{bm} = U_{23}^m U_{12}^m = \frac{1}{2} \begin{pmatrix} \sqrt{2} & \sqrt{2} & 0 \\ -1 & 1 & \sqrt{2} \\ 1 & -1 & \sqrt{2} \end{pmatrix}. \tag{12}$$

Here U_{ij}^m is maximal mixing rotation in the ij-plane. Let us consider two possible QLC scenarios which differ by origin of the bi-maximal mixing and lead to different predictions.

1). QLC1: The bi-maximal mixing is generated by the neutrino mass matrix, presumably due to seesaw. The charged lepton mass matrix produces the CKM mixing as a consequence of the q-l symmetry: $m_l \approx m_d$. Therefore

$$U_{PMNS} = U_{CKM}^\dagger \Gamma_\alpha U_{bm}, \tag{13}$$

where $\Gamma_\alpha \equiv diag(1, 1, e^{i\alpha})$ is the phase matrix which appears in general at diagonalization. In this case exact relation (10) is not realized since the U_{12}^{CKM} rotation matrix should be permuted with U_{23}^m in (13) to reduce (13) to the standard parametrization form (1). As a consequence, the QLC relation is modified:

$$\sin\theta_{12} = \sin(\pi/4 - \theta_C) + 0.5\sin\theta_C(\sqrt{2} - 1 - V_{cb}\cos\alpha). \tag{14}$$

Numerically (without the RGE effects) we find $\sin^2\theta_{12} = 0.3345$ for $\alpha \sim 90°$ and $\sin^2\theta_{12} = 0.330$ for $\alpha = 0$. This is practically indistinguishable from the tri-bimaximal mixing prediction $\sin^2\theta_{12} = 0.3333$. Practically the same predictions for 1-2 mixing are obtained from two different combinations of matrices: $U_{23}^m U_{12}(\arcsin(1/\sqrt{3}))$ and $U_{12}(\theta_C)U_{23}^m U_{12}^m$ which are completely independent. Therefore an equality of the predictions is just accidental coincidence. This means that one of the two approaches (QLC1 or tri-bimaximal mixing) is wrong. To some extend that can be tested by measuring the 1-3 mixing. In the QLC1-scenario one obtains

$$\sin^2\theta_{13} = 0.5\sin^2\theta_C \approx 0.0245, \tag{15}$$

whereas the tri-bimaximal mixing implies $\sin^2\theta_{13} = 0$ unless some corrections are introduced.

2). QLC2: Maximal mixing comes from the charged lepton mass matrix and the CKM mixing originates from the neutrino mass matrix due to the q-l symmetry: $m_D \sim m_u$ (assuming also that in the context of seesaw the RH neutrino mass matrix does not influence mixing). Consequently,

$$U_{PMNS} = U_{bm}\Gamma_\alpha U_{CKM}^\dagger. \tag{16}$$

In this case the QLC relation for 1-2 mixing is satisfied precisely: $\sin\theta_{12} = \sin(\pi/4 - \theta_C)$. Now $\sin^2\theta_{13} \approx \sin^2\theta_{12}V_{cb}^2$ is extremely small.

All three predictions for 1-2 mixing (from QLC1, QLC2 and tri-bimaximal mixing) are within 1σ errors from the b.f. point. The tri-bimaximal and QLC1 predictions almost coincide, the b.f. value is in between the QLC2 and two other predictions: $\theta_{12}(QLC2) < \theta_{12}^{exp} < \theta_{12}(QLC1) \approx \theta_{12}(tbm)$. To disentangle these two possibilities one needs to measure the 1-2 mixing with accuracy $\Delta\theta_{12} \sim 1°$ or $\Delta\sin^2\theta_{12} \sim 0.015$ (5%).

There are two main issues related to the QLC relations:

(1) origin of the bi-maximal mixing;

(2) mechanism of propagation of the CKM mixing from the quark sector to the lepton one. The problem here is big difference of mass ratios of the quarks and leptons: $m_e/m_\mu = 0.0047$, $m_d/m_s = 0.04 - 0.06$, as well as difference of masses of muon and s-quark at the GU scale. This means that mixing should weakly depend on or be independent of masses.

So, if not accidental, the QLC relation may have the following implications: (i) the quark-lepton symmetry, (ii) existence of some additional structure which produces the bi-maximal mixing, (iii) mass matrices with weak correlation of the mixing angles and mass eigenvalues. Alternatively, it may imply certain flavor physics with $\sin\theta_C$ being the "quantum" of this physics.

EFFECTS OF NEW NEUTRINO STATES

Effects of new neutrino states (singlets of the SM symmetry group) depend on their masses. Superheavy new states essentially decouple. These states are not produced in laboratory experiments, but they can lead to indirect effects: 1) modify the mass matrix of active neutrinos; 2) violate universality of the weak interactions, *etc.*.

For relatively small masses, say $M_S \ll m_W$, these new states can be produced in reactions thus leading to direct effects but also they modify the mass matrix of active neutrinos. Light new states with $m_S \sim m_\nu$ can lead to non-trivial oscillation effects.

Here we consider two applications of possible existence of new neutrinos states. They realize an idea that these states play the role of additional structures which lead to substantial difference of quark and lepton properties.

Screening of Dirac structure

Let us introduce one heavy neutral state S for each generation and consider mass matrix in the basis (v, N^c, S) of the following form

$$m = \begin{pmatrix} 0 & m_D & 0 \\ m_D^T & 0 & M_D^T \\ 0 & M_D & M_S \end{pmatrix}. \tag{17}$$

Here M_S is the Majorana mass matrix of new fermions. Such a structure can be formed by a lepton number violated in the M_S and some additional symmetry which forbids also 13-element.

For $m_D \ll M_D \ll M_S$ the matrix leads to the double (cascade) seesaw mechanism [25]:

$$m_v = m_D^T M_D^{-1T} M_S M_D^{-1} m_D, \tag{18}$$

and the mass matrix of RH neutrinos becomes $M_R = -M_D M_S^{-1} M_D^T$. If two Dirac mass matrices are proportional each other,

$$M_D = A^{-1} m_D, \quad A \equiv v_{EW}/V_{GU}, \tag{19}$$

they cancel in (18) and we obtain

$$m_v = A^2 M_S. \tag{20}$$

That is, the structure of light neutrino mass matrix is determined by M_S immediately and does not depend on the Dirac mass matrix (the later is screened). The seesaw mechanism provides scale of neutrino masses but not the flavor structure of the mass matrix.

Notice that screening does not depend on the scale of M_S and in fact $M_S \ll M_D$ is also possible. However it is natural to assume that M_D is at the GUT scale, and M_S is at the Planck scale M_{Pl} which leads to correct values of the light neutrino masses. It can be shown that at least in SUSY version the radiative corrections do not destroy screening [26]. The relation (19) can be a consequence of Grand Unification with extended gauge group or/and certain flavor symmetry [26, 27].

Structure of the light neutrino mass matrix depends now on M_S which can be related to some physics at the Planck scale, and consequently, lead to "unusual" properties of neutrinos. In particular,

(i) certain symmetry of M_S can be the origin of "neutrino" symmetry;

(ii) the matrix $M_S \propto I$ leads to the quasi-degenerate mass spectrum;

(iii) M_S can be the origin of bi-maximal mixing thus leading to the QLC relations, if the charged lepton mass matrix generates the CKM rotation.

New states and induced mass matrix

Suppose the active neutrinos acquire (*e.g.*, via seesaw) the Majorana mass matrix m_a. Consider one sterile neutrino, S, with Majorana mass m_S and mixing with active neutrinos characterized by "vector" of masses $\bar{m}_S \equiv (m_{eS}, m_{\mu S}, m_{\tau S})$. Essentially in the basis (ν, N^c, S) this corresponds to the mass matrix of the form

$$m = \begin{pmatrix} 0 & m_D & \bar{m}_S \\ m_D^T & M_R & 0 \\ \bar{m}_S & 0 & m_S \end{pmatrix}. \tag{21}$$

If $m_S \gg m_{iS}$, then after decoupling of S the mass matrix of active neutrinos becomes

$$m_\nu = m_a + m_I, \tag{22}$$

where the last term is the matrix induced by S:

$$m_I = \frac{1}{m_S} \bar{m}_S^T \bar{m}_S. \tag{23}$$

The induced matrix has zero determinant and therefore can be an origin of singular structures. Introducing the active-sterile mixing angle θ_S as $\sin\theta_S = \bar{m}_S/m_S$, we can rewrite the elements of induced matrix as

$$m_I \sim \sin^2\theta_S m_S. \tag{24}$$

The induced matrix may turn out to be the "missed" element which leads to the difference of mixings of quarks and leptons. Let us consider several possibilities.

1). Suppose $\bar{m}_S \propto (0, 1, 1)$, then the induced matrix reproduces the dominant block of the active neutrino mass matrix for the normal mass hierarchy:

$$m_\nu = \frac{\sqrt{\Delta m_{32}^2}}{2} \begin{pmatrix} \cdots & \cdots & \cdots \\ \cdots & 1 & 1 \\ \cdots & 1 & 1 \end{pmatrix}, \tag{25}$$

where "dots" denote small parameters. In this case one can realize a possibility that the original active neutrino mass matrix, m_a, has hierarchical structure with small mixings being similar to the quark mass matrices. From eqs. (25) and (24) we find $\sin^2\theta_S m_S = \frac{1}{2}\sqrt{\Delta m_{32}^2} \sim 0.025$ eV.

2). Let us assume that couplings of S with active neutrinos are universal - flavor "blind": $\bar{m}_S \propto (1, 1, 1)$. Then the induced matrix has form: $m_I \propto D$, where D is the democratic matrix - the second matrix in (5). Suppose that the original active neutrino mass matrix has structure of the first matrix in (5). Then the sum, $m_\nu = m_a + m_I$, reproduces the mass matrix for the tri-bimaximal mixing (5). In this case, according

105

to (5), the parameters of S should satisfy relation $\sin^2 \theta_S m_S = \frac{1}{3}\sqrt{\Delta m_{21}^2} \sim 0.003$ eV. With two sterile neutrinos whole structure (5) can be obtained.

3). New neutrino states are irrelevant if $m_{iS}m_{jS}/m_S \ll (m_a)_{ij}$ or $\sin^2 \theta_S m_S < 0.001$ eV.

Clearly, the presence of induced contribution changes implications of the neutrino results [28, 12]. Since S is beyond the SM structure extended by RH neutrinos, it may be easier to realize "neutrino" symmetries as a consequence of certain symmetry of S couplings with active neutrinos.

Strong bound on θ_S and m_S follow from various cosmological, astrophysical and laboratory bounds on the parameters of new neutrino states (see ref.[12] for details). Still two regions are allowed:

1). Small masses window: $m_S \sim (0.5 - 1)$ eV and $\sin^2 \theta_S = 0.001 - 0.1$, where direct and indirect effects are comparable. This window is disfavored by results of recent analysis of cosmological data [4], and it is closed if the Big Bang nucleosynthesis bound on the effective number of neutrino species $N_v < 4$ is taken.

Notice that there are various ways to avoid the cosmological bounds which however imply an existence of additional physics beyond the Standard model [12].

2). Large masses range: $m_S > 300$ MeV and $\sin^2 \theta_S < 10^{-9}$. Here direct mixing effects are negligible and the presence of new states can not be verified.

SUMMARY

Comparison of the properties of the quarks and leptons shows similar gauge characteristics and strong difference of mass and mixing patterns.

There are several observations which (if not accidental) can strongly influence implications of the results. Those include possible presence of special leptonic (neutrino) symmetries; particular (tri-bimaximal) form of neutrino mixing matrix; quark-lepton complementarity relations. These features may indicate that quarks and leptons are fundamentally different and some new structures of theory exist beyond the seesaw.

Mixing with new neutrino states can play the role of this additional structure. In particular, it can

- produce screening of the Dirac structure;
- generate the induced matrix of active neutrinos with certain symmetry properties. The induced matrix can lead to enhancement of lepton mixings, to generation of the dominant block of the mass matrix in the case of normal mass hierarchy, or to various subdominant structures, e.g., for the tri-bimaximal mixing.

Still the approximate quark-lepton universality can be realized. In this case, the dominant mass or mixing matrices are the same for all fermions and small (of the order $\sin \theta_C$) corrections can produce whole difference. The seesaw mechanism plays the key role in getting of large lepton mixing.

ACKNOWLEDGEMENTS

I am grateful to Prof. S. Khalil for hospitality during my stay in Cairo.

REFERENCES

1. G. L. Fogli et al, hep-ph/0506083.
2. SNO Collaboration (B. Aharmim et al.). *Phys. Rev.* C **72**, 055502 (2005).
3. A. Strumia, F. Vissani, *Nucl. Phys.* B **726**, 294 (2005).
4. U. Seljak, A. Slosar and P. McDonald, arXiv:astro-ph/0604335.
5. H. V. Klapdor-Kleingrothaus, et al, *Phys. Lett.* B **586**, 198 (2004).
6. L. Wolfenstein, *Phys. Rev.* D **18**, 958 (1978); P. F. Harrison, D. H. Perkins and W. G. Scott, *Phys. Lett.* B **458**, 79 (1999), *Phys. Lett.* B **530**, 167 (2002).
7. E. Ma, *Mod. Phys. Lett.* A **17**, 2361 (2002); E. Ma, G. Rajasekaran, *Phys. Rev.* D **64** 113012, (2001); K.S. Babu, E. Ma, J.W.F. Valle, *Phys. Lett.* B **552**, 207 (2003); W. Grimus and L. Lavoura, *JHEP* **0508**, 013 (2005); K. S. Babu and X. G. He, hep-ph/0507217; E. Ma, *Mod. Phys. Lett.* A **20**, 2601 (2005). G. Altarelli and F. Feruglio, hep-ph/0507217; H. G. He, Yong-Yeon Keum and R. R. Volkas, hep-ph/0601001.
8. A. Yu. Smirnov, hep-ph/0402264; M. Raidal, *Phys. Rev. Lett.* **93**, 161801 (2004).
9. H. Minakata, A. Yu. Smirnov, *Phys. Rev.* D **70**, 073009 (2004).
10. T. Fukuyama and H. Nishiura, hep-ph/9702253; R. N. Mohapatra and S. Nussinov, *Phys. Rev.* D **60**, 013002 (1999); E. Ma and M. Raidal, *Phys. Rev. Lett.* **87**, 011802 (2001); C. S. Lam, *Phys. Lett.* B **507**, 214 (2001).
11. A. S. Joshipura, arXiv:hep-ph/0512252.
12. A. Y. Smirnov and R. Zukanovich Funchal, arXiv:hep-ph/0603009.
13. P. Minkowski, *Phys. Lett.* B **67** 421 (1977); T. Yanagida, in *Proc. of Workshop on Unified Theory and Baryon number in the Universe*, eds. O. Sawada and A. Sugamoto, KEK, Tsukuba, (1979); M. Gell-Mann, P. Ramond and R. Slansky, in *Supergravity*, eds P. van Niewenhuizen and D. Z. Freedman (North Holland, Amsterdam 1980); P. Ramond, *Sanibel talk*, retroprinted as hep-ph/9809459; S. L. Glashow, in *Quarks and Leptons*, Cargèse lectures, eds M. Lévy, (Plenum, 1980, New York) p. 707; R. N. Mohapatra and G. Senjanović, *Phys. Rev. Lett.* **44**, 912 (1980).
14. J. C. Pati and A. Salam, *Phys. Rev.* D **10**, 275 (1974).
15. H. Georgi, *In Coral Gables 1979 Proceeding, Theory and experiment in high energy physics*, New York 1975, 329 and H. Fritzsch and P. Minkowski, Annals Phys. **93** 193 (1975).
16. K. S. Babu and S. M. Barr, Phys. Lett. B **381** (1996) 202.
17. C. H. Albright, K. S. Babu and S. M. Barr, Phys. Rev. Lett. **81** (1998) 1167.
18. B. Bajc, G. Senjanovic and F. Vissani, Phys. Rev. Lett. **90** (2003) 051802.
19. E. K. Akhmedov, et al., *Phys. Lett.* B **498**, 237 (2001); R. Dermisek, *Phys. Rev.* D **70**, 033007 (2004).
20. I. Dorsner, A.Yu. Smirnov, *Nucl. Phys.* B **698**, 386 (2004).
21. C. D. Froggatt and H. B. Nielsen, *Nucl. Phys.* B **147**, 277 (1979).
22. A. Yu. Smirnov, *Phys. Rev.* D **48**, 3264 (1993).
23. A. S. Joshipura and A. Y. Smirnov, arXiv:hep-ph/0512024.
24. F. Vissani, hep-ph/9708483; V. D. Barger, et al, *Phys. Lett.* B **437**, 107 (1998).
25. R. N. Mohapatra, *Phys. Rev. Lett.* **56**, 561 (1986); R. N. Mohapatra and J. W. F. Valle, *Phys. Rev.* D **34**, 1642 (1986).
26. M. Lindner, M. A. Schmidt, A. Yu. Smirnov, *JHEP* **0507**, 048 (2005).
27. O. Vives, hep-ph/0504079; J. E. Kim and J. C. Park, hep-ph/0512130.
28. K.R.S. Balaji, A. Perez-Lorenzana, A.Yu. Smirnov, *Phys. Lett.* B **509**, 111 (2001).

Neutrino Masses and Interactions
in a Model with Nambu-Goldstone Bosons

E.A. Paschos

Institut für Physik, Universität Dortmund, D-44221 Dortmund, Germany

Abstract.

A natural scenario for the generation of neutrino masses is the see-saw mechanism, in which a large right-handed neutrino mass makes the left-handed neutrinos light. We review a special case when the Majorana masses originate from spontaneous breaking of a global U(1)XU(1) symmetry. The interactions of the right-handed with the left-handed neutrinos at the electorweak scale further break the global symmetry giving mass to one pseudo Nambu-Goldstone boson (pNGB). The pNGB can then generate a long-range force. Leptogenesis occurs through decays of heavy neutrinos into the light ones and Higgs particles. The pNGB can become the acceleron field and the neutrino masses vary with the value of the scalar field[1].

---

[1] Invited Talk presented at the Second International Conference on High Energy Physics (CICHEP II), January 14-17,2006; Cairo, Egypt. The talk is a brief preview of the results in reference [11].

CP881, *Cairo International Conference on High Energy Physics (CICHEP II)*,
edited by S. Khalil

INTRODUCTION

The standard model of electroweak theory is succesful in explaining most of the phenomena we observed till now. The neutrino mass is the only indication we have that requires physics beyond the standard model. It has been established that the neutrino sector must be extended to accommodate the observed mass-squared difference for the neutrinos in the atmospheric, solar, as well as in the Laboratory experiments.

The observed neutrino masses are too small to treat them at par with charged fermions. The most natural explanation for the smallness of the neutrino masses comes from the see-saw mechanism [1]. One introduces the right-handed neutrinos in the model that allows the usual Dirac masses for the neutrinos to be of the order of charged lepton masses. Since the right-handed neutrinos are singlets, it is now possible to allow very heavy Majorana masses $M_1, M_2, ...,$ for right-handed neutrinos. This introduces the scale of lepton number violation in the model and in turn induces tiny Majorana masses for the left-handed neutrinos.

When Majorana masses of the right-handed neutrinos are introduced, they break lepton number explicitly. However, it is more natural to associate any new scale of the theory with a symmetry whose breaking produces neutrino masses. In a forthcoming article we propose [11], that the Majorana masses of the right-handed neutrinos originate from the spontaneous breaking of lepton number, which leads to a massless Goldstone boson, the Majoron. However, since the residual global symmetry is not exact, one Goldstone boson picks up small mass producing ultra low mass pseudo Nambu-Goldstone bosons (pNGB). An early proposal, along these lines, was the axion, which was introduced to solve the strong CP problem through the Peccei-Quinn symmetry [2, 3]. The axion is still being searched for as a candidate for dark matter [4]. In strong interactions, it is the breaking of symmetries within QCD that generates a large part of the proton's mass, which is relatively large compared to the masses of current quarks. The result is the appearance of low mass pions – the pNGB's of the spontaneously broken chiral symmetry. This concept of pNGB was applied [5] to the general formalism to a long-range force and phase transitions [6] occurring at a late epoch of the universe, thus developing quintessence models contemporaneously with other authors [7]. Neutrino physics has also found applications of quintessence model, in which the quintessence or the acceleron are formed as condensates and couple to neutrinos varying their masses [8]. This scenario has some interesting consequences [9, 10]. In this review we point out that the pNGB in models of Majorana neutrinos can play the role of the acceleron, introducing an exponential potential, and giving rise to a long-range force that may be of of phenomenological importance [11]. Finally, the model generates a lepton asymmetry through the decay of heavy neutrinos to light ones and Higgs particles.

TWO GENERATION PNGB MODEL

We demonstrate here the idea behind the pNGB model of Majorana neutrinos with a two generation example. The theory possesses a global $U(1)_A \times U(1)_B$ symmetry. The breaking of this symmetry spontaneously sets a large scale in the theory far above the scale of electroweak symmetry breaking. There are two right-handed neutrinos N_1 and N_2, which are singlets under the standard model gauge groups and transform under the global symmetry as $(1,0)$ and $(0,1)$ respectively. The global symmetry prevents any Majorana mass of the right-handed neutrinos. We also introduce two singlet scalar fields $\Phi_1(x)$ and $\Phi_2(x)$ transforming under the global symmetry as $(2,0)$ and $(0,2)$ respectively, whose interactions with the right-handed neutrinos are given by

$$\mathscr{L}_M = \frac{1}{2}\alpha_1 \bar{N}_1 N_1^c \Phi_1 + \frac{1}{2}\alpha_2 \bar{N}_2 N_2^c \Phi_2. \tag{1}$$

The vacuum expectation values (*vevs*) of these scalars will give Majorana masses to the right-handed neutrinos and there will be two Goldstone bosons. As we shall demonstrate below, the Dirac masses of the neutrinos at the electroweak scale will break one combination of the global symmetry softly giving a small mass to one of the Goldstone bosons, which now becomes a pseudo Nambu-Goldstone boson (pNGB) of the model. The other Goldstone boson, the Majoron, remains massless. Since the Majoron couples only to singlet fields, the theory is consistent with phenomenology of known particles.

After the fields $\Phi_i, i = 1, 2$ acquire *vevs*, we can express them in terms of their *vevs* σ_i and decay constants f (we assume their decay constants to be same, $f_1 \sim f_2 \sim f$) and the Nambu-Goldstone bosons ϕ_i as

$$\alpha_i \Phi_i \rightarrow \alpha_i \langle \Phi_i \rangle e^{2i\phi_i/f_i} = \alpha_i \sigma_i e^{2i\phi_i/f_i} = M_i e^{2i\phi_i/f_i}. \tag{2}$$

At this stage it is possible to make phase transformations to the fields N_i that eliminate the ϕ_i, implying that both Goldstone bosons are massless. The self interactions of the Goldstone bosons, given by

$$\mathscr{L}_\Phi = \frac{1}{2}M_\Phi^2 \Phi^\dagger \Phi + \frac{1}{4}\lambda(\Phi^\dagger \Phi)^2 + \partial_\mu \Phi^\dagger \partial^\mu \Phi = \frac{\sigma^2}{f^2}\partial_\mu \phi \partial^\mu \phi + f(\Phi^\dagger \Phi)$$

also do not contain any term that can give masses to the Goldstone bosons.

We now write down all the Yukawa terms involving the neutrinos, including the Dirac mass terms

$$\begin{aligned}
\mathscr{L}_{mass} &= \frac{1}{2}M_1 \bar{N}_1 N_1^c e^{2i\phi_1/f} + \frac{1}{2}M_2 \bar{N}_2 N_2^c e^{2i\phi_2/f} + m e^{i\alpha}\bar{N}_1 v_1 + m\varepsilon e^{i\beta}\bar{N}_1 v_2 \\
&\quad + \lambda m\varepsilon' e^{i\gamma}\bar{N}_2 v_1 + \lambda m e^{i\xi}\bar{N}_2 v_2.
\end{aligned} \tag{3}$$

We introduced the Dirac mass m and some scaling parameters $\lambda, \varepsilon, \varepsilon'$ in order to write down the Dirac mass terms that break the $U(1)_{A-B}$ global symmetry softly. We included all the phases which contribute to CP violation. Rephasing of the fields

$$N_i \rightarrow e^{i\phi_2/f} N_i \quad \text{and} \quad v_i \rightarrow e^{i\phi_2/f} v_i \tag{4}$$

and rephasing of the CP phases, leads to the full mass matrix

$$\begin{aligned}
\mathscr{L}_\mu &= \frac{1}{2} M_1 \bar{N}_1 N_1^c e^{2i\phi/f} + \frac{1}{2} M_2 \bar{N}_2 N_2^c + m \bar{N}_1 v_1 + m\varepsilon e^{i\eta} \bar{N}_1 v_2 \\
&\quad + \lambda m\varepsilon' e^{i\eta} \bar{N}_2 v_1 + \lambda m \bar{N}_2 v_2 + H.c.,
\end{aligned} \tag{5}$$

where $2\eta = \gamma - \alpha + \beta - \xi$. Thus we finally have one CP phase η and one combination of the fields $\phi = \phi_1 - \phi_2$, which becomes the pNGB. The other combination of fields $\phi_1 + \phi_2$ correspond to the invisible singlet Majoron and remains massless by decoupling from the theory. This conclusion is independent of the choice of phase transformation.

We shall now explicitly demonstrate how such $U(1)_A \times U(1)_B$ global symmetry breaking soft Dirac mass term may appear in the theory after the electroweak symmetry breaking. If we assign the $U(1)_A \times U(1)_B$ quantum numbers

$$\ell_1 \equiv \begin{pmatrix} v_e \\ e^- \end{pmatrix} \equiv (1,0) \quad \text{and} \quad \ell_2 \equiv \begin{pmatrix} v_\mu \\ \mu^- \end{pmatrix} \equiv (0,1),$$

then the usual Higgs doublet H with the assignment $(0,0)$ (for this to give masses to the charged fermions) can give only the diagonal Dirac mass terms. We need two more Higgs doublets $H_1 \equiv (-1,+1)$ and $H_2 \equiv (+1,-1)$ in order to generate the complete Dirac mass we discussed above. The mass term of the Lagrangian now becomes

$$\begin{aligned}
\mathscr{L}_{mass} &= \frac{1}{2} M_1 \bar{N}_1 N_1^c e^{2i\phi_1/f} + \frac{1}{2} M_2 \bar{N}_2 N_2^c e^{2i\phi_2/f} + f_{11} \bar{N}_1 v_1 H + f_{12} \bar{N}_1 v_2 H_1 \\
&\quad + f_{21} \bar{N}_2 v_1 H_2 + f_{22} \bar{N}_2 v_2 H.
\end{aligned} \tag{6}$$

These dimension-4 terms do not break the global symmetry. However, after the electroweak symmetry breaking, when all the doublets H and H_i acquire vevs, the global symmetry is broken. The part of the Lagrangian with all neutrino mass terms is:

$$\begin{aligned}
-\mathscr{L}_{mass} &= \frac{1}{2} M_1 \bar{N}_1 N_1^c e^{2i\phi_1/f} + \frac{1}{2} M_2 \bar{N}_2 N_2^c e^{2i\phi_2/f} + m e^{i\alpha} \bar{N}_1 v_1 + m\varepsilon e^{i\beta} \bar{N}_1 v_2 \\
&\quad + \lambda m\varepsilon' e^{i\gamma} \bar{N}_2 v_1 + \lambda m e^{i\xi} \bar{N}_2 v_2.
\end{aligned} \tag{7}$$

After rephasing of the fields it reduces to equation (5).

To find out the mass of the pNGB, we consider the effective potential generated by the interactions of the scalar field through the mass terms. The Colemann-Weinberg potential for ϕ is computed through the leading loop in Fig. 1. It has the remarkable property that the symmetry

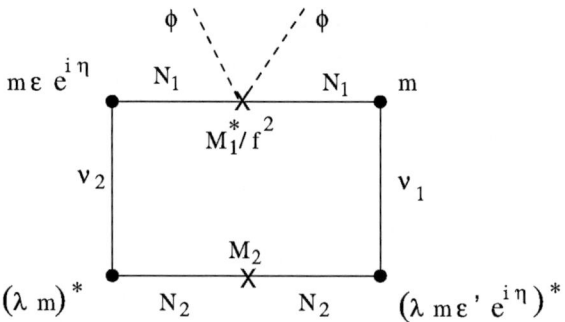

FIGURE 1. Loop diagram for the effective potential

structure of the theory makes the loop finite. The reason is that the ϕ field could be eliminated if any of the vertices is set to zero. This diagram is also invariant under any rephasing of the fields, which can be confirmed by observing that the most general phase transformation of the neutrinos,

$$N_i \rightarrow e^{ip_i}N_i \quad \text{and} \quad v_i \rightarrow e^{iq_i}, \tag{8}$$

transforms

$$
\begin{aligned}
m &\rightarrow e^{i(q_1-p_1)}\, m; & m\varepsilon &\rightarrow e^{i(q_2-p_1)} \\
m\lambda\varepsilon' &\rightarrow e^{i(q_1-p_2)}\, m\lambda\varepsilon'; & \lambda m &\rightarrow e^{i(q_2-p_2)}\, \lambda m; \\
M_1 &\rightarrow e^{-i2p_1}\, M_1; & M_2 &\rightarrow e^{-i2p_2}\, M_2;
\end{aligned}
\tag{9}
$$

so that

$$(m) \cdot (M_1^*/f^2) \cdot (m\,\varepsilon\,e^{i\eta}) \cdot (\lambda\,m)^* \cdot (M_2) \cdot (\lambda\,m\,\varepsilon'\,e^{i\eta})^*$$

and hence the diagram remains invariant.

The explicit calculation gives

$$V_{eff}(\phi^2) = -\frac{m^4\lambda^2\varepsilon\varepsilon'}{4\pi^2}\frac{M_1 M_2 \log\left(\frac{M_1^2}{M_2^2}\right)}{M_1^2 - M_2^2}\cos\left(\frac{2\phi}{f}\right), \tag{10}$$

which has the minima at $\phi = 0, \pi f, 2\pi f, \cdots$. Expanding around one of the minima, we can write down a mass term

$$V_{eff}(\phi) = \frac{m^4\lambda^2\varepsilon\varepsilon'}{2\pi^2}\frac{M_1 M_2 \log\left(\frac{M_1^2}{M_2^2}\right)}{M_1^2 - M_2^2}\frac{\phi^2}{f^2} \tag{11}$$

Thus the mass of the ϕ field is now

$$m_\phi = \frac{m^2 \lambda \sqrt{\varepsilon \varepsilon'}}{\pi f} \frac{M_1 M_2 \log\left(\frac{M_1^2}{M_2^2}\right)}{M_1^2 - M_2^2}. \tag{12}$$

The symmetry at the scale of $f \sim M_i$ protects the mass of the pseudo Nambu-Goldstone boson, so an explicit soft breaking of the symmetry at the scale m can generate a mass of the order of m^2/f. This light pNGB ($\mathscr{A} = i\phi$) can generate a long-range force and also become the acceleron field to explain the smallness of dark energy.

NEUTRINO MASSES

We consider next the structure of neutrino masses in this model. After the global symmetries are broken by the *vevs* of the scalars, the model reduces to the usual see-saw models [1], except for the pNGB which generates a new long-range force because of its small mass. The interactions required for the usual leptogenesis [12, 13, 14] through the decays of right-handed neutrino are also present and will be discussed in the next section.

We shall first consider the time evolution of the light neutrino states without including the effects of the pNGB, which is determined by the matrix

$$-\mathscr{L}_{eff} = m_{ij}^T M_i^{-1} m_{ij} \, \nu_i \nu_j = \frac{m^2}{M} \left[(\nu_1 \ \ \nu_2) \begin{pmatrix} 1 + (\lambda \varepsilon')^2 e^{2i\eta} & e^{i\eta}(\varepsilon + \lambda \varepsilon') \\ e^{i\eta}(\varepsilon + \lambda \varepsilon') & \lambda^2 + \varepsilon^2 e^{2i\eta} \end{pmatrix} \begin{pmatrix} \nu_1 \\ \nu_2 \end{pmatrix} \right]. \tag{13}$$

This gives a mixing angle

$$\tan 2\theta = 2 \frac{\varepsilon + \lambda \varepsilon'}{(1 - \lambda^2)e - i\eta + (\lambda^2 \varepsilon'^2 - \varepsilon^2)ei\eta} \tag{14}$$

which is large for $\varepsilon, \varepsilon' \ll 1$ and $\lambda \approx 1$. This mass matrix can be diagonalized to

$$M^{diag} = \begin{pmatrix} m_1 e^{i\theta_1} & 0 \\ 0 & m_2 e^{i\theta_2} \end{pmatrix} \tag{15}$$

with

$$m_{1,2} = \frac{m^2}{M}[1 \pm (\varepsilon + K) + O(\varepsilon^2)], \quad \text{and} \quad \tan\theta_{1,2} = \frac{1 \pm (\varepsilon + K)\cos\eta}{\pm(\varepsilon + K)\sin\eta},$$

where $K = \lambda \varepsilon'$. This part of the discussion is the same as in any other model with a see-saw mechanism. It can account for the large mixing between two generations.

We shall now consider the effect of the light scalar field, the pNGB in the model. We obtain the interactions of the light neutrinos with the scalar field by keeping only the terms linear in pNGB, given by

$$\mathcal{L} = \frac{m^2}{2M} \overline{\Psi^c} \left\{ \begin{pmatrix} m_1 & 0 \\ 0 & m_2 \end{pmatrix} - \frac{\phi}{f} e^{i\eta} \begin{pmatrix} (\varepsilon - K)e^{-i\theta_1} & 0 \\ 0 & (K - \varepsilon)e^{-i\theta_2} \end{pmatrix} \right\} \Psi + H.C. \qquad (16)$$

The long range force introduced by the pNGB in this model will have direct consequences in neutrino oscillation experiments [15]. The extension of the model to three generations and the implications of the new force will be discussed elsewhere [11].

LEPTON ASYMMETRY OF THE UNIVERSE

The direct connection between the neutrino masses and the baryon asymmetry of the universe is now well established. In our case, lepton number violating couplings were introduced in order to give masses to the neutrinos. These couplings generate a lepton asymmetry [12, 13, 14], which will be converted to a baryon number asymmetry as the universe expands and approaches the electroweak phase transition.

In general, the generation of the lepton asymmetry of the universe requires three ingredients:

1. Lepton number violation, which is also the source of neutrino masses;
2. CP violation, which comes from the interference of tree level and one loop diagrams
3. Departure of the lepton number violating interactions from equilibrium

The decays of the right-handed neutrinos violate lepton number and the Yukawa couplings contain a phase η which can give CP-even and CP-odd amplitudes. If the right-handed neutrino mass can now satisfy the out-of-equilibrium condition, this model will be able to generate the required baryon asymmetry of the universe.

In our model lepton number is violated at tree-level in decays of the right-handed neutrinos

$$\begin{aligned} N_i &\rightarrow \ell_j + H_a^\dagger \\ &\rightarrow \ell_j^c + H_a \end{aligned}$$

The one loop diagrams, like vertex corrections and self energies [12, 13] will contain CP-even and -odd amplitudes. The structure of the Lagrangian in eqs (5) and (6) has loops only when the Higgs particles mix with each other. Let us denote by Vo1 the mixing between H_0 and H_1 and similarly by V_{02} the mixing between H_0 amd H_2. Such mixings are generated by the quartic couplings of the potential when the scalar acquire vevs. Here we do not restrict ourselves to a specific form of the potential and represent the generic mixings by V_{0i}. The interference between tree and vertex diagrams gives the asymmetry

$$\delta = \frac{\Gamma(N_1 \rightarrow \ell H^\dagger) - \Gamma(N_1 \rightarrow \ell^c H)}{\Gamma(N_1 \rightarrow \ell H^\dagger) + \Gamma(N_1 \rightarrow \ell^c H)}$$

$$= \frac{3}{16\pi} \frac{M_1}{M_2} \frac{m^2 \lambda^2}{v_1^2 + v_0^2 \varepsilon^2} \left[|V_{02}|^2 \frac{v_1^2}{v_2^2} \varepsilon'^2 - |V_{01}|^2 \varepsilon^2 \right] \sin 2\eta \tag{17}$$

where $v_a = \langle H_a \rangle, a = 0, 1, 2$ and we assumed $M_2 >> M_1$, so that the asymmetry is generated by the decays of N_1. In the context of leptogenesis the lightest N_1 is naturally out of thermal equilibrium when it decays. The dilusion factor depends on the relative magnitude of

$$\Gamma_{N_1} = \frac{|f_{1i}|^2}{16\pi} M_1 \tag{18}$$

relative to the Hubble constant

$$H(M_1) = 1.7 \sqrt{g_*} \frac{T^2}{M_P} \qquad \text{at } T = M_1. \tag{19}$$

When the ratio $\kappa = \Gamma/H(M_1)$ is small the lepton excess is large. On the other hand, for large κ the time of expansion is large relative to the lifetime of N_1 so that many decays and recombinations can take place in the cosmological scale $1/H$ giving a smaller excess of leptons. Finally, the sphaleron interactions convert a fraction of the produced $(B - L)$ asymmetry to a baryon asymmetry

$$\frac{n_B}{s} = \frac{24 + 4n_H}{66 + 13n_H} \frac{n_{B-L}}{s}. \tag{20}$$

This conversion takes place from the time of leptogenesis down to the time of the electroweak phase transition. Here n_H is the number of Higgs doublets in our model. In summary, it is evident that the model can generate the required amount of baryon asymmetry of the universe.

ORIGIN OF DARK ENERGY

One of the most challenging questions at present is why the cosmological constant is very small but nonvanishing. If the dark energy were due to the presence of a cosmological constant, then one expects

$$\rho_{DE} = \rho_{vac} = E_0^4$$

where E_0 is the energy associated with a particle or a field. The observed cosmological constant coresponds to an $E_0 \sim 2 \times 10^{-3}$ eV, which is very small for the scales of most elementary particles. It is of the order of neutrino masses and it has recently been proposed that the dark energy tracks the neutrino density [8]. If the neutrino masses vary as functions of a light scalar field (called the acceleron) the dark energy density may track the neutrino masses. This would then explain why the scale of cosmological constant is the same as the scale of neutrino masses. In the present model, the pNGB introduces a spatial dependence into the Majorana mass M1

$$\mathcal{L}_\mu = \frac{1}{2} M_1(\mathscr{A}) \bar{N}_1^c N_1 + \frac{1}{2} M_2 \bar{N}_2^c N_2 + m \bar{N}_1 v_1 + m \varepsilon \bar{N}_1 v_2$$
$$+ \lambda m \varepsilon' \bar{N}_2 v_1 + \lambda m \bar{N}_2 v_2 + H.c. \tag{21}$$

The first term $M_1(\mathscr{A})$ has an exponential functional dependence on the scalar field. The effective neutrino mass then varies explicitly with the acceleron field $m_\nu(\mathscr{A}) = m^T M^{-1}(\mathscr{A})m$, as required by models of mass varying neutrinos [8].

SUMMARY

A large Majorana scale is a necessary ingredient of the see-saw mechanism. At this very high energy there may exist global symmetries which are broken in order to create the Majorana masses. Remnants of the global symmetry survive in the low energy interactions. In a two generation model developed with my collaborators Hill, Mocioiu and Sarkar [11], the breaking of the symmetry produces two Nambu-Goldstone bosons. The introduction of Dirac masses at the electroweak scale further breaks the symmetry softly giving a small and finite mass to one of the Nambu-Goldstone bosons and leaves the other scalar massless. The exchange of the scalar particle introduces a new long-range force between neutrinos.

Finally, the model has additional attractive properties. It describes correctly the masses and mixings of the light neutrinos. The decays of the heavy neutrinos generate a lepton asymmetry consistent with leptogenesis. The neutrino masses have an explicit dependence on the scalar field, which may bring density effects with the pNGB playing the role of the "acceleron".

ACKNOWLEDGEMENT

I wish to thank my collaborators Drs. C.T. Hill, I. Mocioiu and U. Sarkar for a pleasant and fruitful collaboration. An extensive article mentioned in ref. [11] is in preparation. I also thank Dr. Y.-F. Zhou and Mr. A. Karavtsev for working with me on extensions of the model. The financial support of BMBF, Bonn under contract 05HT 4PEA/9 is gratefully ackonwledged. Finally, I thank the DFG for a travelling grant to participate at the conference.

REFERENCES

1. M. Gell-Mann, P. Ramond, and R. Slansky, in *Supergravity*, edited by P. van Nieuwenhuizen and D. Z. Freedman (North-Holland, Amsterdam, 1979), p. 315; T. Yanagida, in *Proceedings of the Workshop on the Unified Theory and the Baryon Number in the Universe*, edited by O. Sawada and A. Sugamoto (KEK Report No. 79-18, Tsukuba, Japan, 1979), p. 95;
    P. Minkowski, Phys. Lett. **B 67**, 421 (1977);
    R. N. Mohapatra and G. Senjanovic, Phys. Rev. Lett. **44**, 912 (1980);
    J. Schechter and J.W.F. Valle, Phys. Rev. **D 22**, 2227 (1980); Phys. Rev. **D 25**, 774 (1982).
2. R.D. Peccei and H.R. Quinn, Phys. Rev. Lett. **38**, 1440 (1977); Phys. Rev. **D 16**, 1791 (1977).
3. S. Weinberg, Phys. Rev. Lett. **40**, 223 (1978);
    F. Wilczek, Phys. Rev. Lett. **40**, 279 (1978).

4. K. Zioutas et al., Phys. Rev. Lett. **94**, 121301 (2005).
5. C.T. Hill and G.G. Ross, Phys. Lett. **B 203**, 125 (1988); Nucl. Phys. **B 311**, 253 (1998).
6. C.T. Hill, D.N. Schramm and J.N. Fry, Nucl. Part. Phys. **19**, 25 (1989).
7. C. Wetterich, Nucl. Phys. **B302**, 668 (1988);
    P. J. E. Peebles and B. Ratra, Astrophys. J. **325**, L17 (1988).
8. R. Fardon, A. E. Nelson, and N. Weiner, JCAP **0410**, 005 (2004);
    see also P. Q. Hung, hep-ph/0010126;
    P. Gu, X. Wang, and X. Zhang, Phys. Rev. **D68**, 087301 (2003).
9. D. B. Kaplan, A. E. Nelson, and N. Weiner, Phys. Rev. Lett. **93**, 091801 (2004);
    G. Dvali, Nature **432**, 567 (2004).
10. X.-J. Bi, P. Gu, X. Wang, and X. Zhang, Phys. Rev. **D69**, 113007 (2004);
    P.-H. Gu and X.-J. Bi, Phys. Rev. **D70**, 063511 (2004).
11. C.T. Hill, I. Mocioiu, E.A. Paschos and U. Sarkar, *in preparation*.
12. M. Fukugita and T. Yanagida, Phys. Lett. **B174**, 45 (1986).
13. M.A. Luty, Phys. Rev. **D45**, 455 (1992);
    W. Buchmuller, P. Di Bari and M. Plumacher, Annals Phys. **315**, 305 (2005).
14. M. Flanz, E.A. Paschos and U. Sarkar, Phys. Lett. **B 345**, 248 (1995);
    M. Flanz, E.A. Paschos, U. Sarkar and J. Weiss, Phys. Lett. **B 389**, 693 (1996);
15. V. Barger, P. Huber, and D. Marfatia, Phys. Rev. Lett. **95**, 211802 (2005);
    M. Cirelli, M. C. Gonzalez-Garcia, and C. Pena-Garay, Nucl. Phys. **B719**, 219 (2005).

Reconstructing see-saw models

Alejandro Ibarra

Instituto de Física Teórica, CSIC/UAM, C-XVI
Cantoblanco, 28049 Madrid, Spain.

Abstract. In this talk we discuss the prospects to reconstruct the high-energy see-saw Lagrangian from low energy experiments in supersymmetric scenarios. We show that the model with three right-handed neutrinos could be reconstructed in theory, but not in practice. Then, we discuss the prospects to reconstruct the model with two right-handed neutrinos, which is the minimal see-saw model able to accommodate neutrino observations. We identify the relevant processes to achieve this goal, and comment on the sensitivity of future experiments to them. We find the prospects much more promising and we emphasize in particular the importance of the observation of rare leptonic decays for the reconstruction of the right-handed neutrino masses.

Keywords: Supersymmetric Standard Model, Neutrino Physics, Beyond Standard Model
PACS: 12.60.Jv,14.60.St,12.60.-i

INTRODUCTION

Neutrino observations can be well described by adding a dimension five operator to the Standard Model Lagrangian [1]:

$$\mathscr{L}_{eff} = \mathscr{L}_{SM} - \frac{1}{2}\kappa_{ij}(L_i \cdot H)^T(L_j \cdot H) + h.c. \tag{1}$$

so that after the electroweak symmetry breaking a Majorana mass term for the neutrinos is generated:

$$\mathscr{M}_\nu = \kappa_{ij}\langle H^0\rangle^2. \tag{2}$$

One of the most pressing questions in neutrino physics is to determine the origin of this effective operator. Among all the proposals, the type I see-saw mechanism is perhaps the most plausible. It consists on adding to the Standard Model particle content three right-handed neutrino fields, so that the most general Lagrangian compatible with the Standard Model gauge symmetry reads:

$$\mathscr{L}_{eff} = \mathscr{L}_{SM} + \nu_R^{c\,T}\mathbf{Y}_\nu L \cdot H - \frac{1}{2}\nu_R^{c\,T}\mathbf{M}_\nu\,\nu_R^c + h.c. \tag{3}$$

If the right-handed neutrino masses are much larger than the scale of electroweak symmetry breaking, the right-handed neutrinos decouple at low energies and the theory can be well described by the effective Lagrangian given by eq.(1), with

$$\kappa_{ij} = \mathbf{Y}_\nu^T\mathbf{M}_\nu^{-1}\mathbf{Y}_\nu, \tag{4}$$

which is naturally suppressed by the large mass scale of the right-handed neutrinos, thus providing a natural explanation for the small neutrino masses.

CP881, *Cairo International Conference on High Energy Physics (CICHEP II)*,
edited by S. Khalil

In this work we would like to discuss the possibility of reconstructing the complete see-saw Lagrangian, eq.(3), using just low energy experiments [2, 3]. There are several motivations to study this problem. Although the effective theory provides a good description of neutrino experiments, we would like to have a deeper understanding of the observed neutrino parameters. In particular, we would like to understand why neutrino masses are so small compared to the rest of the fermion masses, or why there are two large mixing angles, in stark contrast to the quark sector. Clearly, the answers to these questions necessarily lie in the fundamental theory and not in the effective theory.

As a more ambitious goal we would like to unravel the so-called flavour puzzle, namely why fermion masses and mixing angles have the structure they present. The non-trivial flavour structure observed in the leptonic sector indeed provides additional clues to the solution of the flavour puzzle, but again, these clues lie in the fundamental theory rather than in the effective theory. If the fundamental theory responsible for the leptonic flavour physics could be disentangled, one could look for patterns in the eigenvalues of the neutrino Yukawa coupling or in the right-handed neutrino masses, analogous to the intriguing relations observed in the quark sector, $m_u : m_c : m_t \sim \lambda^8 : \lambda^4 : 1$ and $m_d : m_s : m_b \sim \lambda^4 : \lambda^2 : 1$, being λ the Cabibbo angle. In addition, one could look in the fundamental neutrino parameters for similarities with the charged lepton sector or with the quark sector, in the quest for hints of some possible underlying Grand Unified Theory. Needless to say, the ultimate solution to the flavour puzzle would require an understanding of the origin of Yukawa couplings, which is still lacking.

Finally, it would be interesting to determine the scale of new physics, namely the masses of the right-handed neutrinos. Apart from the intrinsic interest of determining parameters of the complete Lagrangian, determining the scale of new physics could have implications for some models beyond the Standard Model, that are intimately related to some particular mass scale. For example, some Grand Unified Theories predict masses for the right-handed neutrinos close to the gauge unification scale, $M_X \sim 10^{15-16}$ GeV. The determination of masses for the right handed neutrinos close to this scale would undoubtedly give strong support to this class of models. Also, the mechanism of leptogenesis to produce the observed baryon asymmetry of the Universe heavily relies on the mass of the lightest right-handed neutrino. The determination of this mass in the range favoured by leptogenesis would also give strong support to this mechanism.

APPROACHES TO DETERMINE SEE-SAW PARAMETERS

The approaches to reconstruct the high-energy see-saw Lagrangian, can be roughly classified into two main classes: top-down approach and bottom-up approach.

The top-down approach consists on selecting a particular model at high energies (a Grand Unified Model, a Froggatt-Nielsen model, a superstring model...), compute the low energy predictions and compare with the experiments. This approach is completely motivated by theory and has the advantage that the analysis is restricted just to scenarios that are very well motivated from the theoretical point of view. Nevertheless, there is still a huge freedom and many different possibilities arise. And what is more frustrating, despite these scenarios are *a priori* the best candidates as fundamental see-saw models, the simplest ideas do not seem to work, indicating perhaps that we are being mislead by our theoretical prejudices.

A radically different approach is the bottom-up approach, that consists on exploiting all the information available at low energies on the leptonic sector, in order to reconstruct the high-energy theory. Contrary to the top-down approach, the bottom-up approach is completely phenomenological and it is impossible to get mislead by aesthetics. However, it is very difficult to realize in practice.

In particular, following this approach in the Standard Model is hopeless. In the basis where the charged lepton Yukawa coupling and the right-handed mass matrix are diagonal, the complete Lagrangian depends on three independent parameters in the right-handed mass matrix (the three right-handed masses), and fifteen independent parameters in the neutrino Yukawa coupling, of which nine are moduli and six are phases. On the other hand, in the effective theory the neutrino mass matrix depends just on nine parameters: three masses, three mixing angles, and three phases. Clearly, part of the information about the complete Lagrangian has been lost in the decoupling process and cannot be recovered just from neutrino experiments, to be precise six real parameters and three phases.

On the other hand, in the Minimal Supersymmetric Standard Model (MSSM) there is additional information about the high-energy see-saw Lagrangian. Namely, radiative corrections on slepton parameters carry information about the combination $\mathbf{Y}_\nu^\dagger \mathbf{Y}_\nu \equiv P$. This information could disentangled from measurements in the slepton sector, provided the mechanism of supersymmetry breaking is specified. Assuming that the slepton mass matrices are proportional to the identity at the high energy scale, quantum effects induced by the right-handed neutrinos would yield at low energies a left-handed slepton mass matrix with a non-trivial structure, whose measurement would provide additional information about the see-saw parameters. To be more specific, the low energy left-handed slepton mass matrices read, in the leading-log approximation [4]:

$$(m_{\tilde{\ell},\tilde{\nu}}^2)_{ij} \simeq (\text{diagonal part})_{\tilde{\ell},\tilde{\nu}} - \frac{1}{8\pi^2}(3m_0^2 + A_0^2)(\mathbf{Y}_\nu^\dagger \mathbf{Y}_\nu)_{ij} \log \frac{M_X}{M}, \tag{5}$$

where m_0 and A_0 are respectively the soft scalar mass and the soft trilinear term at the cut-off scale, M_X, and M is the mass scale of the right-handed neutrinos. In this formula, "diagonal-part" includes the tree level soft mass matrix, the radiative corrections from gauge and charged lepton Yukawa interactions, and the mass contributions from F- and D-terms (that are different for charged sleptons and sneutrinos). Therefore, the measurement at low energies of rare lepton decays, electric dipole moments and slepton mass splittings would allow the determination of $m_{\tilde{\ell},\tilde{\nu}}^2$, and consequently would provide information about the combination $\mathbf{Y}_\nu^\dagger \mathbf{Y}_\nu \equiv P$.

It is remarkable that the matrix P encodes precisely the necessary information to reconstruct the high-energy see-saw parameters [2]. This can be easily understood from the following procedure. Using the singular value decomposition $\mathbf{Y}_\nu = V_R D_Y V_L^\dagger$, with V_R and V_L unitary matrices and $D_Y = \text{diag}(y_1, y_2, y_3)$ the diagonal matrix of the Yukawa eigenvalues, the matrix P reads:

$$P \equiv \mathbf{Y}_\nu^\dagger \mathbf{Y}_\nu = V_L^\dagger D_Y^2 V_L. \tag{6}$$

Then, from the diagonalization of P, the matrices V_L and D_Y could be straightforwardly determined. On the other hand, from $\mathcal{M}_\nu = \mathbf{Y}_\nu^T D_{\mathbf{M}_\nu}^{-1} \mathbf{Y}_\nu \langle H_u^0 \rangle^2$ and the singular value decomposition of $\mathbf{Y}_\nu$, it follows that

$$D_Y^{-1} V_L^* \mathcal{M}_\nu V_L^\dagger D_Y^{-1} = V_R^* D_{\mathbf{M}_\nu}^{-1} V_R^\dagger, \tag{7}$$

where the left hand side of this equation is known ($\mathcal{M}_\nu$ is one of our inputs, and V_L and D_Y were obtained from eq. (6)). Therefore, V_R and $D_{\mathbf{M}_\nu}$ can also be determined. This simple procedure shows that starting from the low energy observables $\mathcal{M}_\nu$ and P it is possible to determine *uniquely* the matrices $D_{\mathbf{M}_\nu}$ and $\mathbf{Y}_\nu = V_R D_Y V_L^\dagger$.

In general, although the reconstruction of the high energy see-saw parameters from low energy observables is possible in theory, in practice it could be very difficult, if not impossible. The reconstruction procedure requires the measurement of all the parameters in $\mathcal{M}_\nu$ and P, which is not feasible at least with the present and proposed experiments. In particular, the measurement of the Majorana phases in the neutrino mass matrix, the mass splitting between the lightest sleptons, and the electric dipole moments of the muon and the tau do not seem possible even with the next round of experiments.

In the view of the limitations of the previous approaches, one might try to pursue a more humble approach, taking elements from both the top-down and the bottom-up approaches. This hybrid approach would consist on a bottom-up approach with some well-motivated hypotheses about the high-energy theory. For example, we could assume that the neutrino Yukawa coupling is symmetric, or that the neutrino Yukawa eigenvalues are in the relation $m_u : m_c : m_t$, inspired in $SO(10)$ Grand Unified Theories. Another possibility is to assume that the Yukawa couplings present texture zeros, inspired by the success of the Gatto-Sartori-Tonin relation in the quark sector. Finally we could assume that the Standard Model is extended with just two right-handed neutrinos instead of three, the so-called two right-handed neutrino (2RHN) model. In the rest of the talk we will concentrate in this last hypothesis: the two right-handed neutrino model.

THE TWO RIGHT-HANDED NEUTRINO MODEL

The phenomenological motivation to consider this model is that oscillation experiments indicate that two new mass scales have to be introduced, in order to account for the solar and atmospheric mass splittings. These two mass scales could be associated to the masses of two right-handed neutrinos, therefore a see-saw model with just two right-handed neutrinos can already accommodate all the observations, being the third one in principle not necessary.

Another motivation to consider the 2RHN model is that it corresponds to interesting limits of the complete three right-handed neutrino (3RHN) model, that is undoubtedly more plausible from the theoretical point of view. Let us discuss first the situations where the 3RHN model can be well approximated by the 2RHN model from the point of view of the neutrino mass matrix, and then the situations where it can be approximated also from the point of view of the radiative corrections, parametrized by the matrix P [3]. In the basis where the right-handed neutrino mass matrix is diagonal, the low energy effective neutrino mass matrix reads:

$$(\mathscr{M}_v)_{ij} = \frac{y_{1i}y_{1j}}{M_1} + \frac{y_{2i}y_{2j}}{M_2} + \frac{y_{3i}y_{3j}}{M_3}, \tag{8}$$

where $y_{ij} = (\mathbf{Y}_v)_{ij}$. Two right-handed neutrinos dominate the see-saw when

$$\frac{y_{1i}y_{1j}}{M_1} \ll \frac{y_{2i}y_{2j}}{M_2}, \frac{y_{3i}y_{3j}}{M_3} \quad \text{or}$$
$$\frac{y_{2i}y_{2j}}{M_2} \ll \frac{y_{1i}y_{1j}}{M_1}, \frac{y_{3i}y_{3j}}{M_3} \quad \text{or}$$
$$\frac{y_{3i}y_{3j}}{M_3} \ll \frac{y_{1i}y_{1j}}{M_1}, \frac{y_{2i}y_{2j}}{M_2} \quad \text{for all } i,j = 1,2,3. \tag{9}$$

The most interesting cases are the first and the third. The first one corresponds to the case in which the Yukawa couplings for the first generation of right handed neutrinos are tiny, $y_{1i} \ll y_{2i}, y_{3i}$, for $i = 1,2,3$. If this is the case, the radiative corrections are also dominated by the same two right-handed neutrinos, the two heaviest ones. Therefore, in this case the 3RHN model can be well approximated by a 2RHN model, both from the point of view of neutrino masses as of radiative corrections. Since the two relevant right-handed neutrinos are the two heaviest ones, the corresponding Yukawa couplings could be large, and the radiative corrections could be sizable.

The third case corresponds to the situation where the mass of the heaviest right-handed neutrino is much larger than the mass of the other two, $M_3 \gg M_1, M_2$. However, in general the heaviest right-handed neutrino will produce sizable contributions to the radiative corrections. If this is the case, the 3RHN model could be reduced to a 2RHN model only from the point of view of neutrino masses, but not from the point of view of the radiative corrections. Nevertheless, there are some circumstances in which the heaviest right-handed neutrino indeed does not contribute to the radiative corrections and does not leave any imprint in P, so that this matrix is only determined by the Yukawa couplings of the two lightest generations of right-handed neutrinos. If this occurs, the 3RHN model would also be well approximated by a 2RHN model from the point of view of the radiative corrections. This situation arises for example when the mass of the heaviest right-handed neutrino is very close to the Planck mass, although this possibility seems a bit contrived.

A more plausible situation arises in models with gauge mediated supersymmetry breaking. So far, we have implicitly assumed that the boundary conditions for the soft breaking terms are set at the Planck scale. However, if the mass of the messenger particles involved in the supersymmetry breaking mechanism is smaller than M_3 but larger than M_2, then the heaviest right-handed neutrino would decouple at an energy larger than the energy at which supersymmetry breaking is communicated to the observable sector. Consequently, it would not participate in the radiative corrections of the parameters of the Lagrangian. If this is the case, only the two lightest right-handed neutrinos would contribute to the radiative corrections and to the neutrino mass generation, and therefore the 3RHN model could be well approximated by a 2RHN model.

PARAMETRIZATIONS OF THE 2RHN MODEL

For the purposes of this study the most interesting feature of the 2RHN model is that it depends on many less parameters than the complete 3RHN model, while is still capable to accommodate neutrino observations. In the basis where the right-handed mass matrix is diagonal, the model is described by two right-handed neutrino masses and a 2×3 neutrino Yukawa coupling, that depends on six independent real parameters and there phases. In consequence, the complete Lagrangian depends on eleven parameters, of which eight are moduli and three are phases. On the other hand at low energies, in principle experiments could measure a total of eighteen parameters, of which twelve are moduli and six are phases. We find then that in the 2RHN model many predictions arise, both in the neutrino mass matrix and in the radiative effects [3].

To be specific, the effective neutrino mass matrix in the 2RHN model is rank 2, and therefore the lightest neutrino mass automatically vanishes and only two possible spectra may arise:

- Normal hierarchy: $m_1 = 0$, $m_2 = \sqrt{\Delta m_{sol}^2}$, $m_3 = \sqrt{\Delta m_{atm}^2}$
- Inverted hierarchy: $m_1 = \sqrt{\Delta m_{atm}^2 - \Delta m_{sol}^2}$, $m_2 = \sqrt{\Delta m_{atm}^2}$, $m_3 = 0$

The only Majorana phase corresponds to the phase difference between the two non-vanishing mass eigenvalues. Therefore, the number of unmeasurable parameters in the 2RHN model is reduced to three moduli and one phase.

To derive the predictions on the radiative effects it is convenient to express the Yukawa coupling in the following form [5]

$$\mathbf{Y}_v = D_{\sqrt{M_v}} R D_{\sqrt{m}} U^\dagger / \langle H_u^0 \rangle, \tag{10}$$

where $D_{\sqrt{m}} = \text{diag}(\sqrt{m_1}, \sqrt{m_2}, \sqrt{m_3})$ is the diagonal matrix of the square roots of the light neutrino mass eigenvalues (which has $m_1 = 0$ for the normal hierarchy and $m_3 = 0$ for the inverted hierarchy), $D_{\sqrt{M_v}} = \text{diag}(\sqrt{M_1}, \sqrt{M_2})$ is the diagonal matrix of the square roots of the right handed neutrino masses, U is the leptonic mixing matrix, and R is an orthogonal matrix that in the 2RHN model with normal hierarchy has the following structure (in what follows we will concentrate on the case with normal hierarchy, since the analysis for the case with inverted hierarchy is completely analogous):

$$R = \begin{pmatrix} 0 & \cos z & \xi \sin z \\ 0 & -\sin z & \xi \cos z \end{pmatrix}, \tag{11}$$

with z a complex parameter and $\xi = \pm 1$ a discrete parameter that accounts for a discrete indeterminacy in R. In consequence, the elements of the neutrino Yukawa matrix read:

$$
\begin{aligned}
(\mathbf{Y}_v)_{1\alpha} &= \sqrt{M_1}(\sqrt{m_2}\cos z\, U_{\alpha 2}^* + \xi \sqrt{m_3}\sin z\, U_{\alpha 3}^*)/\langle H_u^0 \rangle, \\
(\mathbf{Y}_v)_{2\alpha} &= \sqrt{M_2}(-\sqrt{m_2}\sin z\, U_{\alpha 2}^* + \xi \sqrt{m_3}\cos z\, U_{\alpha 3}^*)/\langle H_u^0 \rangle,
\end{aligned}
\tag{12}
$$

The three moduli and the phase that are not determined by low energy experiments are identified in this parametrization with the two right-handed masses M_1 and M_2, and the complex parameter z.

Notice that we have included all the low energy phases in the definition of the matrix U, *i.e.* we have written the leptonic mixing matrix in the form $U = V \, \mathrm{diag}(1, e^{-i\phi/2}, 1)$, where ϕ is the Majorana phase and V has the form of the CKM matrix:

$$V = \begin{pmatrix} c_{13}c_{12} & c_{13}s_{12} & s_{13}e^{-i\delta} \\ -c_{23}s_{12} - s_{23}s_{13}c_{12}e^{i\delta} & c_{23}c_{12} - s_{23}s_{13}s_{12}e^{i\delta} & s_{23}c_{13} \\ s_{23}s_{12} - c_{23}s_{13}c_{12}e^{i\delta} & -s_{23}c_{12} - c_{23}s_{13}s_{12}e^{i\delta} & c_{23}c_{13} \end{pmatrix}, \tag{13}$$

so that the neutrino mass matrix is $\mathcal{M}_\nu = U^* \mathrm{diag}(m_1, m_2, m_3) U^\dagger$. It is straightforward to check that the Yukawa coupling eq.(10) indeed satisfies the see-saw formula $\mathcal{M}_\nu = Y_\nu^T \mathrm{diag}(M_1^{-1}, M_2^{-1}) Y_\nu \langle H_u^0 \rangle^2$.

The Yukawa coupling affects the renormalization group equation of the slepton parameters through the combination $P = Y_\nu^\dagger Y_\nu$, that depends in general on six moduli and three phases. Since the Yukawa coupling depends in the 2RHN model on only three unknown moduli and one phase, so does P, and consequently it is possible to obtain predictions on the moduli of three P-matrix elements and the phases of two P-matrix elements. Namely, from eq.(10) one obtains that:

$$U^\dagger P U = U^\dagger Y_\nu^\dagger Y_\nu U = D_{\sqrt{m}} R^\dagger D_{M_\nu} R D_{\sqrt{m}} / \langle H_u^0 \rangle^2. \tag{14}$$

In the case with normal hierarchy $m_1 = 0$, from where it follows that $(U^\dagger P U)_{1i} = 0$, for $i = 1, 2, 3$, leading to three relations among the elements in P. For instance, one could derive the diagonal elements in P in terms of the off-diagonal elements:

$$P_{11} = -\frac{P_{12}^* U_{21}^* + P_{13}^* U_{31}^*}{U_{11}^*},$$

$$P_{22} = -\frac{P_{12} U_{11}^* + P_{23}^* U_{31}^*}{U_{21}^*},$$

$$P_{33} = -\frac{P_{13} U_{11}^* + P_{23} U_{21}^*}{U_{31}^*}. \tag{15}$$

The observation of these correlations would be non-trivial tests of the 2RHN model.

The relations for the phases arise from the hermicity of P, since the diagonal elements in P have to be real. Taking as the independent phase the argument of P_{12}, one can derive from eq.(15) the arguments of the remaining elements:

$$e^{i\mathrm{arg}P_{13}} = \frac{-i \, \mathrm{Im}(P_{12}U_{21}U_{11}^*) \pm \sqrt{|P_{13}|^2 |U_{11}|^2 |U_{31}|^2 - [\mathrm{Im}(P_{12}U_{21}U_{11}^*)]^2}}{|P_{13}||U_{31}U_{11}^*|},$$

$$e^{i\mathrm{arg}P_{23}} = \frac{i \, \mathrm{Im}(P_{12}U_{21}U_{11}^*) \pm \sqrt{|P_{23}|^2 |U_{21}|^2 |U_{31}|^2 - [\mathrm{Im}(P_{12}U_{21}U_{11}^*)]^2}}{|P_{23}||U_{31}U_{21}^*|}, \tag{16}$$

where the $\pm$ sign has to be chosen so that the eigenvalues of P are positive. It is important to remark that the hermicity of P is not guaranteed for any value of P_{12}, $|P_{13}|$, $|P_{23}|$; only some particular ranges for the parameters are allowed, corresponding to the values for which the arguments of the square roots in eq.(16) are positive.

We conclude then that the P-matrix parameters P_{12}, $|P_{13}|$ and $|P_{23}|$ can be regarded as independent and can be used in an alternative parametrization of the 2RHN model. Together with the five moduli and the two phases of the neutrino mass matrix, sum up to the eight moduli and the three phases necessary to reconstruct the high-energy Lagrangian of the 2RHN model. Interestingly enough, most of the low energy parameters required to reconstruct the 2RHN model have good prospects to be measured or at least to be further constrained by future experiments. Concerning the parameters of the neutrino mass matrix, two mixing angles and the three masses have already been determined, under the assumption of normal hierarchy. On the other hand θ_{13} is already constrained to be small, $\sin^2 \theta_{13} < 0.046$ at 3σ [1], and will be further constrained by future experiments, the phase δ could be measured with superbeams, and the only parameter that will remain poorly determined is the Majorana phase. Finally, the parameters from the matrix P could be determined or constrained by measurements of rare lepton decays and the electric dipole moment of the electron, for which there are good perspectives in the next few years.

THE RECONSTRUCTION PROCEDURE

In Section 2 we have discussed that the complete Lagrangian can be written in terms of the five moduli and two phases of the neutrino mass matrix, and the three independent moduli and two phases of the matrix P, that is involved in the radiative corrections of the slepton parameters. In this section we will derive *exact* formulas for the high energy parameters in terms of these low energy parameters.

To this end, we will use the parametrization of the Yukawa couplings in eq.(10), so that all our ignorance of the high energy theory is encoded in the right-handed neutrino masses, M_1 and M_2, and the complex parameter in the matrix R, z. Let us define the hermitian matrix $Q \equiv U^\dagger P U$, that depends *exclusively* on parameters that in principle could be measured in low energy experiments. The first row and column vanish and yield the relations among the P-matrix elements already presented in eq.(15). On the other hand, the remaining elements Q_{22}, Q_{23}, Q_{33}, can be written in terms of the high-energy parameters M_1, M_2 and z. Therefore, one can invert the equations to derive exact expressions for the high-energy parameters in terms of the low energy parameters in Q. These expressions are [3] :

$$
M_1 = \frac{1}{2}\left[\sqrt{\left(\frac{Q_{33}}{m_3}+\frac{Q_{22}}{m_2}\right)^2 + \frac{(Q_{23}-Q_{23}^*)^2}{m_2 m_3}} - \sqrt{\left(\frac{Q_{33}}{m_3}-\frac{Q_{22}}{m_2}\right)^2 + \frac{(Q_{23}+Q_{23}^*)^2}{m_2 m_3}}\right]\langle H_u^0\rangle^2,
$$

$$
M_2 = \frac{1}{2}\left[\sqrt{\left(\frac{Q_{33}}{m_3}+\frac{Q_{22}}{m_2}\right)^2 + \frac{(Q_{23}-Q_{23}^*)^2}{m_2 m_3}} + \sqrt{\left(\frac{Q_{33}}{m_3}-\frac{Q_{22}}{m_2}\right)^2 + \frac{(Q_{23}+Q_{23}^*)^2}{m_2 m_3}}\right]\langle H_u^0\rangle^2,
$$

$$
\cos 2z = \left(\frac{Q_{33}^2}{m_3^2}-\frac{Q_{22}^2}{m_2^2}+\frac{(Q_{23}+Q_{23}^*)(Q_{23}-Q_{23}^*)}{m_2 m_3}\right)\frac{\langle H_u^0\rangle^4}{M_2^2 - M_1^2}. \tag{17}
$$

To complete the reconstruction procedure, the Yukawa coupling would be derived from these parameters using eq.(10) and where the discrete parameter ξ in eq.(11) is deter-

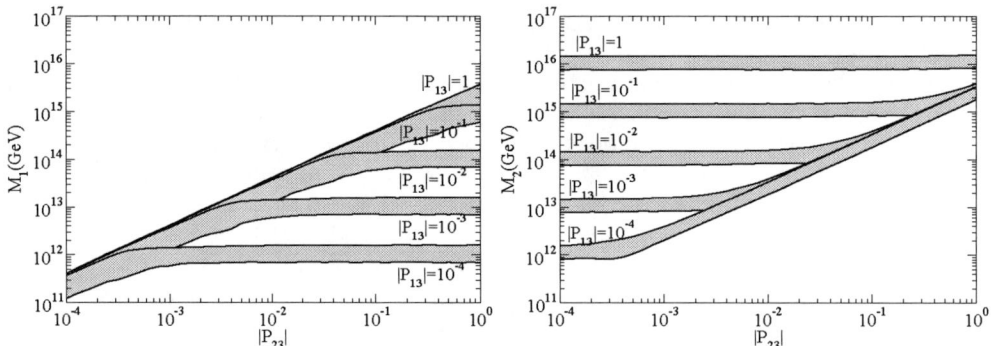

FIGURE 1. Reconstructed right-handed neutrino masses for different values of $|P_{13}|$ and $|P_{23}|$, in the limit $|P_{12}| \ll |P_{13}|, |P_{23}|$. Here, we have taken $\theta_{23} \simeq \pi/4$ and $\theta_{12} \simeq \pi/6$, and we have sampled over different values for ϕ and δ for $\theta_{13} = 0.1$. The shaded areas represent the regions at 2σ from the main value.

mined by:

$$\xi = \frac{\sqrt{m_2 m_3}}{Q_{23} \langle H_u^0 \rangle^2} (M_1 \sin z \cos z^* - M_2 \cos z \sin z^*). \tag{18}$$

We would like to illustrate now the reconstruction procedure for a particular choice of the matrix P, namely when $|P_{12}| \ll |P_{13}|, |P_{23}|$. Physically, this limit would correspond to a situation where $BR(\mu \rightarrow e\gamma) \ll BR(\tau \rightarrow e\gamma), BR(\tau \rightarrow \mu\gamma)$. Note that in this limit the only independent phase in P is irrelevant, therefore, the only independent parameters in P are $|P_{13}|$ and $|P_{23}|$, that are related to the rate for $\tau \rightarrow e\gamma$ and $\tau \rightarrow \mu\gamma$, respectively. On the other hand, in the neutrino mass matrix we know all the parameters except θ_{13}, that is constrained by present experiments to be small, and the phases δ and ϕ. We can then reconstruct the high energy parameters in terms of the low energy data, including an indeterminacy stemming from our ignorance on the phases and the angle θ_{13}. The result is shown in Fig.1, where we show the regions at 2σ from the main value for the reconstructed right-handed masses as a function of $|P_{23}|$, for different values of $|P_{13}|$. It is apparent from this figure that in this limit, $|P_{12}| \ll |P_{13}|, |P_{23}|$, the observation of rare lepton decays could allow the reconstruction of the right-handed masses up to a factor of three.

As θ_{13} becomes smaller the indeterminacy in the reconstruction process is reduced, and some analytical formulas can be derived for the reconstructed parameters in different limits:

- $|P_{12}| \ll |P_{13}| \ll |P_{23}|$ (or $BR(\mu \rightarrow e\gamma) \ll BR(\tau \rightarrow e\gamma) \ll BR(\tau \rightarrow \mu\gamma)$)

$$M_1 \simeq 2\sqrt{\frac{2}{3} \frac{|P_{13}|}{m_2}} \langle H_u^0 \rangle^2,$$

$$M_2 \simeq \frac{2|P_{23}|}{m_3} \langle H_u^0 \rangle^2,$$

126

$$\mathbf{Y}_\nu \simeq \sqrt{|P_{23}|} \left(\begin{array}{ccc} \sqrt{\dfrac{|P_{13}|}{\sqrt{6}|P_{23}|}}e^{i\phi/2} & \sqrt{\sqrt{\dfrac{3}{8}}\dfrac{|P_{13}|}{|P_{23}|}}e^{i\phi/2} & -\sqrt{\sqrt{\dfrac{3}{8}}\dfrac{|P_{13}|}{|P_{23}|}}e^{i\phi/2} \\[2ex] -\dfrac{|P_{13}|}{2|P_{23}|} & 1 & 1 \end{array} \right). \quad (19)$$

Note that in this limit the lightest right-handed mass is essentially determined by the rate for the process $\tau \to e\gamma$, while the heaviest one, by the process $\tau \to \mu\gamma$.

- $|P_{12}| \ll ||P_{23}| \ll |P_{13}|$ (or $BR(\mu \to e\gamma) \ll BR(\tau \to \mu\gamma) \ll BR(\tau \to e\gamma)$)

$$M_1 \simeq \frac{8|P_{23}|\langle H_u^0 \rangle^2}{\sqrt{9m_2^2 + 16m_3^2 + 24m_2 m_3 \cos\phi}},$$

$$M_2 \simeq \frac{|P_{13}|\sqrt{9m_2^2 + 16m_3^2 + 24m_2 m_3 \cos\phi}}{\sqrt{6}m_2 m_3}\langle H_u^0 \rangle^2,$$

$$\mathbf{Y}_\nu \simeq \sqrt{\sqrt{6}\frac{\Delta}{\Delta^*}|P_{13}|} \left(\begin{array}{ccc} \sqrt{\dfrac{\sqrt{6}|P_{23}|}{|P_{13}|}\dfrac{m_2 e^{i\phi}}{|\Delta|^2}} & \sqrt{\dfrac{|P_{23}|}{\sqrt{6}|P_{13}|}\dfrac{\Delta^*}{\Delta}} & \sqrt{\dfrac{|P_{23}|}{\sqrt{6}|P_{13}|}\dfrac{-3m_2 e^{i\phi}+4m_3}{|\Delta|^2}} \\[2ex] -\dfrac{e^{i\phi/2}}{\sqrt{6}} & \dfrac{\sqrt{6}m_2 e^{-i\phi/2}}{\Delta^2}\dfrac{|P_{23}|}{|P_{13}|} & e^{i\phi/2} \end{array} \right) \quad (20)$$

where $\Delta = \sqrt{3m_2 e^{-i\phi} + 4m_3}$. Contrary to the previous limit, the lightest right-handed mass is essentially determined by the process $\tau \to \mu\gamma$, and the heaviest by $\tau \to e\gamma$.

Finally, in the limit in which all the parameters are real, the expressions are further simplified, by substituting $\phi = 0$ or π in these formulas, to account for the cases where light neutrinos have the same or opposite CP parities.

CONCLUSIONS

In this work we have studied the possibility of reconstructing the high energy see-saw Lagrangian in terms of low energy observables. We have discussed that the reconstruction of the complete see-saw model with three right-handed neutrinos could be possible in theory, but probably not in practice. However, the reconstruction of the model with two right-handed neutrinos could indeed be possible. We have also stressed the fact that the model with two right-handed neutrinos is not just a toy model for neutrino masses, but it also corresponds to interesting and physically plausible limits of the three right-handed neutrino model.

REFERENCES

1. For a review on neutrino physics, see M. C. Gonzalez-Garcia and Y. Nir, Rev. Mod. Phys. **75** (2003) 345 [arXiv:hep-ph/0202058].
2. S. Davidson and A. Ibarra, JHEP **0109** (2001) 013,
3. A. Ibarra, JHEP **0601** (2006) 064
4. F. Borzumati and A. Masiero, Phys. Rev. Lett. **57** (1986) 961;
5. J. A. Casas and A. Ibarra, Nucl. Phys. B **618** (2001) 171.

About Mass, CP and Extra Dimensions

J.-M. Frère

Physique Théorique CP 225, ULB, B-1050 Bruxelles Belgium

Abstract. We discuss the notion of mass, mostly for fermions, and its relation to the breaking of CP invariance, the natural symmetry of gauge interactions. In a first model, we show how compactification on a Vortex in 2 extra dimensions leads to a replication of generations in 3+1, with challenging mass patterns, and testable consequences in flavour-changing neutral currents (family-number conserving), both at low energies and at future colliders. In different model, we show how CP violation can result from compactification from 4+1 to 3+1 dimensions.

Keywords: Mass generation, extra dimensions, flavour changing neutral currents, CP violation
PACS: PACS 11.10.Kk, 12.10.Kt, 11.30.Er, 11.30.Hv, 12.15.Mm ,12.

INTRODUCTION

While gauge boson masses are intimately linked to broken local symmetry, no such logical connection occurs in general for fermions. In the Standard Model, at least some of the mass must be linked to symmetry breaking, but this is due to the fact that left- and right-handed fermions are not in the same representations. Hypothetical vectorlike fermions could indeed have masses without symmetry breaking. Another situation is known for the effective, sometimes called "constituent" mass. It is usually interpreted as a result of strong interactions leading to a breaking of chiral symmetry, and used as a model for "dynamical symmetry breaking". Quite another possible origin for masses is from extra dimensions. Here the problem is almost the opposite. If the scales associated with extra dimensions are heavy, it is now a matter of preserving some light states. This is often realized through some form of localisation on topological singularities: domain walls in one, or vortices for two extradimensions.

In this presentation, we review two topics relating fermion masses and extra dimensions. The first aspect may seem at first a variant of fermion "multilocalisation", [1] leading to the generation of mass patterns from overlap of wave functions in extra dimensions. The scheme considered here is however much more constrained and deterministic, as the fermions are all localized in the same way on a unique topological singularity, and the overlap functions are precisely determined by the dynamics of this singularity. This work is based on the series of papers [3],[4],[5], [6],with predictions for colliders in [7]and recent conjectures about a light Brout-Englert-Higgs particle in [8].

The second part deals with another fundamental question, namely: "How does CP (or T) violation enter in a fundamental theory?". The main point is that CP is the natural symmetry of gauge interactions, and therefore attempts at unifying scalar couplings (effective or not) with gauge interactions pose the problem of breaking CP symmetry. We prove that this can occur through dimensional reduction, by the inclusion of "Hosotani loops" (in this case, the equivalent of Wilson loops, but looping around the extra dimen-

CP881, *Cairo International Conference on High Energy Physics (CICHEP II)*,
edited by S. Khalil

sion). In a simple example, we generate CP violation from a 4+1 theory with only real couplings. This work is based on reference [9], and its generalization to include grand unified structures [10].

THREE FAMILIES IN ONE

In this section, we describe in general terms a model evolved in collaboration with Serguey Troitsky and Maxim Libanov. Full mathematical details can be found in the original papers, and we will here mainly list the salient results.

It may still be useful to start with a little history of the model. It was initially introduced by S. Troitsky and M. Libanov in a different form, possibly lighter in field content, but with only approximate symmetry. [2], using a vortex of winding number 3. The following version, still formulated for flat extra dimensions, simulated this vortex with a scalar field of winding number one (with cubic coupling to fermions, and avoided any explicit breaking of symmetry by introducing supplementary scalar fields.[3]

In both cases, the first rationale for the use of a topological defect (the vortex) was to confine the fermions to a small region of the extra-dimensional space, taken otherwise to be flat and infinite. The vortex being built out of a scalar field Φ and an auxiliary gauge field A^μ (both unrelated to the Standard Model fields, these fields don't enter directly the currently observable phenomenology), it is possible to apply the Index theorem, and to conclude that for an effective winding number n exactly n massless chiral fermion modes persist in the remaining 3+1 dimensions for each fermion field coupling to the 6-d structure.

We have thus developed the phenomenology from this onset, but soon realized that the flat and unbounded extra dimensions represented a liability, with the need of confining the observed gauge fields, which otherwise could couple to fermionic modes outside the structure. A more involved version, largely with the same characteristics, was thus developed with the extra dimensions now assuming spherical topology.

Why maintain the topological singularity?

Part of the answer lies of course in the counting of light (massless at compactification scale) fermion modes, and the resulting replication as an origin to the very notion of particle families.

Obviously, the number of light fermions is not a direct prediction of the model in its present state, as we have to put in by hand the winding number, but the mass hierarchy of fermions, and a number of phenomenological implications are quite generic. We will detail them somewhat now.

The Model - fermion wave functions and mass hierarchies

While the latest formulation has the extra dimensions on a sphere, we will for peda-gogical reasons describe the situation for a plane. The outcome in terms of fermion wave functions, selection rules and masses is very similar. For the plane, the extra variables are chosen as (r, φ), while for the sphere (of radius R) we use $(R\theta, \varphi)$. Assuming the

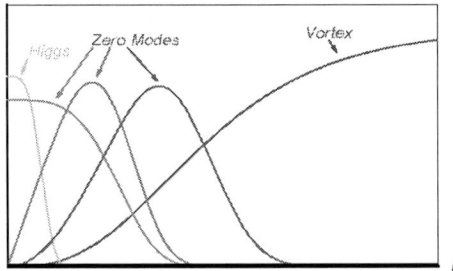

FIGURE 1. Profiles of the relevant wave functions in the radial coordinate

above-mentioned vortex structure is described by a field Φ such that

$$\Phi(x, r, \varphi) = \Phi(x)\, \Phi_1(r)\, e^{i\varphi} \tag{1}$$

This of course has only winding number 1, but the winding number 3 structure is achieved by coupling all the fermion fields according to:

$$\overline{\Psi}\Phi^3\Psi \tag{2}$$

This coupling may look surprising, as it is obviously non-renormalisable in 4 dimensions. We are not considering renormalisability of the theory here, as the 6 dimensional context is most likely an effective model, but even in this case it might be suitable that the 4-dim reduction be renormalisable. In this context, the reason we proceed as above is mostly simplicity: we could obviously replace this structure with a winding-3 solution for a fundamental field $\widetilde{\Phi}$, and find similar solutions for the couplings which will follow. This would however unnecessarily clutter the discussion, and we opt at this point for the simplicity of the field content.

The Yukawa coupling in equation (2) is only responsible for linking the fermions to the vortex. For each type of fermion field Ψ_T introduced. Here the type T represents the various quarks, $U_R, D_R, Q_L = (U_L, D_L)$ or the leptons and the indices L, R refer to the 4-dim chirality of the associated particles after dimensional reduction. Only one field is introduced to represent all the "up" quarks, for instance. The vortex structure leads, through the index theorem to 3 massless localized chiral solutions in 3+1 dimensions, which we will associate with the family replication. Typically, we have

$$\Psi_{Tn} \sim f_{Tn}(r)e^{i(3-n)\varphi} \tag{3}$$

What is important here is that each "family" of solutions has a different "winding" behaviour in the variable φ. The labeling of the modes is quite arbitrary; it was chosen here so that in the simplest case the larger mass is associated to the third generation. Rather than analytic expressions, we give a pictorial representation of the radial dependence of the corresponding functions $f_{Tn}(r)$. The fermionic mode 3 is non-vanishing at $r = 0$, with the modes behaving near the origin as:

$$f_{Tn}(r) \sim r^{3-n} \quad r \to 0 \tag{4}$$

We have thus realized a situation similar to, but much more constrained than the multilocalisation in 5 dimensions. Here, we don't have the freedom to place our various families at arbitrary spacing, but are faced instead with non-trivial overlaps fixed by the vortex properties.

This far, we only dealt with the "background" of the problem, and generated essentially massless states in 3+1 dimensions. Now comes the time to generate the "usual" fermion masses, and to break the $SU(3) \times SU(2) \times U(1)$ symmetry. This is done in the usual way, through the Brout-Englert-Higgs mechanism, at the price of introducing a scalar doublet, which we will name H. In fact, for the purpose of separating the various quantum numbers, we write, instead of the usual coupling:

$$\mathscr{L}_{Yukawa} = \lambda_U \Psi_{U_R} H^\dagger X \Psi_{Q_L} + \varepsilon \lambda_U \Psi_{U_R} H^\dagger \Phi \Psi_{Q_L} + h.c. \tag{5}$$

where Φ is the background field already described, while X is a singlet with non-vanishing value at $r = 0$ and winding number 0. The reason for introducing these couplings and X itself, is to simplify the breaking structure; it would be perfectly possible to introduce single fields to replace the combinations $H\Phi$ and HX. Obviously, similar terms obtain for the other fermion types.

The H field also couples to Φ, and this results in a non-trivial profile, centered at the origin, and exemplified in figure 1. Note that $H^\dagger X$ and $H^\dagger \varphi$

$$H^\dagger X \sim e^{i \, 0 \, \varphi} \tag{6}$$

$$H^\dagger \varphi \sim e^{i\varphi} \tag{7}$$

Mixing of families and alternatives

The reduction to 4 dimensions involves integration over φ and r (or $\theta = r/R$ in the case of a sphere) and generates the mass terms (we just give the "up" quark as an example,and neglect the term in ε for the time being):

$$m_{U,ij} = \lambda \int dr \int d\varphi f_{U_R,i}^\dagger f_{Q_L,j} X(r) H(r) e^{i(j-i)\varphi} \tag{8}$$

Clearly, the integration over φ guarantees that only diagonal entries occur, and the ratio of the masses is given by the overlap between the fermion and scalar wave functions in the coordinate r. If, for instance, the profile of the scalar combination is concentrated close to the origin (as suggested in figure 1), we get $m_3 \gg m_2 \gg m_1$ (more on this later).

Mixing is however needed, and it can be introduced by the term proportional to $H^\dagger \Phi$ in equation (5). This results in a mass matrix

$$m_{U,ij} = \begin{pmatrix} m_1 & \varepsilon \, m_{12} & 0 \\ 0 & m_2 & \varepsilon \, m_{23} \\ 0 & 0 & m_3 \end{pmatrix} \tag{9}$$

The mass eigenstates are obtained by diagonalization, but the Kobayashi-Maskawa matrix itself finds its origin in the differential rotation of the "up" and "down" quarks. That these rotations are different can result from different choices of λ, ε.

131

The possibility also exists to use completely different patterns, for instance choosing the $H^\dagger\Phi$ contribution as leading (which we are free to do, since the Yukawa couplings are arbitrary) leads to a completely off-diagonal mass matrix, a feature which can come in handy when dealing with the large rotations between neutrinos and charged leptons.

Family number near-conservation and Kaluza-Klein modes; Experimental constraints

We turn now to the gauge interactions, and to their effect in 3+1 diemnsions. As usual, we will need to face a number of Kaluza-Klein excitations for each gauge boson introduced. In the present case, each such "tower" carries a double label, since it must refer both to radial and angular excitations. Generically, (A stands for any gauge boson, γ, W, Z, gluons):

$$A_{lm}(r,\varphi) = a_{lm}(r)e^{im\varphi} \tag{10}$$

The lowest state (massless before symmetry breaking) has a flat profile in both variables, which guarantees charge universality on one hand, and diagonal couplings (even in the mass basis) for the neutral bosons. Mass relations between the lm modes depend on the topology of the model.

We will focus in this section on a particular feature of our approach, namely the presence of flavour-changing neutral currents, but with approximate conservation of "family number". Indeed, even if we neglect the Kobayashi-Maskawa mixing, the $m \neq 0$ modes generate transitions between families. Winding number (φ dependence) acts as an effective "family number", which is conserved at both vertices where an excitation A_{lm} connects to the fermions. For instance, the Z boson excitations $Z' \sim e^{\pm i\varphi}$, allow the following transitions:

$$s+\bar{d} \Rightarrow Z' \Rightarrow s+\bar{d}, \tag{11}$$
$$s+\bar{d} \Rightarrow Z' \Rightarrow \mu+\bar{e}, \tag{12}$$
$$s+\bar{d} \Rightarrow Z' \Rightarrow \tau+\bar{\mu} \tag{13}$$

As discussed in [6], this leads to powefull constraints. In particular, the second process above induces the transition $K \Rightarrow \mu+\bar{e}$, which is severely constrained experimentally. The strength of the interaction is given by the usual gauge coupling constant, the mass of the Z' excitation, and an extra factor due to the overlap of the wave functions for fermions of the first and second generation, and the radial dependence of the Z': $\kappa_{1,2} = \int dr\, f^\dagger_{U_R,1}\, f_{Q_L,2}\, z_{0,1}(r)$, leading to the limit:

$$M_{Z'} \geq \kappa_{12} \cdot 100\text{TeV} \tag{14}$$

Future tests of this model will benefit much from currently planned precision lepton flavour violation experiments at moderate energies, as well as collider experiments trying to produce directly the Kaluza-Klein excitations.

Expectations at colliders

We expect mainly two types manifestations at colliders:

- the Brout-Englert-Higgs boson mass
- production of the Kaluza-Klein excitations.

The first issue is tackled in [8], where a connection between the mass of H and the extension of the scalar wave function in the extra coordinate r. As already specified, large mass ratios $m_3/m_2, m_2/m_1$ require that the wave function be quite concentrated close to the origin. For the simplest scheme discussed above, [8] gets an upper bound close to the current limit.

The second issue deals with direct production of Kaluza-Klein excitations. As seen from equation (14), there is here a trade-off between the overlap factor and the actual mass of the new gauge boson. Constraints from K meson decays put the mass range above in the (unlikely) case that the overlap factor κ is close to 1. This is clearly out of reach of LHC, but a smaller overlap may bring the scale down considerably, albeit at the cost of production efficiency (thus requiring the full luminosity for detection). We have studied this production as a function of κ, assuming the bound on M to be saturated in each case; in [7], we see that for an integrated luminosity of $100 fb^{-1}$ and $\sqrt{s} = 14 TeV$, a few events can be expected if $M < 3 TeV$ in the $\mu^+ e^-$ final mode (the charge-conjugate channel is less sensitive). Note however that this is an exceptionally clean final state to look for!

Considerably more events are expected if we deal with the Kaluza-Klein excitations of the gluons instead, but there, even if the family-number near conservation still obtains, the extraction of the signal will be considerably more difficult, and detailed modeling is requested.

CP VIOLATION FROM EXTRA DIMENSIONS

A toy model

The motivation for this work was given already in the introduction: if a fundamental theory has all couplings related to gauge interactions, how can it violate the natural symmetry of those, i.e., CP ? One possibility is to have *spontaneous* CP violation, involving several, relatively complex vacuum expectation values. This is certainly possible, requires at least 3 such vev's, but can appear somewhat contrived. We look here for a more structural approach.

We start with a simple example, showing how a complex mass matrix can arise from a purely real Lagrangian in 5+1 dimensions.

P,CP and CPT in extra dimensions

Basically two choices are possible for the definition of Parity: either flip all the spatial coordinates, (central inversion), or only one (specular reflection). For an odd number of spatial dimensions, they are equivalent (up to a rotation), but not for an even number. In this case, the central inversion is part of the rotation group, while the specular reflection stays an extra operation. It turns out that the symmetry we are really concerned with here (and which enters the CPT theorem, with a corresponding definition of C) is the latter.

A somewhat surprising fact is that, in 4+1 dimensions, the simple Lagrangian:

$$\mathcal{L} = i\bar{\psi}\slashed{D}\psi - M\bar{\psi}\psi \tag{15}$$

where $D^a = d/dx^a + iA^a$ is already P-violating. This is unfamiliar, and will need a word of explanation, which will prove the key to the CP breaking mechanism.

If we go back to 3+1 dimensions, we note that the two expressions

$$\bar{\psi}\psi = \bar{\psi}_L\psi_R + \bar{\psi}_R\psi_L \tag{16}$$
$$\bar{\psi}i\gamma_5\psi = i(\bar{\psi}_L\psi_R - \bar{\psi}_R\psi_L) \tag{17}$$

are related by a chiral transformation, or more simply by the allowed change

$$\psi_R \to i\psi_R \tag{18}$$

There is no fundamental difference here between scalar and pseudoscalr couplings, and the simultaneous presence of both "scalar" and "pseudoscalar" independent terms is requested to have parity violation in (3+1) dim.

This is not the case in 4+1 dimensions, since we can no longer consider independently ψ_L, and ψ_R, but need to deal with the full 4-component spinor grouping both. Seen otherwise, the presence of $\gamma_4 = i\gamma_5$ in the kinetic term effectively forbids the transformation (18).

First example

Returning to (15), we note that afer singling out one extra dimension (typically, x_4), we will generate terms of the type:

$$\bar{\psi}(M + i\gamma_5 X_4)\psi \tag{19}$$

where X_4 may originate either from the derivative or the A_4 term.

Such terms lead directly to a complex mass, even if the original scalar couplings (M) were all real.

Now, of course two objections are in order. The first deals with the observability of this complex coupling, namely would it imply CP violation in 3+1 dimensions? We know the answer to this question is negative, as a chiral transformation can be used to rotate the phase away. In presence of other interactions, like QCD, this would however contribute to the θ term. We will return to this in the next section, as the obvious answer is to enlarge the gauge group we consider (what we have here corresponds only to QED!). The second objection deals with the gauge invariance of the procedure. If indeed X_4 is related to the vacuum expectation value of a field, like $X_4 = <A_4>$, the statement is not gauge invariant.

The answer to the second objection is of course straightforward, and one has to introduce either the line integral (for unbounded space) or the loop integral (for compact space):

$$X_4 = \int dy A_4(y) \tag{20}$$

where y is the extra dimension. In the case of a closed loop, this is just a Wilson loop. It can be thought of (if we imbed the loop in an unphysical plane) as the flux of A across the (unphysical) surface spanned by the loop.

Alternatively, if we are dealing with a segment or orbifold structure, A_4 can still be gauged away, but at the cost of introducing non-(anti-)periodical boundary conditions induced by:

$$\psi'(y) = e^{-i \int_0^y dy A_4(y)} \psi(y) \tag{21}$$

The use of such a line integral to introduce symmetry breaking has been investigated extensively by Hosotani [11], [12]in the framework of dynamical symmetry breaking.

A working example

We now turn to the simplest possible case of a working CP violation model along the lines sketched above. For this purpose, we use a fermion doublet, interacting with an $SU(2)$ gauge group according to :

$$\bar{\Psi} i(\partial^A - iW_a^A \tau^a) \gamma_A \Psi + M \bar{\Psi} \Psi \tag{22}$$

and assume the Hosotani loop (seen here as a kind of boundary condition):

$$\langle W_4 \rangle = \int dy\, W_4(y) = \begin{pmatrix} w & \\ & -w \end{pmatrix} \tag{23}$$

This generates a mass matrix

$$\mathcal{M} = \begin{pmatrix} M + iw\gamma_5 & \\ & M - iw\gamma_5 \end{pmatrix}.$$

The phases can be rotated away by a gauge transformation, but this leads to complex coefficients at the charged W vertices, leading to an effective "W_3-dipole moment" through the graph shown in figure 2.

Towards realistic models - chiral examples

The above models have shown how to reach effective CP violation, but are still a far cry from being realistic. For one thing, we have not discussed how to obtain chiral fermi-

135

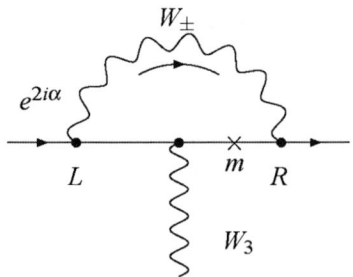

FIGURE 2. example of induced edm

ons: indeed spinors in 4+1 dimensions contain both chiralities when naively reduced to 3+1 dimensions. The obvious trick is to use a topological defect (like a domain wall, for instance). In this case, the index theorem shows that for each 4+1-d fermion, we get only one localized chiral mode in 3+1; the chirality (L or R) depending on the sign of the coupling between the scalar representing the domain wall (the presence of the latter is ensured by suitable boundary conditions). By choosing the scalar responsible for the topological singularity to be part of a larger gauge group, we can in fact obtain both chiralities from a single representation: imagine for instance an octet of such background scalars under SU(3) breaking with vev in the direction λ_8, and coupled to a triplet of fermions.

The details of such mechanism become quite intricate, and can be found in [9] where we propose semi-realistic models, endowed with non-trivial mass matrices and chiral fermions.

A complete picture finds its natural expression in the context of grand unified theories. In this case, we have shown that the first usable unification group is $SO(11)$ (rather than the usual $SU(5)$ or $SO(10)$). The source of this extension is precisely in the need to localize both left- and right-handed fermions.

CONCLUSION

While I did not want to go into the technical details, I hope this presentation has shown interesting and novel ways to address fundamental questions like the family replication and the origin of CP violation in the context of extra dimensions.

ACKNOWLEDGMENTS

I wish to thank very warmly my colleagues Serguey Troitsky, Maxim Libanov and Emin Nugaev for our ongoing collaboration on the 6-dimensional mode, as well as Nicolas Cosme and Laura Lopez-Honorez for the study of CP violation in extra dimensions. I thank of course the organisers of the CICHEP meeting in Cairo for a challenging conference, and acknowledge support from IISN and the belgian science policy (IAP V/27).

REFERENCES

1. N. Arkani-Hamed and M. Schmaltz, Phys. Rev. **D61** (2000) 033005 [hep-ph/9903417];
    G. Dvali and M. Shifman, Phys. Lett. **B475** (2000) 295 [hep-ph/0001072];
    T. Gherghetta and A. Pomarol, hep-ph/0003129;
    D. E. Kaplan and T. M. Tait, JHEP **0006** (2000) 020 [hep-ph/0004200];
    S. J. Huber and Q. Shafi, hep-ph/0010195.
2. M. V. Libanov and S. V. Troitsky, Nucl. Phys. B **599**, 319 (2001) [arXiv:hep-ph/0011095].
3. J. M. Frere, M. V. Libanov and S. V. Troitsky, Phys. Lett. B **512**, 169 (2001) [arXiv:hep-ph/0012306].
4. J. M. Frere, M. V. Libanov and S. V. Troitsky, JHEP **0111**, 025 (2001) [arXiv:hep-ph/0110045].
5. J. M. Frere, M. V. Libanov, E. Y. Nugaev and S. V. Troitsky, JHEP **0306**, 009 (2003) [arXiv:hep-ph/0304117].
6. J. M. Frere, M. V. Libanov, E. Y. Nugaev and S. V. Troitsky, JHEP **0403**, 001 (2004) [arXiv:hep-ph/0309014].
7. J. M. Frere, M. V. Libanov, E. Y. Nugaev and S. V. Troitsky, JETP Lett. **79**, 598 (2004) [Pisma Zh. Eksp. Teor. Fiz. **79**, 734 (2004)] [arXiv:hep-ph/0404139].
8. M. V. Libanov and E. Y. Nugaev, arXiv:hep-ph/0512223.
9. N. Cosme, J. M. Frere and L. Lopez Honorez, Phys. Rev. D **68**, 096001 (2003) [arXiv:hep-ph/0207024]
10. N. Cosme and J. M. Frere, Phys. Rev. D **69**, 036003 (2004) [arXiv:hep-ph/0303037]
11. Y. Hosotani, Phys. Lett. B **126** (1983) 309.
12. Y. Hosotani, Phys. Lett. B **129** (1983) 193.

$\mathrm{E_6SSM}$

S. F. King, S. Moretti and R. Nevzorov

School of Physics and Astronomy, University of Southampton,
Southampton, SO17 1BJ, UK

Abstract. In this talk we discuss an E_6 inspired supersymmetric (SUSY) model with an extra $U(1)_N$ gauge symmetry under which right-handed neutrinos have zero charge. In this exceptional supersymmetric standard model ($\mathrm{E_6SSM}$) the μ–term is generated dynamically after the electroweak symmetry breaking. We specify the particle content of the model and argue that the presence of a Z' and exotic particles predicted by $\mathrm{E_6SSM}$ allows the lightest Higgs boson to be significantly heavier than in the MSSM and NMSSM. Other possible manifestations of $\mathrm{E_6SSM}$ at the LHC are also discussed.

Keywords: Supersymmetry; Extra gauge groups; Electroweak symmetry breaking.
PACS: 12.60.Jv, 12.60.Cn, 12.60.Fr

INTRODUCTION

The cancellation of quadratic divergences in the supersymmetric models does not allow to solve the hierarchy problem of the standard model (SM) entirely. Indeed, the superpotential of the simplest supersymmetric extension of the SM — minimal supersymmetric (SUSY) standard model (MSSM) contains a bilinear term $\mu H_d H_u$. In order to get the correct pattern of electroweak (EW) symmetry breaking the parameter μ is required to be of the order of electroweak or SUSY breaking scale. At the same time the incorporation of the MSSM into supergravity or Grand Unified theories (GUT) results in $\mu \sim M_X - M_{Pl}$, where M_X and M_{Pl} are GUT or Planck scales respectively. This is the so–called μ–problem.

An elegant solution to the μ problem naturally arises in the framework of superstring inspired E_6 models. At the string scale E_6 can be broken directly to the rank-6 subgroup $SU(3)_C \times SU(2)_L \times U(1)_Y \times U(1)_\psi \times U(1)_\chi$. Two anomaly-free $U(1)_\psi$ and $U(1)_\chi$ symmetries of the rank-6 model are defined by: $E_6 \to SO(10) \times U(1)_\psi$, $SO(10) \to SU(5) \times U(1)_\chi$. Near the string scale the rank-6 model can be reduced further to an effective rank–5 model with only one extra $U(1)'$ gauge symmetry. The extra $U(1)'$ gauge symmetry forbids an elementary μ term but allows interaction $\lambda S H_d H_u$ in the superpotential. The scalar component of the SM singlet superfield S acquires a non-zero vacuum expectation value (VEV) breaking $U(1)'$ and giving rise to an effective μ term. Here we review a particular E_6 inspired supersymmetric model with an extra $U(1)_N$ gauge symmetry in which right handed neutrinos do not participate in the gauge interactions [1].

At collider energies the gauge group is:

$$SU(3)_C \times SU(2)_L \times U(1)_Y \times U(1)_N \tag{1}$$

CP881, *Cairo International Conference on High Energy Physics (CICHEP II)*,
edited by S. Khalil

where the Standard Model is augmented by an additional $U(1)_N$ gauge group which is defined so that right-handed neutrinos are neutral under it and can be superheavy. This gauge group is supposed to descend from an E_6 GUT gauge group which is broken at the GUT scale. The $U(1)_N$ gauge group (a particular case of $U(1)'$) is broken near the TeV energy scale giving rise to a massive Z' gauge boson which can be discovered at the LHC.

To ensure anomaly cancellation the low energy particle content of the E_6SSM must include complete fundamental 27 representations of E_6. Thus in addition to the three families of SM quarks and leptons we predict three families of exotic quark states $D_i, \overline{D}_i$ which carry a $B - L$ charge $\left(\pm\frac{2}{3}\right)$, and singlet fields S_i which carry non-zero $U(1)_N$ charges and therefore survive down to the EW scale. We also predict three families of states H_{1i} and H_{2i} which have the quantum numbers of Higgs doublets. We also require a further pair H' and $\overline{H}'$ from incomplete extra $27'$ and $\overline{27}'$ representations to survive to low energies in order to ensure gauge coupling unification. Thus in addition to a Z' the E_6SSM involves extra matter beyond the MSSM.

THE SUPERPOTENTIAL AND PARAMETER COUNTING

The superpotential of the E_6SSM involves a lot of new Yukawa couplings in comparison to the SM. In general these new interactions violate baryon number conservation and induce non-diagonal flavour transitions. To suppress baryon number violating and flavour changing processes one can postulate a Z_2^H symmetry under which all superfields except one pair of H_{1i} and H_{2i} (say $H_d \equiv H_{13}$ and $H_u \equiv H_{23}$) and one SM-type singlet field $(S \equiv S_3)$ are odd. The Z_2^H symmetry reduces the structure of the Yukawa interactions to:

$$W_{E_6\text{SSM}} \simeq \lambda_i S(H_{1i}H_{2i}) + \kappa_i S(D_i\overline{D}_i) + f_{\alpha\beta}S_\alpha(H_d H_{2\beta}) +$$
$$\tilde{f}_{\alpha\beta}S_\alpha(H_{1\beta}H_u) + W_{MSSM}(\mu = 0), \tag{2}$$

where $\alpha, \beta = 1, 2$ and $i = 1, 2, 3$. In Eq. (2) we ignore H' and $\overline{H}'$ for simplicity. Here we define $\lambda \equiv \lambda_3$. The $SU(2)$ doublets H_u and H_d play the role of Higgs fields generating the masses of quarks and leptons after electroweak symmetry breaking (EWSB). Therefore it is natural to assume that only S, H_u and H_d acquire non-zero VEVs. If λ or κ_i are large at the grand unification (GUT) scale M_X they affect the evolution of the soft scalar mass m_S^2 of the singlet field S rather strongly resulting in negative values of m_S^2 at low energies that trigger the breakdown of the $U(1)_N$ symmetry. To guarantee that only H_u, H_d and S acquire a VEV we impose a certain hierarchy between the couplings H_{1i} and H_{2i} to the SM-type singlet superfields S_i: $\lambda \gg \lambda_{1,2}, f_{\alpha\beta}$ and $\tilde{f}_{\alpha\beta}$.

The masses of the fermion components of H' and $\overline{H}'$ are induced by the term $\mu' H'\overline{H}'$ in the superpotential. The corresponding mass term is not involved in the process of EWSB. Therefore parameter μ' remains arbitrary. Gauge coupling unification requires μ' to be within 100 TeV. The masses of scalar components of H' and $\overline{H}'$ are determined by the soft masses $m_{H'}^2$ and $m_{\overline{H}'}^2$ as well as by μ' and the corresponding bilinear scalar coupling B' in the scalar potential. Because μ' and B' can be complex the spectrum of survival components of $27'$ and $\overline{27}'$ is determined by six parameters.

The superpotential (2) contains 14 new Yukawa couplings as compared to the MSSM with $\mu = 0$. They are accompanied by 14 trilinear scalar couplings in the SUSY scalar potential. In addition the scalar potential of E_6SSM includes 13 soft SUSY masses: six masses of exotic squarks $m_{\tilde{D}_i}$ and $m_{\tilde{\overline{D}}_i}$, four masses of non–Higgs fields $m_{\tilde{H}_{1,\alpha}}$ and $m_{\tilde{H}_{2,\alpha}}$ ($\alpha = 1, 2$) and three masses of SM singlet scalar fields $m_{S_i}^2$. Because Yukawa and trilinear scalar couplings can be complex the Z_2^H–symmetric E_6SSM involves 75 new parameters in comparison to the MSSM with $\mu = 0$ which determine masses and couplings of extra fields. Thirty of them are phases. Some of these phases can be eliminated by the appropriate redefinition of new superfields.

Although Z_2^H eliminates any problem related with baryon number violation and non-diagonal flavour transitions it also forbids all Yukawa interactions that would allow the exotic quarks to decay. Since models with stable charged exotic particles are ruled out by different experiments [2] the Z_2^H symmetry must be broken. But the breakdown of Z_2^H should not give rise to the operators leading to rapid proton decay. There are two ways to overcome this problem. The resulting Lagrangian has to be invariant either with respect to Z_2^L symmetry, under which all superfields except lepton ones are even, or with respect to Z_2^B discrete symmetry, which implies that exotic quark and lepton superfields are odd whereas the others remain even. The terms in the superpotential which permit exotic quarks to decay and are allowed by the E_6 symmetry can be written in the following alternative forms, depending on which discrete symmetry is imposed. If Z_2^L is imposed then the following couplings are allowed:

$$W_1 = g_{ijk}^Q D_i(Q_j Q_k) + g_{ijk}^q \overline{D}_i d_j^c u_k^c, \qquad (3)$$

which implies that exotic quarks are diquarks. If Z_2^B is imposed then the following couplings are allowed:

$$W_2 = g_{ijk}^E e_i^c D_j u_k^c + g_{ijk}^D (Q_i L_j) \overline{D}_k. \qquad (4)$$

which implies that exotic quarks are leptoquarks. We assume that the violation of the Z_2^H symmetry in the E_6SSM is mainly caused by the Yukawa couplings of the exotic particles to the quarks and leptons of the third generations. This assumption results in three (six) extra Yukawa couplings if exotic quarks are diquarks (leptoquarks). Thus together with the trilinear scalar couplings this would increase the total number of independent parameters by 12 (24) degrees of freedom.

PHENOMENOLOGICAL IMPLICATIONS

The E_6SSM Higgs sector includes two Higgs doublets H_u and H_d as well as a SM–like singlet field S. After the breakdown of the gauge symmetry two CP-odd and two charged Goldstone modes in the Higgs sector are absorbed by the Z, Z' and $W^{\pm}$ gauge bosons so that only six physical degrees of freedom are left. They represent three CP-even (as in the NMSSM), one CP-odd and two charged Higgs states (as in the MSSM).

As in any other SUSY model the mass of the lightest CP–even Higgs boson m_h in the E_6SSM is limited from above. In Fig. 1 we plot the two-loop upper bounds on the mass of the lightest Higgs particle in the MSSM, NMSSM and E_6SSM as a function of $\tan\beta$. At moderate values of $\tan\beta$ ($\tan\beta = 1.6 - 3.5$) the upper limit on the lightest Higgs boson mass in the E_6SSM is considerably higher than in the MSSM and NMSSM. It reaches the maximum value $150 - 155\,$GeV at $\tan\beta = 1.5 - 2$ [1]. The main reason for the increased Higgs mass in the E_6SSM is due to the increased upper limit on the coupling λ (caused by the extra exotic states) which controls the important F-term contribution to m_h. At large $\tan\beta > 10$ the theoretical restriction on m_h in the E_6SSM is $4 - 5\,$GeV larger than the one in the MSSM and NMSSM because of the $U(1)_N$ D-term contribution to m_h^2. The discovery at future colliders of a relatively heavy SM-like Higgs boson with mass $140 - 155\,$GeV will permit to distinguish the E_6SSM from the MSSM and NMSSM.

Other possible manifestations of our exceptional SUSY model at the LHC are related with the presence of a Z' and of exotic multiplets of matter. For instance, a relatively light Z' will lead to enhanced production of l^+l^- pairs ($l = e, \mu$). Fig. 2 shows the differential distribution in invariant mass of the lepton pair l^+l^- (for one species of lepton $l = e, \mu$) in Drell–Yan production at the LHC with and without light exotic quarks with representative masses of exotic quarks $\mu_{D_i} = 250$ GeV for all three generations and with $M_{Z'} = 1.2\,$TeV. This distribution is promptly measurable at the CERN collider with a high resolution and would enable one to not only confirm the existence of a Z' state but also to establish the possible presence of additional exotic matter, by simply fitting to the data the width of the Z' resonance. The analysis performed in [3] revealed that a Z' boson in E_6 inspired models can be discovered at the LHC if its mass is less than $4 - 4.5\,$TeV. At the same time the determination of its couplings should be possible up to $M_{Z'} \sim 2 - 2.5\,$TeV [4].

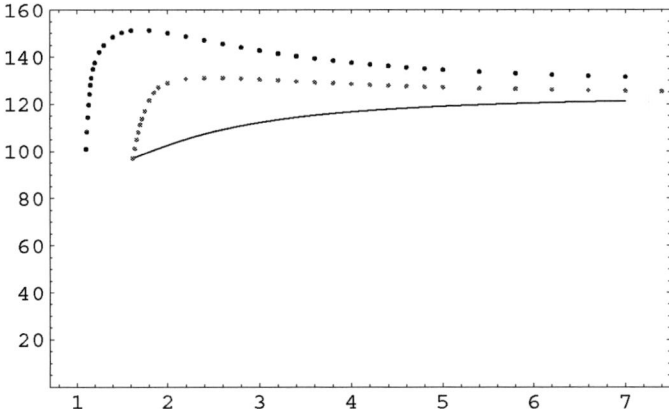

FIGURE 1. Two-loop upper bound on the lightest Higgs mass versus $\tan\beta$. The solid, lower and upper dotted lines correspond to the theoretical restrictions on the lightest Higgs mass in the MSSM, NMSSM and E_6SSM respectively.

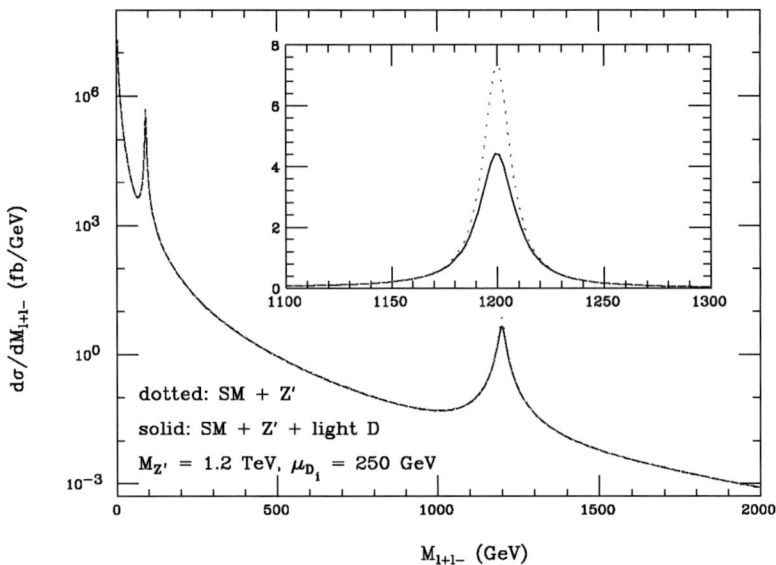

FIGURE 2. Differential cross section in the final state invariant mass, denoted by $M_{l^+l^-}$, at the LHC for DY production ($l = e$ or μ only) in presence of a Z' with and without the (separate) contribution of exotic D-quarks with $\mu_{Di} = 250\,\text{GeV}$ for $M_{Z'} = 1.2\,\text{TeV}$.

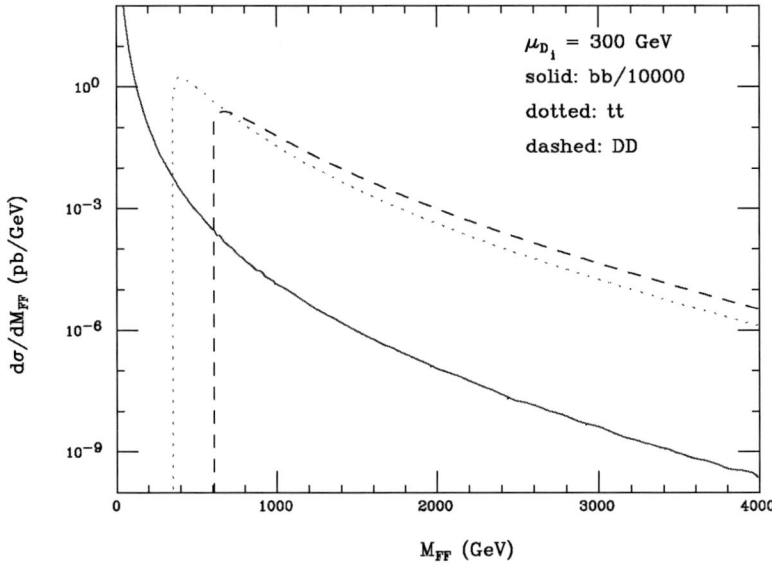

FIGURE 3. Cross section at the LHC for pair production of exotic D-quarks as a function of the invariant mass of $D\bar{D}$ pair. Similar cross sections of $t\bar{t}$ and $b\bar{b}$ production are also included for comparison.

The exotic quarks can be also relatively light in the E_6SSM since their masses are set by the Yukawa couplings κ_i and λ_i that may be small. Then the production cross section of exotic quark pairs at the LHC can be comparable with the cross section of $t\bar{t}$ production (see Fig. 3). Since we have assumed that Z_2^H is mainly broken by operators involving quarks and leptons of the third generation the lightest exotic quarks decay into either two heavy quarks QQ or a heavy quark and a lepton $Q\tau(\nu_\tau)$, where Q is either a b- or t-quark. This results in the growth of the cross section of either $pp \to Q\bar{Q}Q^{(\prime)}\bar{Q}^{(\prime)}+X$ or $pp \to Q\bar{Q}l^+l^-+X$. The discovery of the Z' and exotic quarks predicted by the E_6SSM would represent a possible indirect signature of an underlying E_6 gauge structure at high energies and provide a window into string theory.

ACKNOWLEDGMENTS

The authors would like to to acknowledge support from the PPARC grant PPA/G/S/2003/00096, the NATO grant PST.CLG.980066 and the EU network MRTN 2004-503369.

REFERENCES

1. S. F. King, S. Moretti and R. Nevzorov, *Phys. Rev.* **D73**, 035009 (2006); S. F. King, S. Moretti and R. Nevzorov, *Phys. Lett.* **B634**, 278–284 (2006).
2. T. K. Hemmick et al., *Phys. Rev.* **D41** 2074–2080 (1990).
3. J. Kang and P. Langacker, *Phys. Rev.* **D71** 035014 (2005).
4. M. Dittmar, A. Djouadi, and A.-S. Nicollerat, *Phys. Lett.* **B583** 111–120 (2004).

Modular Inflation and the Curvaton

George Lazarides

*Physics Division, School of Technology, Aristotle University of Thessaloniki,
GR-54124 Thessaloniki, Greece*

Abstract. Supersymmetric Peccei-Quinn models which provide a suitable candidate for the curvaton field are studied. These models also solve the μ problem, while generating the Peccei-Quinn scale dynamically. The curvaton is a pseudo Nambu-Goldstone boson corresponding to an angular degree of freedom orthogonal to the axion. Its order parameter increases substantially following a phase transition during inflation. This results in a drastic amplification of the curvaton perturbations. Consequently, these models are able to accommodate low-scale inflation with Hubble parameter at the TeV scale such as modular inflation. We find that modular inflation with the orthogonal axion as curvaton can indeed account for the observations for natural values of the parameters. In particular, the spectral index can easily be made adequately lower than unity in accord with the recent data.

Keywords: Inflation, curvaton
PACS: 98.80.Cq

INTRODUCTION

The precise cosmological observations of the last decade have established inflation (for a review on inflation, see e.g. [1]) as an essential extension of the hot big bang model. However, the case of slow-roll inflation (i.e. the case with inflaton mass $\ll H_*$, the Hubble parameter at the time when the cosmological scales exit the horizon during inflation) suffers from the fact that, typically, supergravity (SUGRA) introduces [2, 3] corrections to the inflaton mass of order H_* during inflation. One way to keep the inflaton mass under control is to use as inflaton a pseudo Nambu-Goldstone boson (PNGB) field, since the flatness of the potential of such a field is protected by a global U(1) symmetry. Such candidates are [4] the string axions, which are the imaginary parts of string moduli fields with the flatness of their potential lifted only by (soft) supersymmetry (SUSY) breaking. This results in inflaton masses $\lesssim H_*$. The resulting modular inflation can be of the fast-roll type [5]. Fast-roll inflation lasts only a limited number of e-foldings, which, however, can be enough to solve the horizon and flatness problems.

In modular inflation, $H_* \sim 1$ TeV and, thus, the inflationary energy scale is much lower than the grand unified theory (GUT) scale. As a result, the perturbations of the inflaton are not sufficiently large to account for the required density perturbations for explaining the large scale structure in the universe and the temperature perturbations in the cosmic microwave background radiation (CMBR). (For a low-scale inflation model where the inflaton perturbations are adequate, see [6].) Thus, a curvaton [7] (see also [8]), i.e. another "light" field during inflation, is necessary to provide the observed curvature perturbation. However, even the curvaton cannot [9] generically help us to reduce the inflationary scale to energies low enough for modular inflation. This is possible only in certain curvaton models which amplify [10, 11] additionally the curvaton perturbations.

CP881, *Cairo International Conference on High Energy Physics (CICHEP II)*,
edited by S. Khalil

We will construct [12] a class of SUSY Peccei-Quinn (PQ) models [13] which possess such an amplification mechanism and also solve naturally the strong CP and μ problems. We use as curvaton an angular degree of freedom orthogonal to the QCD axion, which we will call orthogonal axion (the radial PQ field was used as curvaton in [14]). We study the characteristics of the scalar potential in this class of models. We then focus on curvaton physics and derive a number of important constraints necessary for the model to be a successful curvaton model. Finally, we concentrate on a concrete example of this class of models and show that it can indeed work for natural values of parameters.

MODULAR INFLATION

After gravity mediated soft SUSY breaking, the potential of the inflaton s, which is a canonically normalized string axion, is [4]

$$V(s) = V_{\mathrm{m}} - \frac{1}{2}m_s^2 s^2 + \cdots, \tag{1}$$

where $V_{\mathrm{m}} \sim (m_{3/2}M_{\mathrm{P}})^2$ and $m_s \sim m_{3/2}$ with $m_{3/2} \sim 1$ TeV and $M_{\mathrm{P}} \simeq 2.44 \times 10^{18}$ GeV being the gravitino mass and the reduced Planck mass respectively. The ellipsis in (1) denotes terms which are expected to stabilize $V(s)$ at $s \sim M_{\mathrm{P}}$. The inflationary potential V_* at the time when the cosmological scales exit the horizon is of intermediate scale:

$$V_*^{\frac{1}{4}} \sim \sqrt{m_{3/2}M_{\mathrm{P}}} \sim 10^{10.5} \text{ GeV} \tag{2}$$

for which $H_* \sim m_{3/2}$.

In this model, inflation can be of the fast-roll type, where

$$s = s_{\mathrm{i}}\exp(F_s\Delta N) \quad \text{with} \quad F_s \equiv \frac{3}{2}\left(\sqrt{1 + \frac{4c}{9}} - 1\right), \quad c \equiv \left(\frac{m_s}{H_*}\right)^2 \sim 1. \tag{3}$$

Here, ΔN is the number of the elapsed e-foldings and s_{i} the initial value of the inflaton field s. From the above, one can obtain the inflation potential N e-foldings before the end of inflation as

$$V(N) \simeq V_{\mathrm{m}}\left(1 - e^{-2F_sN}\right). \tag{4}$$

Even in the fast-roll case, modular inflation keeps the Hubble parameter H rather rigid. Indeed, the slow-roll parameter

$$\varepsilon_H \equiv -\frac{\dot{H}}{H^2} = \frac{1}{2}c^2\left(\frac{s}{M_{\mathrm{P}}}\right)^2 \ll 1, \tag{5}$$

because $c \sim 1$ and $s \ll M_{\mathrm{P}}$ during inflation (the overdot denotes derivative with respect to the cosmic time t).

The initial conditions for the inflaton field are determined by the quantum fluctuations which send the field off the top of the potential hill. Hence, we expect that the initial value for the inflaton is

$$s_i \simeq \frac{H_m}{2\pi}, \quad \text{where} \quad H_m \simeq \frac{\sqrt{V_m}}{\sqrt{3}M_P}. \tag{6}$$

Assuming that the final value of s is close to its vacuum expectation value (VEV) $\langle s \rangle \sim M_P$, we find, from (3), that the total number of e-foldings is

$$N_{tot} \simeq \frac{1}{F_s} \ln \left(\frac{M_P}{m_{3/2}} \right), \tag{7}$$

where we took into account that $H_m \sim m_{3/2}$.

AMPLIFICATION OF THE CURVATON PERTURBATIONS

We consider a PNGB curvaton σ whose order parameter $v = v(t)$ (determined by the values of the radial fields in the model) takes [10, 11] a different (larger) expectation value in the vacuum than during inflation and, in particular, when the cosmological scales exit the horizon. The potential for the real canonically normalized field σ is

$$V(\sigma) = (v\tilde{m}_\sigma)^2 \left[1 - \cos \left(\frac{\sigma}{v} \right) \right] \quad \Rightarrow \quad V(|\sigma| \ll v) \simeq \frac{1}{2}\tilde{m}_\sigma^2 \sigma^2, \tag{8}$$

where $\tilde{m}_\sigma = \tilde{m}_\sigma(v)$ is the variable mass of σ. In the vacuum, $v = v_0$ and $\tilde{m}_\sigma = m_\sigma$.

In the curvaton scenario, the curvature perturbation observed by the cosmic microwave background explorer (COBE) [15], $\zeta \simeq 2 \times 10^{-5}$, is given by

$$\zeta \sim \Omega_{dec}\zeta_\sigma, \tag{9}$$

where ζ_σ is the partial curvature perturbation of the curvaton and Ω_{dec} is the ratio of the curvaton energy density ρ_σ to the total energy density of the universe ρ at the time of the decay of the curvaton:

$$10^{-2} \lesssim \Omega_{dec} \equiv \left. \frac{\rho_\sigma}{\rho} \right|_{dec} \leq 1. \tag{10}$$

The lower bound originates (see [16]) from the 95% confidence level bound on the possible non-Gaussian component of the curvature perturbation from the CMBR data obtained by the Wilkinson microwave anisotropy probe (WMAP) satellite [17]. The partial curvature perturbation of the curvaton when the latter oscillates in a quadratic potential is given [18] by

$$\zeta_\sigma \sim \left. \frac{\delta\sigma}{\sigma} \right|_{dec} \sim \left. \frac{\delta\sigma}{\sigma} \right|_{osc}, \tag{11}$$

where "osc" denotes the onset of curvaton oscillations at a time given by $H_{osc} \sim m_\sigma$ (we assume that $\tilde{m}_\sigma$ had reached its vacuum value before the onset of oscillations).

The phase $\theta \equiv \sigma/v$ corresponding to σ remains frozen until the oscillations begin:

$$\theta_{osc} \simeq \theta_*, \quad \delta\theta_{osc} \simeq \delta\theta_*, \tag{12}$$

where star denotes the values of quantities at the time when the cosmological scales exit the inflationary horizon and $\delta\theta$ is the perturbation in θ. This implies that

$$\left.\frac{\delta\sigma}{\sigma}\right|_{osc} = \left.\frac{\delta\theta}{\theta}\right|_{osc} \simeq \left.\frac{\delta\theta}{\theta}\right|_* = \left.\frac{\delta\sigma}{\sigma}\right|_*. \tag{13}$$

From the curvaton perturbation during inflation $\delta\sigma_* = H_*/2\pi$, we then find [12] that

$$\delta\sigma_{osc} \simeq \frac{H_*}{2\pi\varepsilon}, \quad \text{where} \quad \varepsilon \equiv \frac{v_*}{v_0} \ll 1. \tag{14}$$

So, after the end of inflation when v assumes its vacuum value, *the curvaton perturbation is amplified by a factor* ε^{-1}. Finally, one can show [12] that

$$\sigma_{osc} \sim \frac{H_*\Omega_{dec}}{\pi\varepsilon\zeta} \quad \text{and} \quad \varepsilon \geq \varepsilon_{min} \equiv \frac{H_*}{2\pi v_0}, \tag{15}$$

where we used the relations $\delta\sigma_*/\sigma_* \leq 1$ and $\sigma_{osc} \lesssim v_0$.

CAN WE CONSTRUCT PQ MODELS WITH A PNGB CURVATON?

In the SUGRA extension of the minimal supersymmetric standard model (MSSM), there exist D- and F-flat directions in field space which can generate intermediate scales

$$M_I \sim (m_{3/2}M_P^n)^{\frac{1}{n+1}}, \tag{16}$$

where n is a positive integer. It is natural to identify M_I with the symmetry breaking scale f_a of the PQ symmetry $U(1)_{PQ}$, such that a μ term is generated with $\mu \sim f_a^{n+1}/M_P^n \sim m_{3/2}$ [19]. This would simultaneously resolve the strong CP and μ problems of MSSM.

For this, we need a non-renormalizable superpotential term

$$\lambda P^{n+1}h_1 h_2/M_P^n, \tag{17}$$

where λ is a dimensionless parameter, P is a standard model (SM) singlet superfield with $\langle P \rangle \sim M_I$ and h_1, h_2 are the electroweak Higgs doublets. One can show [12] that P must necessarily carry a non-zero PQ charge. As a consequence, P has a flat potential. To lift the flatness of its potential and generate an intermediate VEV for $P \sim M_I$, we must introduce [20, 21, 22] a second SM singlet Q with non-zero PQ charge having a coupling of the type

$$\xi P^{n+3-k}Q^k/M_P^n, \tag{18}$$

where ξ is a dimensionless parameter and k is a positive integer smaller than $n+3$.

After soft SUSY breaking, the scalar potential possesses [12] non-trivial minima at

$$|P|, \; |Q| \sim (m_{3/2} M_P^n)^{\frac{1}{n+1}}, \tag{19}$$

where $U(1)_{PQ}$ is spontaneously broken and f_a and μ are generated dynamically. The soft masses-squared m_P^2, $m_Q^2 \sim m_{3/2}^2$ of P, Q can have either sign, while the coefficient A of the soft A-term corresponding to the coupling in (18), which is generally complex with $|A| \sim m_{3/2}$, must be large enough for the non-trivial minima to exist if m_P^2, $m_Q^2 > 0$.

To implement our scenario, we need a valley of local minima of the potential which has negative inclination and $|P| \ll |P|_0$ and $|Q| \ll |Q|_0$, where $|P|_0$ and $|Q|_0$ are the vacuum values of $|P|$ and $|Q|$ respectively. If the system slowly rolls down this valley during inflation, the order parameter $v \ll v_0$ and our amplification mechanism for the curvaton perturbations may work. This can be achieved only if one of the masses-squared m_P^2, m_Q^2 is negative. Let us take $m_P^2 < 0$ and $m_Q^2 > 0$. In this case, the scalar potential is [22] unbounded below unless $k = 1$. So, we restrict ourselves to the case $k = 1$. One can show [12] that the orthogonal axion in this case acquires a mass of order $m_{3/2}$ during inflation and, thus, does not qualify as a PNGB curvaton.

The addition of a third SM singlet superfield S, however, with a coupling

$$\xi_q P^{n+3-p-q} Q^p S^q / M_P^n, \tag{20}$$

where p, q are non-negative integers with $p + q \le n + 3$ and $q \ge 3$, can drastically change [12] the situation allowing the implementation of our mechanism. In the next section, we will present a concrete class of models of this category.

PQ MODELS WITH AN AXION-LIKE CURVATON

We consider a class of extensions of MSSM which are based on the SM gauge group, but also possess a global anomalous PQ symmetry $U(1)_{PQ}$, a global non-anomalous R symmetry $U(1)_R$, and a discrete Z_2^P symmetry. Note, in passing, that global continuous symmetries can effectively arise [23] from the discrete symmetry groups of many compactified string theories (see e.g. [24]). In addition to the usual MSSM superfields h_1, h_2 (Higgs $SU(2)_L$ doublets), l_i ($SU(2)_L$ doublet leptons), e_i^c ($SU(2)_L$ singlet charged leptons), q_i ($SU(2)_L$ doublet quarks), and u_i^c, d_i^c ($SU(2)_L$ singlet anti-quarks) with $i = 1$, 2, 3 being the family index, the models contain the SM singlet superfields P, Q, and S. The charges of the superfields under $U(1)_{PQ}$ and $U(1)_R$ are

$$\text{PQ}: \; P(-2), \; Q(2), \; S(0), \; h_1, \; h_2(n+1),$$
$$\text{R}: \; P(\frac{n+3}{2}), \; Q(\frac{n-1}{2}), \; S(\frac{n+1}{2}), \; h_1, \; h_2(0) \tag{21}$$

with the "matter" (quark and lepton) superfields having $\text{PQ} = -(n+1)/2$ and $\text{R} = (n+1)(n+3)/4$. The integer n is taken to be of the form

$$n = 4l + 1, \quad \text{where} \quad l = 0, 1, 2, ..., \tag{22}$$

for reasons to be explained below. Finally, under Z_2^P, P changes sign. Baryon (and lepton) number is [25] automatically conserved to all orders in perturbation theory as a consequence of $U(1)_R$ (and $U(1)_{PQ}$). The Z_2 subgroup of $U(1)_{PQ}$ coincides with the matter parity symmetry Z_2^{mp}, which changes the sign of all matter superfields.

The most general superpotential compatible with these symmetries is

$$
W = y_{eij}l_ih_1e_j^c + y_{uij}q_ih_2u_j^c + y_{dij}q_ih_1d_j^c
$$
$$
+ \lambda P^{n+1}h_1h_2/M_P^n + \sum_{k=0}^{(n+3)/4} \lambda_k S^{n+3-4k}(PQ)^{2k}/M_P^n, \tag{23}
$$

where y_{eij}, y_{uij}, y_{dij} are the usual Yukawa coupling constants, λ, λ_k are complex dimensionless parameters, and summation over the family indices is implied.

The scalar potential

The resulting scalar potential for PQ breaking after soft SUSY breaking is

$$
V = |F_P|^2 + |F_Q|^2 + |F_S|^2 + V_{\text{soft}}, \tag{24}
$$

where

$$
F_P = \sum_{k=1}^{(n+3)/4} 2k\lambda_k \frac{S^{n+3-4k}(PQ)^{2k-1}Q}{M_P^n}, \tag{25}
$$

$$
F_Q = \sum_{k=1}^{(n+3)/4} 2k\lambda_k \frac{S^{n+3-4k}(PQ)^{2k-1}P}{M_P^n}, \tag{26}
$$

and

$$
F_S = \sum_{k=0}^{(n-1)/4} (n+3-4k)\lambda_k \frac{S^{n+2-4k}(PQ)^{2k}}{M_P^n} \tag{27}
$$

are the F-terms, and

$$
V_{\text{soft}} = m_P^2|P|^2 + m_Q^2|Q|^2 + m_S^2|S|^2 + \left[A \sum_{k=0}^{(n+3)/4} \lambda_k \frac{S^{n+3-4k}(PQ)^{2k}}{M_P^n} + \text{h.c.} \right] \tag{28}
$$

the soft SUSY-breaking terms. Here, the soft SUSY-breaking masses-squared m_P^2, m_Q^2, and m_S^2 are of the order of the $m_{3/2}^2$ and can, in principle, have either sign. However, the potential V is [12] bounded below only if m_P^2, and m_Q^2 are positive. For definiteness, we will take these two soft masses-squared to be equal, i.e. we will put $m_P^2 = m_Q^2 \equiv m^2$. Also, for simplicity, we assumed universal soft SUSY-breaking A-terms with $|A| \sim m_{3/2}$.

For reasons which will become clear later, we take $m_S^2 < 0$. Therefore, the origin in field space ($P = Q = S = 0$) is a saddle point of the potential with positive curvature in the P and Q directions and negative in the S direction. We will call it trivial saddle point.

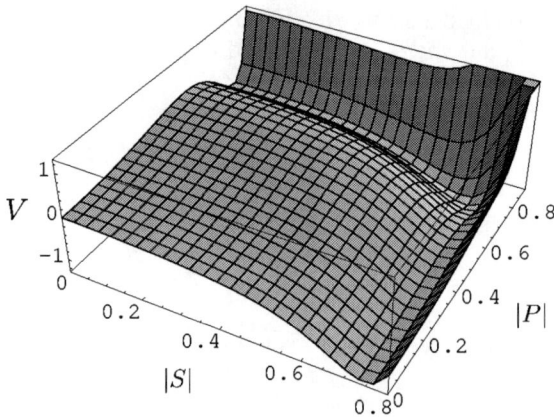

FIGURE 1. Plot of V defined in (24)-(28) in units of $M_\text{I}^{14}/M_\text{P}^{10}$ with respect to $|S|$ and $|P|$, which are in units of M_I. We took $n = 5$, $M_\text{I} \equiv (mM_\text{P}^5)^{1/6}$, $m_P^2 = m_Q^2 = -m_S^2 \equiv m^2$, $A = -9\,m$, and $\lambda_0 = \lambda_1 = \lambda_2 = 1$. Also, $|P| = |Q|$ and $\theta_S = \theta_P = \theta_Q = 0$ (θ_S, θ_P, θ_Q are the phases of S, P, Q) so that the potential is minimized. The trivial and shifted valleys as well as the trivial and non-trivial minima are clearly visible.

One can show [12] that the potential V has a "trivial" valley of local minima which lies on the S axis (i.e. at $P = Q = 0$) and is clearly visible in Figure 1. Moreover, there exists a "trivial" minimum on this valley at $|S| \sim (m_{3/2}M_\text{P}^n)^{1/(n+1)}$, where $U(1)_\text{PQ}$ is unbroken and no μ term is generated. So, we should avoid ending up at this trivial minimum.

The potential V possesses [12] non-trivial minima too with

$$|P| = |Q|, \; |S| \sim (m_{3/2}M_\text{P}^n)^{\frac{1}{n+1}}, \tag{29}$$

where $U(1)_\text{PQ}$ is broken and a μ term is generated. Actually, after soft SUSY breaking (included in V), the only global symmetry surviving in a non-trivial minimum is Z_2^mp.

The shifted valley of minima. We expand V for $|S| \ll |P| \sim |Q|$. The leading term is

$$
\begin{aligned}
V_{(0)} &= \frac{(n+3)^2}{4}|\lambda_{\frac{n+3}{4}}|^2\frac{|PQ|^{n+1}(|P|^2+|Q|^2)}{M_\text{P}^{2n}} + m_P^2|P|^2 + m_Q^2|Q|^2 \\
&\quad - 2|A||\lambda_{\frac{n+3}{4}}|\frac{(|P||Q|)^{\frac{n+3}{2}}}{M_\text{P}^n}\cos\left[\frac{n+3}{2}(\theta_P + \theta_Q)\right],
\end{aligned}
\tag{30}
$$

where θ_P and θ_Q are the phases of P and Q respectively and $A\lambda_{(n+3)/4}$ is taken real and negative by field rephasing.

The potential $V_{(0)}$ is minimized with respect to the phases θ_P and θ_Q for

$$
\frac{n+3}{2}(\theta_P + \theta_Q) = 0 \quad \text{modulo} \quad 2\pi, \quad |P| = |Q|,
$$
$$
\left(\frac{|P|^{n+1}}{M_\text{P}^n}\right)_+ \equiv x_+ = \frac{|A| + \sqrt{|A|^2 - 4(n+2)m^2}}{(n+2)(n+3)|\lambda_{\frac{n+3}{4}}|} \quad \text{for} \quad |A|^2 > 4(n+2)m^2. \tag{31}
$$

The presence of the term $\lambda_{(n+3)/4}(PQ)^{(n+3)/2}/M_{\mathrm{P}}^n$ in the superpotential is vital to the existence of this "shifted" minimum. In view of Z_2^P, however, this term can only exist if $(n+3)/2$ is an even positive integer, which implies the restriction in (22).

For $|S| \ll |P| \sim |Q|$, the shifted minimum of $V_{(0)}$ is also a minimum of V with respect to $|P|$ and $|Q|$ at a practically S-independent position. Thus, for small values of $|S|$, we obtain a "shifted" valley of minima of V at almost constant values of $|P|$ and $|Q|$. This valley, which is clearly visible in Figure 1, has negative inclination for non-zero and small values of $|S|$, due to the negative mass term of S. It starts from the "shifted" saddle point which lies at $|S| = 0$ and $|P|$, $|Q|$ equal to their values at the shifted minimum of $V_{(0)}$. Note, in passing, that a shifted valley of minima was first used in [26] as an inflationary path in order to avoid the overproduction of doubly charged [27] magnetic monopoles at the end of SUSY hybrid inflation [2, 28] in a Pati-Salam [29] GUT model.

The PNGB curvaton. The dominant S-dependent part of V can be expressed as

$$
\begin{aligned}
V_{(1)} &= m_S^2 |S|^2 - 2|A||\lambda_{\frac{n-1}{4}}| \frac{|S|^4(|P||Q|)^{\frac{n-1}{2}}}{M_{\mathrm{P}}^n} \cos\left(4\theta_S + \frac{n-1}{2}(\theta_P + \theta_Q)\right) \\
&\quad + \frac{(n-1)(n+3)}{2} |\lambda_{\frac{n-1}{4}}||\lambda_{\frac{n+3}{4}}||S|^4(|P|^2 + |Q|^2) \frac{(|P||Q|)^{n-1}}{M_{\mathrm{P}}^{2n}} \\
&\quad \times \cos\left(4\theta_S - 2(\theta_P + \theta_Q)\right),
\end{aligned}
\tag{32}
$$

where θ_S is the phase of S and $A\lambda_{(n-1)/4}$ is made real and negative. The potential $V_{(1)}$ is minimized with respect to the phases θ_P, θ_Q, θ_S for $4\theta_S + (n-1)(\theta_P + \theta_Q)/2 = 0$ modulo 2π. Defining the real canonically normalized fields

$$
\phi_P \equiv \sqrt{2}|P|\theta_P, \quad \phi_Q \equiv \sqrt{2}|Q|\theta_Q, \quad \phi_S \equiv \sqrt{2}|S|\theta_S,
\tag{33}
$$

we find [12] three angular mass eigenstates on the shifted valley for $|S| \ll |P| = |Q|$:

1. The axion field $a = (\phi_P - \phi_Q)/\sqrt{2}$, which remains massless to all orders.
2. $\phi_{PQ} \equiv (\phi_P + \phi_Q)/\sqrt{2}$ with mass-squared $m_{PQ}^2 = |A|(n+3)^2|\lambda_{(n+3)/4}|x_+/2 \sim m_{3/2}^2$.
3. ϕ_S, our PNGB curvaton (orthogonal axion) σ with suppressed mass-squared

$$
\tilde{m}_\sigma^2 = \frac{8}{n+2}|\lambda_{\frac{n-1}{4}}|x_+ \frac{|S|^2}{|P|^2}\left[(n+5)|A| - (n-1)\sqrt{|A|^2 - 4(n+2)m^2}\right] \sim \frac{|S|^2}{|P|^2}m_{3/2}^2.
\tag{34}
$$

Note that the Z_2^P symmetry is [12] very important for the PNGB nature of ϕ_S.

SUGRA corrections [2, 3, 30] during inflation can be introduced by simply replacing A, m^2, and m_S^2 by their effective values:

$$
\bar{A} = A + c_A H, \quad \bar{m}^2 = m^2 + c_{PQ}H^2, \quad \bar{m}_S^2 = m_S^2 + c_S H^2 = c_S H^2 - |m_S^2|
\tag{35}
$$

respectively. Here, c_A is a complex parameter of order unity, while c_{PQ} and c_S are real and positive parameters again of order unity. We can arrange the parameters so that

$\bar{m}_S^2 > 0$ during the initial stages of inflation. In this case, the shifted saddle point of V becomes a local minimum and the system may be initially trapped in it. H decreases during inflation and, thus, at some moment of time, this minimum turns into a saddle point and the system starts slowly rolling down the shifted valley. During this slow roll-over, $\phi_S \equiv \sigma$ is an effectively massless PNGB field which can act as curvaton.

CURVATON PHYSICS

The required ε. The phase $\theta = \theta_S \sim 1$ corresponding to the curvaton degree of freedom remains frozen until the onset of the curvaton oscillations. Hence, we expect to have $\sigma_{osc} \sim \theta v_0$. From (15), we then find

$$\varepsilon \sim \frac{\Omega_{dec}}{\pi \zeta \theta} \left(\frac{m_{3/2}}{M_P} \right)^{\frac{n}{n+1}} \gtrsim \varepsilon_{min} \sim \left(\frac{m_{3/2}}{M_P} \right)^{\frac{n}{n+1}}, \tag{36}$$

where we have also used that $H_* \sim m_{3/2}$ and $v_0 \sim (m_{3/2} M_P^n)^{1/(n+1)}$. Assuming that σ decays before big bang nucleosynthesis (BBN), i.e. its decay width $\Gamma_\sigma \sim m_\sigma^3 / v_0^2$ is greater than the Hubble parameter H_{BBN} at BBN, one shows [10, 12] that

$$\varepsilon < \frac{\Omega_{dec}^{\frac{1}{2}}}{\pi \zeta} \left(\frac{M_P}{T_{BBN}} \right)^{\frac{1}{2}} \left(\frac{m_{3/2}}{M_P} \right)^{\frac{5}{4}} \sim 10^{-4} \Omega_{dec}^{\frac{1}{2}}, \tag{37}$$

where $T_{BBN} \approx 1$ MeV is the cosmic temperature at BBN. From (36) and (37), we find [12] that

$$n > \frac{8 + \log(\Omega_{dec}^{\frac{1}{2}}/\theta)}{7 - \log(\Omega_{dec}^{\frac{1}{2}}/\theta)}, \tag{38}$$

which implies that $n \geq 1$ for $\theta \sim 1$. The requirement that $\Gamma_\sigma > H_{BBN}$ yields [12]

$$m_\sigma \gtrsim 10^{\frac{n-9}{n+1}} \text{ TeV} \quad \Rightarrow \quad n \leq \frac{9 + \log(m_\sigma / \text{TeV})}{1 - \log(m_\sigma / \text{TeV})} \tag{39}$$

for $m_\sigma < 10$ TeV. This inequality demands that $n \leq 9$ for $m_\sigma \lesssim 1$ TeV. So, we get $1 \leq n \leq 9$.

Reheating the universe. We will assume that the curvaton decays after dominating the energy density of the universe, i.e. $\Omega_{dec} \simeq 1$, which is [12] crucial for avoiding the overclosure of the universe by axions (see below). The curvaton dominates [12] when $H = H_{dom}$, where

$$H_{dom} \sim \left(\frac{\sigma_{osc}}{M_P} \right)^4 \min\{m_\sigma, \Gamma_{inf}\} \tag{40}$$

with $\Gamma_{inf} \sim g^2 m_{3/2}$ being the inflaton decay width (g is the coupling constant of the inflaton to its decay products). It can be shown [12] that the requirement that $\Gamma_\sigma < H_{dom}$

152

(i.e. σ decays after dominating the energy density of the universe) results in the bound

$$g > \frac{1}{\theta^2}\left(\frac{m_{3/2}}{M_P}\right)^{\frac{n-2}{n+1}} \quad \Rightarrow \quad n > \frac{30 - \log g - 2\log\theta}{15 + \log g + 2\log\theta} \quad \Rightarrow \quad n \geq 2 \quad \text{for} \quad \theta \sim 1. \quad (41)$$

Hot big bang begins after the decay of the curvaton at a reheat temperature

$$T_{\text{REH}} \sim \sqrt{\Gamma_\sigma M_P} \sim m_{3/2}\left(\frac{m_{3/2}}{M_P}\right)^{\frac{1}{2}\left(\frac{n-1}{n+1}\right)} \geq T_{\text{BBN}} \quad \text{for} \quad n \leq 9. \quad (42)$$

Diluting the axions. For $n > 1$, $f_a \approx v_0 \sim (m_{3/2}M_P^n)^{1/(n+1)} \gg 10^{12}$ GeV. This normally leads to axion overproduction overclosing the universe. However, if the curvaton dominates the universe before decaying, the entropy generated [31] during its decay

$$\frac{S_{\text{after}}}{S_{\text{before}}} \sim \left(\frac{H_{\text{dom}}}{\Gamma_\sigma}\right)^{\frac{1}{2}} \sim \frac{g\,\theta^2 v_0^3}{m_{3/2}M_P^2} \sim g\theta^2\left(\frac{M_P}{m_{3/2}}\right)^{\frac{n-2}{n+1}} \quad (43)$$

can adequately dilute the axions (see [32]).

The evolution of v. The order parameter $v \propto |S|$ must be slowly rolling during inflation to preserve the approximate scale invariance of the perturbations (see below). So, it should follow the equation

$$3H|\dot{S}| + \bar{m}_S^2|S| \simeq 0 \quad \Rightarrow \quad \frac{\dot{v}}{v} = \frac{|\dot{S}|}{|S|} = \frac{1}{3}c_S\left(\frac{|m_S^2|}{c_S H^2} - 1\right)H. \quad (44)$$

Using (4) and the fact that $|m_S^2| \equiv c_S H_x^2 \simeq c_S H_m^2(1 - e^{-2F_s N_x})$, this equation becomes

$$\frac{3}{c_S}\frac{d\ln|S|}{dN} = \frac{e^{-2F_s N_x} - e^{-2F_s N}}{1 - e^{-2F_s N}}, \quad (45)$$

where H_x and N_x correspond to the phase transition which changes the sign of $\bar{m}_S^2$ during inflation. The solution of (45) is

$$\frac{6}{c_S}\ln\left(\frac{|S|_*}{|S|_x}\right) = (1 - e^{-2F_s N_x})F_s^{-1}\ln\left(\frac{e^{2F_s N_x} - 1}{e^{2F_s N_*} - 1}\right) - 2(N_x - N_*), \quad (46)$$

where $|S|_* \equiv |S|(N_*) \sim (\varepsilon/\varepsilon_{\text{min}})H_*$ and $|S|_x \equiv |S|(N_x) \sim H_x/2\pi$ from quantum fluctuations at the phase transition. The contribution to the spectral index n_s from the evolution of v during inflation is [12] $-H_*^{-1}(\dot{v}/v)_* \leq 0$. The WMAP bound [17] on n_s then implies

$$\frac{c_S}{3}e^{-2F_s N_*}\left(\frac{1 - e^{-2F_s(N_x - N_*)}}{1 - e^{-2F_s N_*}}\right) \leq 0.04. \quad (47)$$

It is important to note that, in the present case, the negative contribution to n_s from the variation of v during inflation leads naturally to spectral indices which can be adequately smaller than unity in accordance with the recent WMAP results [17].

A CONCRETE EXAMPLE

From (39), and (41) and in view of (22), we see that not many choices for n are allowed. In fact, we can only accept the models with $n = 5$, 9 (i.e. $l = 1$, 2) with the latter case being marginal. Hence, to illustrate the above, we take an example with $n = 5$ (i.e. $l = 1$) and the curvaton assuming a random value after the phase transition, i.e. $\theta \sim 1$.

The bound in (39) suggests that this case is acceptable provided that $m_\sigma \gtrsim 220$ GeV. Using (36), we find that $\varepsilon \sim 10^{-8.5}$ (recall that $\Omega_{\text{dec}} \simeq 1$) and $\varepsilon_{\text{min}} \sim 10^{-12.5}$, which yields $|S|_* \sim 10^4 H_*$. We also estimate [12] N_* to be about 38 for $H_* \sim m_{3/2}$. The reheat temperature turns out to be $T_{\text{REH}} \simeq 10$ MeV, while the entropy production at curvaton decay is given by $S_{\text{after}}/S_{\text{before}} \sim 10^{7.5} g$. From (41), we then conclude that $g > 10^{-4.5}$.

The dilution of axions by the entropy produced when the curvaton decays after dominating the universe may lead [20] to a cosmological disaster. A sizable fraction of the curvaton's decay products consists of sparticles, which eventually turn into stable lightest sparticles (LSPs) in models (such as ours) with an unbroken matter parity. The freeze-out temperature of the LSPs is much higher than T_{REH}. Thus, the LSPs freeze out right after their production and can, subsequently, overclose the universe. This problem can be solved [12] by suppressing the Higgsino components of the lighter neutralinos and charginos below 1% and taking the curvaton adequately light.

One can show [12] that all the requirements mentioned above can be satisfied for

$$c_S,\ c \lesssim O(10^{-4}) \quad \Rightarrow \quad |m_S|,\ m_s \lesssim O(10^{-2}) H_*. \tag{48}$$

The smallness of $|m_S|$ is due to the requirement that $|S|$ is slowly rolling during the relevant part of inflation so that the approximate scale invariance of the density perturbations is preserved, whereas the smallness of m_s to the required value of ε ($\gg \varepsilon_{\text{min}}$), which demands substantial variation of $|S|$ from the phase transition during inflation until the time when the cosmological scales exit the inflationary horizon. Such a variation can be achieved with a large number of e-foldings ($N_x \gtrsim O(10^4)$ and $N_{\text{tot}} \gtrsim O(10^5 - 10^6)$), which means that, in our case, modular inflation is not of the fast-roll type.

CONCLUSIONS

We constructed SUSY PQ models generating the PQ scale and the μ term dynamically. They contain a successful PNGB curvaton whose perturbations are suitably amplified to account for the observed curvature perturbations even in low-scale inflationary models such as modular inflation where the inflaton is unable to generate these perturbations. The spectral index of density perturbations can easily satisfy the recent WMAP bound in contrast to other inflationary models. However, due to the very low value of the reheat temperature, baryogenesis may be achieved only via some exotic mechanism (see [12]).

ACKNOWLEDGMENTS

This work was supported by the European Union under the contract MRTN-CT-2004-503369.

REFERENCES

1. G. Lazarides, *Lect. Notes Phys.* **592**, 351 (2002) (hep-ph/0111328); hep-ph/0607032.
2. E. J. Copeland, A. R. Liddle, D. H. Lyth, E. D. Stewart, and D. Wands, *Phys. Rev.* **D 49**, 6410 (1994).
3. M. Dine, L. Randall, and S. Thomas, *Phys. Rev. Lett.* **75**, 398 (1995); *Nucl. Phys.* **B 458**, 291 (1996).
4. P. Binétruy, and M. K. Gaillard, *Phys. Rev.* **D 34**, 3069 (1986); F. C. Adams, J. R. Bond, K. Freese, J. A. Frieman, and A. V. Olinto, *ibid.* **47**, 426 (1993); T. Banks, M. Berkooz, S. H. Shenker, G. W. Moore, and P. J. Steinhardt, *ibid.* **52**, 3548 (1995); R. Brustein, S. P. De Alwis, and E. G. Novak, *ibid.* **68**, 023517 (2003).
5. A. D. Linde, *J. High Energy Phys.* **11**, 052 (2001).
6. R. Allahverdi, K. Enqvist, J. Garcia-Bellido, and A. Mazumdar, hep-ph/0605035.
7. K. Enqvist, and M. S. Sloth, *Nucl. Phys.* **B 626**, 395 (2002); D. H. Lyth, and D. Wands, *Phys. Lett.* **B 524**, 5 (2002); T. Moroi, and T. Takahashi, *ibid.* **522**, 215 (2001), (E) **539**, 303 (2002).
8. S. Mollerach, *Phys. Rev.* **D 42**, 313 (1990); A. D. Linde, and V. Mukhanov, *ibid.* **56**, 535 (1997).
9. D. H. Lyth, *Phys. Lett.* **B 579**, 239 (2004).
10. K. Dimopoulos, D. H. Lyth, and Y. Rodríguez, *J. High Energy Phys.* **02**, 055 (2005).
11. K. Dimopoulos, *Phys. Lett.* **B 634**, 331 (2006).
12. K. Dimopoulos, and G. Lazarides, *Phys. Rev.* **D 73**, 023525 (2006).
13. R. Peccei, and H. Quinn, *Phys. Rev. Lett.* **38**, 1440 (1977); S. Weinberg, *ibid.* **40**, 223 (1978); F. Wilczek, *ibid.* **40**, 279 (1978).
14. K. Dimopoulos, G. Lazarides, D. H. Lyth, and R. Ruiz de Austri, *J. High Energy Phys.* **05**, 057 (2003).
15. C. L. Bennett *et al.*, *Astrophys. J.* **464**, L1 (1996).
16. D. H. Lyth, C. Ungarelli, and D. Wands, *Phys. Rev.* **D 67**, 023503 (2003).
17. D. N. Spergel *et al.*, astro-ph/0603449.
18. K. Dimopoulos, G. Lazarides, D. H. Lyth, and R. Ruiz de Austri, *Phys. Rev.* **D 68**, 123515 (2003).
19. J. E. Kim, and H. P. Nilles, *Phys. Lett.* **B 138**, 150 (1984).
20. K. Choi, E. J. Chun, and J. E. Kim, *Phys. Lett.* **B 403**, 209 (1997).
21. G. Lazarides, and Q. Shafi, *Phys. Rev.* **D 58**, 071702 (1998).
22. G. Lazarides, and Q. Shafi, *Phys. Lett.* **B 489**, 194 (2000).
23. G. Lazarides, C. Panagiotakopoulos, and Q. Shafi, *Phys. Rev. Lett.* **56**, 432 (1986).
24. N. Ganoulis, G. Lazarides, and Q. Shafi, *Nucl. Phys.* **B 323**, 374 (1989).
25. G. Lazarides, and N. D. Vlachos, *Phys. Lett.* **B 459**, 482 (1999); G. Lazarides, *PoS(trieste99)008*,1999 (hep-ph/9905450).
26. R. Jeannerot, S. Khalil, G. Lazarides, and Q. Shafi, *J. High Energy Phys.* **10**, 012 (2000); G. Lazarides, "Supersymmetric Hybrid Inflation," in *Recent Developments in Particle Physics and Cosmology*, edited by G. C. Branco et al., Kluwer Acad. Pub., Dordrecht, 2001, pp. 399–419 (hep-ph/0011130); R. Jeannerot, S. Khalil, and G. Lazarides, *J. High Energy Phys.* **07**, 069 (2002).
27. G. Lazarides, M. Magg, and Q. Shafi, *Phys. Lett.* **B 97**, 87 (1980).
28. G. R. Dvali, Q. Shafi, and R. K. Schaefer, *Phys. Rev. Lett.* **73**, 1886 (1994); G. Lazarides, R. K. Schaefer, and Q. Shafi, *Phys. Rev.* **D 56**, 1324 (1997).
29. J. C. Pati, and A. Salam, *Phys. Rev.* **D 10**, 275 (1974).
30. M. Dine, W. Fischler, and D. Nemeschansky, *Phys. Lett.* **B 136**, 169 (1984); G. D. Coughlan, R. Holman, P. Ramond, and G. G. Ross, *ibid.* **140**, 44 (1984).
31. R. J. Scherrer, and M. S. Turner, *Phys. Rev.* **D 31**, 681 (1985).
32. P. J. Steinhardt, and M. S. Turner, *Phys. Lett.* **B 129**, 51 (1983); G. Lazarides, C. Panagiotakopoulos, and Q. Shafi, *ibid.* **192**, 323 (1987); G. Lazarides, R. K. Schaefer, D. Seckel, and Q. Shafi, *Nucl. Phys.* **B 346**, 193 (1990).

Clusters of Matter and Antimatter

Walter Greiner

Frankfurt Institute for Advanced Studies,Johann Wolfgang Goethe University, Max-von-Laue-Str. 1, 60438 Frankfurt am Main, Germany

Abstract. We first present the vacuum for the e^+-e^- field of QED and show how it is modified for baryons in the nuclear environment. We then study the possibility of producing a new kind of nuclear systems which in addition to ordinary nuclei contain a few antibaryons ($\bar{B} = \bar{p}, \Lambda$, etc). The properties of such systems are described within the relativistic mean-field model by employing G-parity transformed interactions for antibaryons. Calculations are done for finite nuclei from ^{4}He to ^{208}Pb. It is demonstrated that the presence of a real antibaryon leads to a strong rearrangement of a target nucleus resulting in a significant increase of its binding energy and local compression. Noticeable effects remain even after the antibaryon coupling constants are reduced by a factor of 3-4 compared to G-parity motivated values. We have performed detailed calculations of the antibaryon annihilation rates in the nuclear environment by applying the kinetic approach. It is shown hat due to a significant reduction of the reaction Q-values, the in-medium annihilation rates should be strongly suppressed leading to relatively long-lived antibaryon-nucleus systems. Multi-nucleon annihilation channels are analyzed too. We have also estimated formation probabilities of antiproton+A bound systems in antiproton-A reactions and have found that their observation is feasible at the future GSI antibaryon facility. Several observable signatures are proposed. The possibility of producing multi-quark-antiquark clusters is discussed.

Keywords: antimatter, antiprotons, nuclei, cold dense nuclear matter, relativistic mean-field model
PACS: 25.43.+t, 21.60.-n

INTRODUCTION

It is generally accepted that the physical vacuum has a nontrivial structure. This conclusion was first made by Dirac on the basis of his famous equation for a fermion field which describes simultaneously particles and antiparticles. The Dirac equation in the vacuum has a simple form

$$(i\gamma^\mu \partial_\mu - m)\Psi(x) = 0 , \tag{1}$$

where $\gamma^\mu = (\gamma^0, \boldsymbol{\gamma})$ are Dirac matrices, m is the fermion mass and $\Psi(x)$ is a 4-component spinor field. For a plane wave solution $\Psi(x) = e^{-ipx}u_p$ this equation is written as

$$(\widehat{p} - m)u_p = 0 , \tag{2}$$

where $\widehat{p} = \gamma^0 E - \boldsymbol{\gamma}\mathbf{p}$. Multiplying by $(\widehat{p} + m)$ and requiring that $u_p \neq 0$ one obtains the equation $E^2 - \mathbf{p}^2 - m^2 = 0$ which has two solutions

$$E^{\pm}(\mathbf{p}) = \pm\sqrt{\mathbf{p}^2 + m^2} . \tag{3}$$

Here the $+$ sign corresponds to particles with positive energy $E_N(\mathbf{p}) = E^+(\mathbf{p})$, while the $-$ sign corresponds to solutions with negative energy. To ensure stability of the

CP881, *Cairo International Conference on High Energy Physics (CICHEP II)*,
edited by S. Khalil

physical vacuum Dirac has assumed that these negative-energy states are occupied forming what is called now the Dirac sea. Then the second solution of eq. (3) receives natural interpretation: it describes holes in the Dirac sea. These holes are identifi ed with antiparticles. Their energies are obviously given by $E_{\overline{N}}(\mathbf{p}) = -E^{-}(-\mathbf{p}) = \sqrt{\mathbf{p}^2 + m^2}$. Unfortunately, the Dirac sea brings divergent contributions to physical quantities such as energy density, and one should introduce a proper regularization scheme to get rid off these divergences. This picture has received numerous confi rmations in quantum electrodynamics and other fi elds.

THE QED VACUUM

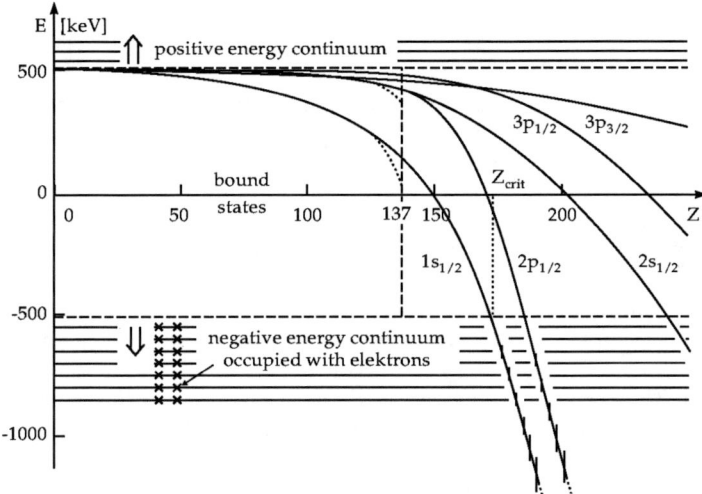

FIGURE 1. Lowest bound states of the Dirac equation for nuclei with charge Z. While the Sommerfeld fine-structure energies (dashed line) for $\xi = 1$ (s states) end at $Z = 137$, the solutions for extended Coulomb potentials (full line) can be traced down to the negative-energy continuum reached at the critical charge Z_{cr} for the $1s$ state. The bound states entering the continuum obtain a spreading width as indicated.

One of the most fascinating aspects is the structure of the vacuum in QED and its change into charged vacuum states under the influence of strong (supercritical) electric fi elds [1]. We shortly remind of this phenomenon.

Fig. 1 shows the diving of the deeply bound states into the lower energy continuum of the Dirac equation.

In the supercritical case the dived state is degenerate with the (occupied) negative electron states. Hence spontaneous $e^{+}e^{-}$ *paircreation* becomes possible, where an electron from the Dirac sea occupies the additional state, leaving a hole in the sea which escapes as a positron while the electron's charge remains near the source. This is a fundamentally new process, whereby the neutral vacuum of QED becomes unstable in supercritical electrical fi elds. It decays within about $10^{-19}s$ into a charged vacuum. The

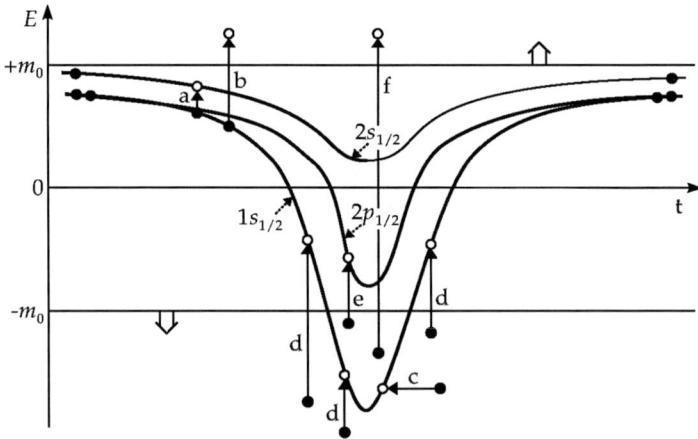

FIGURE 2. Time dependence.of the quasi-molecular energy levels in a supercritical heavy ion collision. The arrows denote various excitation processes which lead to the production of holes and positrons.

charged vacuum is now stable due to the Pauli principle, that is the number of emitted particles remains finite. The vacuum is first charged twice because two electrons with opposite spins can occupy the $1s$ shell. After the $2p_{1/2}$ shell has dived beyond $Z_{cr} = 185$, the vacuum is charged four times, etc. This change of the vacuum structure is not a perturbative effect, as are the radiative QED effects (vacuum polarization, self-energy, etc.).

The time-dependence of the energy levels in a supercritical heavy-ion collision is depicted in Fig. 2. An electron (or hole) which was in a certain molecular eigenstate at the beginning of the collision can be transfered with a certain probability into different states by the dynamics of the collision. This can lead to the hole production in an inner shell by excitation of an electron to a higher state and/or hole production by ionization of an electron to the continuum. Further possibilities are induced positron production by excitation of an electron from the lower continuum to an empty bound level and direct pair production [2].

A comparison of the theoretical predictions and expectations and experimental data is shown in figure 3. Sharp positron peeks can be expected if there were a mechanism in the heavy ion collision leading to a time delay. This may be caused by a pocket in the potential between the two ions. Spontaneous pair production should then be enhanced in supercritical systems. Until now, however, the situation remains inconclusive [2].

THE VACUUM STRUCTURE IN NUCLEAR PHYSICS

It has been noticed already many years ago (see e. g. ref. [3]) that nuclear physics may provide a unique laboratory for investigating the Dirac picture of vacuum. The basis for this is given by relativistic mean-field models which are widely used now for describing nuclear matter and finite nuclei. Within this approach nucleons are described

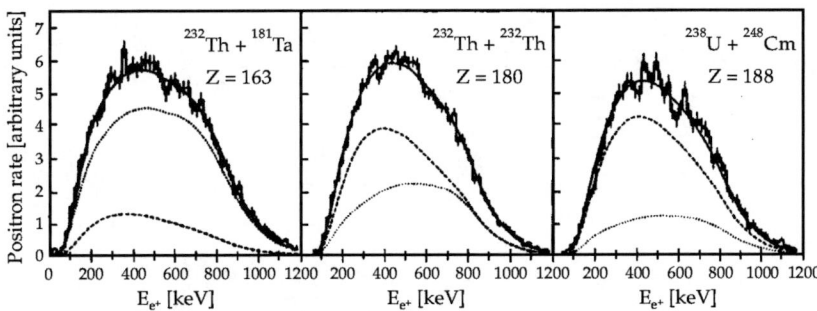

FIGURE 3. Positron energy spectra measured in collisions of Th+Ta, Th+Th, and U+Cm at energies of about 6 MeV per nucleon. The QED predictions (dashed lines) and the experimentally determined background from nuclear pair conversion (dotted lines) add up to the full lines which are in close agreement with experiment.

by the Dirac equation coupled to scalar and vector meson fields. Scalar S and vector V potentials generated by these fields modify plane-wave solutions of the Dirac equation as follows

$$E^{\pm}(\mathbf{p}) = V \pm \sqrt{\mathbf{p}^2 + (m - S)^2} . \tag{4}$$

Again, the $+$ sign corresponds to nucleons with positive energy $E_N(\mathbf{p}) = V + \sqrt{\mathbf{p}^2 + (m - S)^2}$, and the $-$ sign corresponds to antinucleons with energy $E_{\bar{N}}(\mathbf{p}) = -E^{-}(-\mathbf{p}) = -V + \sqrt{\mathbf{p}^2 + (m - S)^2}$. It is remarkable that changing sign of the vector potential for antinucleons is exactly what is expected from the G-parity transformation of the nucleon potential. As follows from eq. (4), in nuclear environment the spectrum of single-particle states of the Dirac equation is modified in two ways. First, the mass gap between positive- and negative-energy states, $2(m - S)$, is reduced due to the scalar potential and second, all states are shifted upwards due to the vector potential. These changes are illustrated in Fig. 4.

It is well known from nuclear phenomenology that good description of nuclear ground state is achieved with $S \simeq 350$ MeV and $V \simeq 300$ MeV so that the net potential for nucleons is $V - S \simeq -50$ MeV. Using the same values one obtains for antinucleons very a deep potential, $-V - S \simeq -650$ MeV. Such a potential would produce many strongly bound states in the Dirac sea. However, if these states are occupied they are hidden from the direct observation. Only creating a hole in this sea, i.e. inserting a real antibaryon into the nucleus, would produce an observable effect. If this picture is correct one should expect the existence of strongly bound states of antinucleons with nuclei. Below we report on our recent study of antibaryon-doped nuclear systems [4, 5].

ANTIBARYONS BOUND IN NUCLEI

Unlike some previous works, we take into account the rearrangement of nuclear structure due to the presence of a real antibaryon. The structure of such systems is calculated

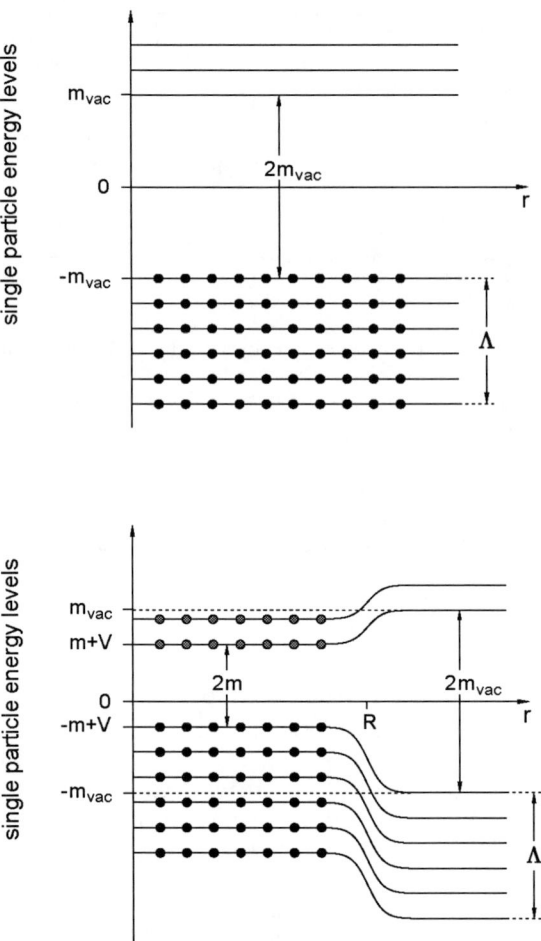

FIGURE 4. Schematic spectrum of Dirac equation in vacuum (upper panel) and in a nucleus of radius R (lower panel). A divergent contribution of negative-energy states is often regularized by introducing a cut-off momentum Λ

using several versions of the relativistic mean–field model (RMF): TM1 [6], NL3 and NL-Z2 [7]. Their parameters were found by fitting binding energies and charge form-factors of spherical nuclei from ^{16}O to ^{208}Pb. The general Lagrangian of the RMF model is written as

$$\mathscr{L} = \sum_{j=B,\overline{B}} \overline{\psi}_j \left(i\gamma^\mu \partial_\mu - m_j \right) \psi_j$$
$$+ \frac{1}{2} \partial^\mu \sigma \partial_\mu \sigma - \frac{1}{2} m_\sigma^2 \sigma^2 - \frac{b}{3} \sigma^3 - \frac{c}{4} \sigma^4$$

160

Sum of proton and neutron densities for ^{16}O (top),
^{16}O with $\bar{\Lambda}$ (bottom left) and ^{16}O with $\bar{p}$ (bottom right)

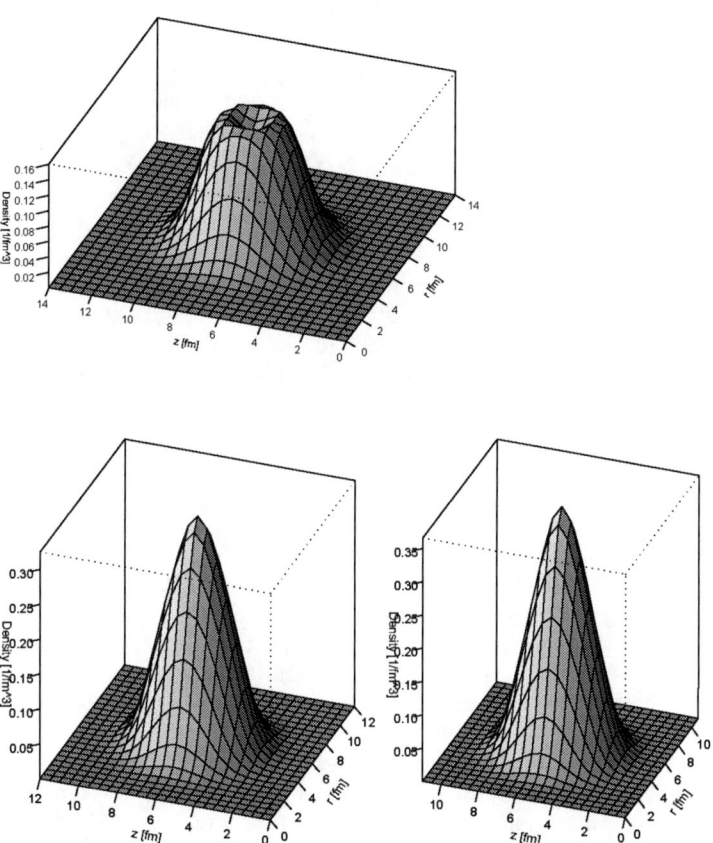

FIGURE 5. Various systems calculated with the parametrization NL-Z2.

$$-\frac{1}{4}\omega^{\mu\nu}\omega_{\mu\nu} + \frac{1}{2}m_\omega^2\omega^\mu\omega_\mu + \frac{d}{4}(\omega^\mu\omega_\mu)^2$$

$$-\frac{1}{4}\vec{\rho}^{\,\mu\nu}\vec{\rho}_{\mu\nu} + \frac{1}{2}m_\rho^2\vec{\rho}^{\,\mu}\vec{\rho}_\mu$$

$$+\sum_{j=B,\bar{B}}\overline{\psi}_j\left(g_{\sigma j}\sigma + g_{\omega j}\omega^\mu\gamma_\mu + g_{\rho j}\vec{\rho}^{\,\mu}\gamma_\mu\vec{\tau}_j\right)\psi_j$$

$$+\text{Coulomb part} \tag{5}$$

Here summation includes valence baryons B, in fact the nucleons forming a nucleus, and valence antibaryons $\bar{B}$ inserted in the nucleus. They are treated as Dirac particles coupled to the scalar-isoscalar (σ), vector-isoscalar (ω) and vector-isovector ($\vec{\rho}$) meson fields.

The calculations are carried out within the mean-field approximation where the meson fields are replaced by their expectation values. Also a "no-sea" approximation is used. This implies that all occupied states of the Dirac sea are "integrated out" so that they do not appear explicitly. It is assumed that their effect is taken into account by nonlinear terms in the meson Lagrangian. Most calculations are done with antibaryon coupling constants which are given by the G-parity transformation ($g_{\sigma \bar{N}} = g_{\sigma N}$, $g_{\omega \bar{N}} = -g_{\omega N}$) and $SU(3)$ flavor symmetry ($g_{\sigma \bar{\Lambda}} = \frac{2}{3} g_{\sigma \bar{N}}$, $g_{\omega \bar{\Lambda}} = \frac{2}{3} g_{\omega \bar{N}}$). In isosymmetric static systems the scalar and vector potentials for nucleons are expressed as $S = g_{\sigma N} \sigma$ and $V = g_{\omega N} \omega^0$.

Following the procedure suggested in Ref. [8] and assuming the axial symmetry of the nuclear system, we solve effective Schrödinger equations for nucleons and an antibaryon together with differential equations for mean meson and Coulomb fields. We explicitly take into account the antibaryon contributions to the scalar and vector densities. It is important that antibaryons give a negative contribution to the vector density, while a positive contribution to the scalar density. This leads to increased attraction and decreased repulsion for surrounding nucleons. To maximize attraction, nucleons move to the center of the nucleus, where the antiproton has its largest occupation probability. This gives rise to a strong local compression of the nucleus and leads to a dramatic rearrangement of its structure.

Results for the ^{16}O nucleus are presented in Fig. 5 which shows 3d plots of nucleon density distributions. The calculations show that inserting an antiproton into the ^{16}O nucleus leads to the increase of central nucleon density by a factor 2–4 depending on the parametrization. Due to a very deep antiproton potential the binding energy of the whole system is increased significantly as compared with 130 MeV for normal ^{16}O. The calculated binding energies of the $\bar{p} - ^{16}$O system are 830, 1050 and 1160 MeV for the NL–Z2, NL3 and TM1, respectively. Due to this anomalous binding we call such systems super bound nuclei (SBN). In the case of antilambdas we rescale the coupling constants with a factor 2/3 that leads to the binding energy of 560÷700 MeV for the $\bar{\Lambda} - ^{16}$O system.

As a second example, we investigate the effect of a single antiproton inserted into the ^{8}Be nucleus. The normal ^{8}Be nucleus is not spherical, exhibiting a clearly visible 2α structure with the ground state deformation $\beta_2 \simeq 1.20$. As seen in Fig. 6, inserting an antiproton in ^{8}Be results in a much less elongated shape ($\beta_2 \simeq 0.23$) and disappearance of its cluster structure. The binding energy increases from 53 MeV to about 700 MeV. Similar, but weaker effects have been predicted [9] for the K^- bound state in the ^{8}Be nucleus.

The calculations have been performed also with reduced antinucleon coupling constants as compared to the G-parity prescription. We have found that the main conclusions about enhanced binding and considerable compression of $\bar{p}$-doped nuclei remain valid even when coupling constants are reduced by factor 3 or so. Now we would like to discuss the structural effect of an antiproton in the doubly magic lead nucleus. A contour plot of the sum of proton and neutron densities is shown in figure 7.

In this case we encounter a quite different scenario: again, the complete system is affected, but not in the sense that the whole nucleus shrinks and becomes very dense. Here, a small and localized region of high density develops within the heavy system. Additionally, the lead nucleus deforms itself. This effect is even larger for

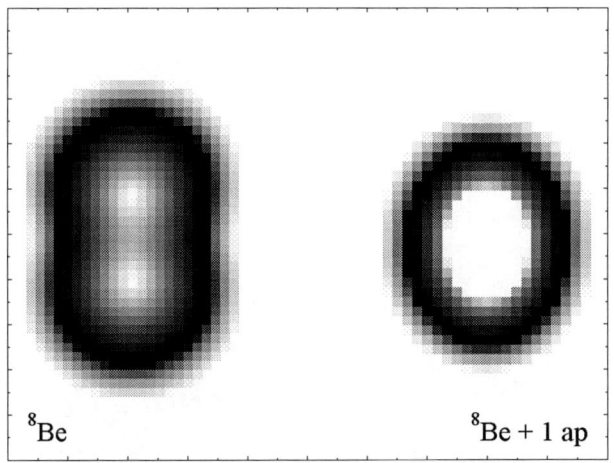

FIGURE 6. Contour plot of nucleon densities for ^{8}Be without (left) and with (right) antiproton calculated with the parametrization NL3.

the case of lead with an $\bar{\alpha}$. The reason for this behaviour can be understood from the properties of the single-particle wavefunctions in the mean-field: In a small region with a deep potential, only states with small angular momenta can be bound deeply. States with higher angular momenta do not have much overlap with the potential. This is exactly what happens here. Basically only the lowest s- and p-states can be bound deeper than for lead without any antiparticles present. Higher lying states do not gain significantly binding or are even lesser bound. The deformation effect probably has two reasons: firstly, a deformation might be energetically favourable to gain some binding for the higher lying states. Secondly, the distortion of the system due to the presence of antiparticles destroys the magicitiy of the system.

LIFETIME ESTIMATES

The crucial question concerning possible observation of the SBNs is their life time. The main decay channel for such states is the annihilation of antibaryons on surrounding nucleons. The energy available for annihilation of a bound antinucleon equals $Q = 2m_N - B_N - B_{\overline{N}}$, where B_N and $B_{\overline{N}}$ are the corresponding binding energies. In our case this energy is at least by a factor 2 smaller as compared with the vacuum value of $2m_N$. This should lead to a significant suppression of the available phase space and thus to a reduced annihilation rate in medium. We have performed detailed calculations assuming that the annihilation rates into different channels are proportional to the available phase space. All intermediate states with heavy mesons like ρ, ω, η as well as multipion channels have been considered. Our conclusion is that decreasing the Q value from 2 GeV to 1 GeV may lead to the reduction of total annihilation rate by factor 20÷30.

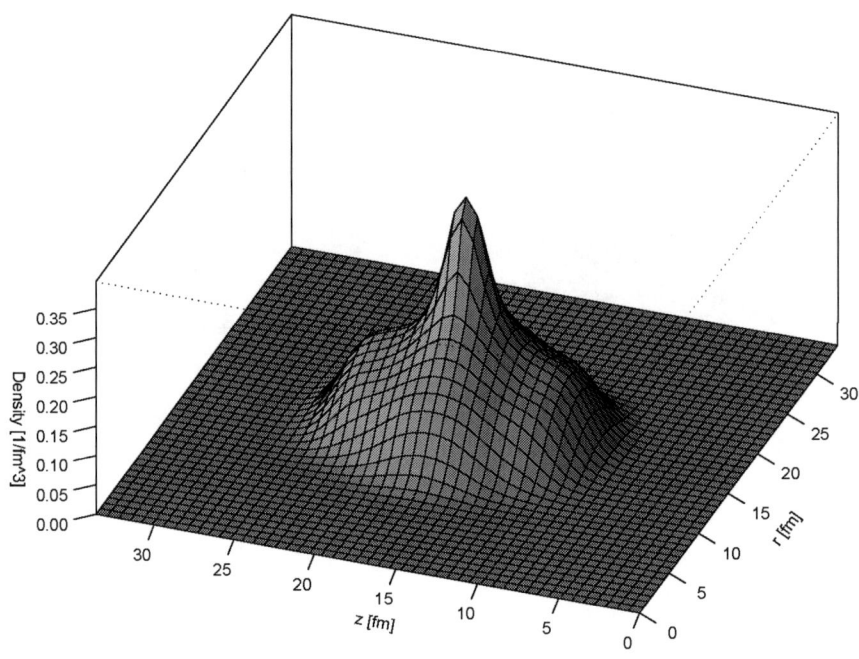

FIGURE 7. Surface plot of the sum of proton and neutron densities for the system ^{208}Pb with $\bar{p}$.

Then we estimate the SBN life times on the level of 5-25 fm/c which makes their observation feasible. This large margin in the life times is mainly caused by uncertainties in the overlap integral between antinucleon and nucleon scalar densities. Longer life times may be expected for SBNs containing antihyperons. The reason is that instead of pions more heavy kaons must be produced in this case. We have also analyzed multi-nucleon annihilation channels (Pontecorvo-like reactions) and have found their contribution to be less than 40% of the single-nucleon annihilation.

We believe that such exotic nuclear states can be produced by using antiproton beams of multi-GeV energy, e.g. at the future GSI facility. It is well known that low-energy antiprotons annihilate on the nuclear periphery (at about 5% of the normal density). Since the annihilation cross section drops signifi cantly with energy, a high-

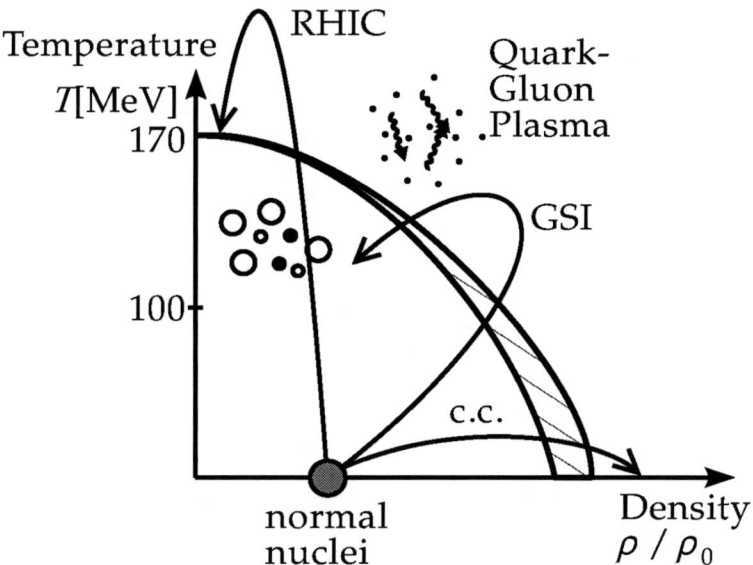

FIGURE 8. The phase diagram of strongly interacting matter. The antiproton-induced cold compression (c.c.) takes place at low temperatures and high densities, a regime that cannot be accessed in heavy-ion collisions. The curves with arrows denote dynamic trajectories of matter in heavy-ion collisions as expected at the future GSI facility and RHIC.

energy antiproton can penetrate deeper into the nuclear interior. Then it can be stopped there in an inelastic collision with a nucleon, e. g. via the reaction $A(\bar{p}, N\pi)_{\bar{p}}A'$, leading to the formation of a $\bar{p}$-doped nucleus. Reactions like $A(\bar{p}, \Lambda)_{\bar{\Lambda}}A'$ can be used to produce a $\bar{\Lambda}$-doped nuclei. Fast nucleons or lambdas can be used for triggering such events. In order to be captured by a target nucleus final antibaryons must be slow in the lab frame. Rough estimates of the SBN formation probability in a central $\bar{p}A$ collision give the values $10^{-5} - 10^{-6}$. With the $\bar{p}$ beam luminocity of $2 \cdot 10^{32}$ cm^{-2}s^{-1} planned at GSI this will correspond to the reaction rate of a few tens of desired events per second.

Several signatures of SBNs can be used for their experimental observation. First, annihilation of a bound antibaryon can proceed via emission of a single photon, pion or kaon with an energy of about 1 GeV (such annihilation channels are forbidden in vacuum). So one may search for relatively sharp lines, with width of $10 \div 40$ MeV, around this energy, emitted isotropically in the SBN rest frame. Another signal may come from explosive disintegration of the compressed nucleus after the antibaryon annihilation. This can be observed by measuring radial collective velocities of nuclear fragments.

COLD COMPRESSION: NUCLEAR AND QUARK MATTER

It is interesting to look at the antibaryon-nucleus system from somewhat different point of view. An antibaryon implanted into a nucleus acts as an attractor for surrounding nucleons. Due to the uncompensated attractive force these nucleons acquire acceleration towards the center. As the result of this inward collective motion the nucleons pile up producing local compression. If this process would be completely elastic it would generate monopole-like oscillations around the compressed SBN state. The maximum compression is reached when the attractive potential energy becomes equal to the compression energy. Simple estimates show that local baryon densities up to 5 times the normal nuclear density may be obtained in this way. It is most likely that the deconfi n-ment transition will occur at this stage and a high-density cloud containing an antibaryon and a few nucleons will appear in the form of a multi-quark-antiquark cluster. One may speculate that the whole ^{4}He or even ^{16}O nucleus can be transformed into the quark phase by this mechanism. As shown in ref. [10], an admixture of antiquarks to cold quark matter is energetically favorable. The problem of annihilation is now transferred to the quark level. But the argument concerning the reduction of available phase space due to the entrance-channel nuclear effects should work in this case too. Thus one may hope to produce relatively cold droplets of the quark phase by the inertial compression of nuclear matter initiated by an antibaryon.

The admixture of antibaryons to fi nite nuclei provides an almost unique doorway to the study of cold compressed nuclear and/or quark matter in the laboratory. This region of the phase diagram of strongly interacting matter, see Fig. 8, is not accessible by collisions of heavy ions, which produce matter which is simultaneously hot and dense.

REFERENCES

1. W. Greiner, B. Müller, J. Rafelksi, *Quantum electrodynamics of strong fields*, Springer Verlag, 2nd edition, December 1985
2. J. Reinhardt and W. Greiner, *Quantum electrodynamics*, Springer Verlag, 3rd edition, February 2003
3. N. Auerbach, A.S. Goldhaber, M.B. Johnson, L.D. Miller, A. Picklesimer, Phys. Lett. **B182** (1986) 221.
4. T. Bürvenich, I.N. Mishustin, L.M. Satarov, H. Stöcker, W. Greiner, Phys. Lett. **B542** (2002) 261.
5. I.N. Mishustin, L.M. Satarov, T.J. Bürvenich, H. Stöcker, and W. Greiner, Phys.Rev. **C 71** (2005) 035201
6. Y. Sugahara and H. Toki, Nucl. Phys. **A579** (1994) 557.
7. M. Bender, K. Rutz, P.–G. Reinhard, J.A. Maruhn, and W. Greiner, Phys. Rev. **C60** (1999) 34304.
8. G. Mao, H. Stöcker, and W. Greiner, Int. J. Mod. Phys. **E8** (1999) 389.
9. Y. Akaishi and T. Yamazaki, Phys. Rev. **C65** (2002) 044005.
10. I.N. Mishustin, L.M. Satarov, H. Stoecker, W. Greiner, Phys. Rev. **C 59** (1999) 3343.

Thermal leptogenesis in a modified cosmic expansion

Nobuchika Okada*,† and Osamu Seto**,‡

* *Theory Division, KEK, Oho 1-1, Tsukuba, Ibaraki 305-0801, Japan*
†*Department of Particle and Nuclear Physics, The Graduate University for Advanced Studies*
(Sokendai), Oho 1-1, Tsukuba, Ibaraki 305-0801, Japan
** *Department of Physics and Astronomy, University of Sussex, Brighton BN1 9QJ, United*
Kingdom
‡*a talk given by O.S.*

Abstract. The thermal leptogenesis in brane world cosmology is studied. In brane world cosmology, the expansion law is modified from the four-dimensional standard cosmological one at high temperature regime in the early universe. As a result, the well-known upper bound on the lightest light neutrino mass induced by the condition for the out-of-equilibrium decay of the lightest heavy neutrino, $\tilde{m}_1 \lesssim 10^{-3}$ eV, can be moderated to be $\tilde{m}_1 \lesssim 10^{-3}$ eV $\times (M_1/T_t)^2$ in the case of $T_t \leq M_1$ with the lightest heavy neutrino mass (M_1) and the "transition temperature" (T_t), at which the modified expansion law in brane world cosmology is smoothly connecting with the standard one. This implies that the degenerate mass spectrum of the light neutrinos can be consistent with the thermal leptogenesis scenario. On the other hand, if the brane world effect is too effective, thermal equilibrium state can not realize. This fact leads to the lower bound on the five-dimensional Planck mass and limits compatible inflation models.

Keywords: Baryon asymmetry, leptogenesis, brane world
PACS: 98.80.Cq, 04.50.+h, 14.60.Pq

1. INTRODUCTION

Cosmological observations have revealed the major ingredients and their abundance in our Universe. The origin of the cosmological baryon asymmetry is still one of open questions in particle physics as well as in cosmology. The asymmetry must have been generated at the very early universe through the processes satisfying the following three conditions, *i*) the existence of baryon number violating interactions, *ii*) C and CP violations and *iii*) the departure from thermal equilibrium [1].

Among various mechanisms of baryogenesis, leptogenesis [2] is attractive because of its simplicity and the connection to neutrino physics. Particularly, the simplest scenario, namely thermal leptogenesis, requires only the thermal excitation of heavy Majorana neutrinos which induce tiny neutrino masses via the seesaw mechanism [3] and provides several implications for the light neutrino mass spectrum [4]. In leptogenesis, the first condition is satisfied by the Majorana nature of heavy neutrinos and the sphaleron effect in the standard model (SM) at the high temperature [5], while the second condition is provided by their CP violating decay. The departure from thermal equilibrium is provided by the expansion of the universe.

CP881, *Cairo International Conference on High Energy Physics (CICHEP II)*,
edited by S. Khalil

The out-of-equilibrium decay is realized if the decay rate is smaller than the expansion rate of the universe,

$$\Gamma_{N_1} < H|_{T=M_1},\tag{1}$$

where Γ_{N_1} and M_1 are the decay rate and the mass of the lightest heavy neutrino, respectively, and H denotes the Hubble parameter. Note that the expansion law is governed by the gravitational theory. Therefore, if general relativity is replaced by another theory at a high energy scale, the universe would undergo non-standard expansion. One of such examples is the brane world cosmology [6]. The following discussion is based on the so-called "RS II" model first proposed by Randall and Sundrum [7]. In the model, the Friedmann equation for a spatially flat spacetime is given by

$$H^2 = \frac{8\pi G_N}{3}\rho\left(1 + \frac{\rho}{\rho_0}\right)\tag{2}$$

where

$$\rho = \frac{\pi^2}{30}g_* T^4\tag{3}$$

is the energy density of the radiation with g_* being the effective degrees of freedom of relativistic particles,

$$\rho_0 = 96\pi G_N M_5^6,\tag{4}$$

with M_5 being the five dimensional Planck mass, and the four dimensional cosmological constant has been tuned to be zero. The second term proportional to ρ^2 is a new ingredient in the brane world cosmology and leads to a non-standard expansion law.

Note that according to Eq. (2) the evolution of the early universe can be divided into two eras. At the era where $\rho/\rho_0 \gg 1$ the second term dominates and the expansion law is non-standard (brane world cosmology era), while at the era $\rho/\rho_0 \ll 1$ the first term dominates and the expansion of the universe obeys the standard expansion law (standard cosmology era). In the following, we call a temperature defined as $\rho(T_t)/\rho_0 = 1$ "transition temperature", at which the evolution of the early universe changes from the brane world cosmology era into the standard one. The transition temperature T_t is determined as

$$T_t \simeq 1.6 \times 10^7 \left(\frac{100}{g_*}\right)^{1/4}\left(\frac{M_5}{10^{11}\,\text{GeV}}\right)^{3/2}\text{GeV},\tag{5}$$

once M_5 is given. Using the transition temperature, we rewrite Eq. (2) into the form,

$$H^2 = \frac{8\pi G_N}{3}\rho\left[1 + \left(\frac{T}{T_t}\right)^4\right] = H_{st}^2\left[1 + \left(\frac{T}{T_t}\right)^4\right],\tag{6}$$

where H_{st} denotes the Hubble parameter in the standard Friedmann equation;

$$H_{st}^2 = \frac{8\pi G_N}{3}\rho.\tag{7}$$

This modification of the expansion law at a high temperature ($T > T_t$) leads some drastic changes for several cosmological issues. In fact, some interesting consequences in the brane world cosmology such as the enhancement of the dark matter relic density [8] and the suppression of the overproduction of gravitino [9] have been pointed out. Here, we show how the modified expansion law in the brane world cosmology affects the thermal leptogenesis [10].

2. A BRIEF OVERVIEW OF THERMAL LEPTOGENESIS

In the seesaw model, the smallness of the neutrino masses can be naturally explained by the small mixings between left-handed neutrinos and heavy right-handed Majorana neutrinos N_i. The basic part of the Lagrangian in the SM with right-handed neutrinos is described as

$$\mathscr{L}_N = -h_{ij}\overline{l_{L,i}}HN_j - \frac{1}{2}\sum_i M_i \overline{N_i^C}N_i + h.c., \tag{8}$$

where $i,j = 1,2,3$ denote the generation indices, h is the Yukawa coupling, l_L and H are the lepton and the Higgs doublets, respectively, and M_i is the lepton-number-violating mass term of the right-handed neutrino N_i in the basis of their mass eigenstates. Hereafter, we assume the hierarchical mass spectrum for the heavy neutrinos, $M_1 \ll M_2, M_3$, for simplicity.

In the case of the hierarchical mass spectrum for the heavy neutrinos, the lepton asymmetry in the universe is generated dominantly by CP-violating out-of-equilibrium decay of the lightest heavy neutrino, $N_1 \to l_L H^*$ and $N_1 \to \overline{l}_L H$. The leading contribution is given by the interference between the tree level and the one-loop level decay amplitudes. The CP-violating parameter defined by

$$\varepsilon \equiv \frac{\Gamma(N_1 \to H + \overline{l}_j) - \Gamma(N_1 \to H^* + l_j)}{\Gamma(N_1 \to H + l_j) + \Gamma(N_1 \to H^* + l_j)}. \tag{9}$$

is approximated as

$$\varepsilon \simeq \frac{3}{16\pi}\frac{M_1 m_3}{v^2}\sin\delta \simeq 10^{-6}\left(\frac{M_1}{10^{10}\text{GeV}}\right)\left(\frac{m_3}{0.05\text{eV}}\right)\sin\delta, \tag{10}$$

through the relations of the seesaw mechanism. Here m_3 is the heaviest light neutrino mass, $v = 174$ GeV is the vacuum expectation value (VEV) of Higgs and $\sin\delta$ is an effective CP phase. In addition, we have normalized m_3 by 0.05 eV which is a preferable value in recent atmospheric neutrino oscillation data $\sqrt{\Delta m_\oplus^2} \simeq 0.05$ eV [11]. Using the above ε, the resultant baryon asymmetry generated via thermal leptogenesis is described as

$$\frac{n_b}{s} \simeq \frac{\varepsilon}{g_*}d, \tag{11}$$

where $g_* \sim 100$ is the effective degrees of freedom in the universe at $T \sim M_1$, and $d \leq 1$ is the so-called dilution factor. This factor parameterizes how the naively expected value

$n_b/s \simeq \varepsilon/g_*$ is reduced due to wash-out processes. To evaluate the resultant baryon asymmetry precisely, numerical calculations are necessary, and the lower bound on the lightest heavy neutrino mass in order to obtain the realistic baryon asymmetry in the present universe $n_b/s \simeq 10^{-10}$ for fixed $\varepsilon/g_* \sin \delta \simeq 10^{-10}$ has been found to be $M_1 \gtrsim 10^9$ GeV [12].

The condition of the out-of-equilibrium decay in Eq. (1) to provide sufficient lepton asymmetry without dilution $d \sim 1$ is rewritten as

$$\tilde{m}_1 \equiv \sum_j (h_{1j}h_{1j}^\dagger) \frac{v^2}{M_1} < \frac{8\pi v^2}{M_1^2} H|_{T=M_1} \equiv m_* \simeq 1 \times 10^{-3} \text{eV}, \tag{12}$$

with the decay width of the lightest heavy neutrinos

$$\Gamma_{N_1} = \sum_j \frac{h_{1j}h_{1j}^\dagger}{8\pi} M_1, \tag{13}$$

and the Friedmann equation (7). This condition can be regarded as the upper bound on the lightest neutrino mass m_1, since the inequality $m_1 \leq \tilde{m}_1$ can be shown [13, 14]. Considering Eq. (10), this upper bound is normally interpreted as an implication that thermal leptogenesis cannot generate sufficient baryon asymmetry in the case of the degenerate mass spectrum of light neutrinos [14, 15].

3. LEPTOGENESIS IN BRANE WORLD COSMOLOGY

Let us consider the case where the lightest heavy neutrinos decay in the brane world cosmology era, namely $M_1 > T_t$. Accordingly, the condition for the out-of-equilibrium decay of the heavy neutrino is modified. From Eqs. (1), (6) and (13), now we obtain the upper bound on the lightest neutrino mass in the brane world cosmology such that

$$\tilde{m}_1 < m_* \left(\frac{M_1}{T_t}\right)^2 \simeq 1 \times 10^{-3} \text{eV} \times \left(\frac{M_1}{T_t}\right)^2, \tag{14}$$

instead of Eq. (12). Note that the upper bound has been moderated due to the enhancement factor $(M_1/T_t)^2$ for $M_1 > T_t$. This result implies that, if T_t is low enough, the thermal leptogenesis scenario is successful even in the case of the degenerate light neutrino mass spectrum.

Note that the mass dependence of the dilution factor d would be slightly modified in brane world cosmology. Indeed, although this fact is confirmed with small differences in some numerical values by solving Boltzmann equations numerically [16], these results roughly justify our naive estimation here in the same manner as the one in the standard cosmology.

In the above discussion, we have implicitly assumed that the lightest heavy neutrino as well as other SM model particles is in thermal equilibrium at a high temperature $T > M_1$. In the following, let us verify whether this situation can be realized in the brane world cosmology.

For most of SM particles with gauge interactions, the interaction rate is estimated as

$$\Gamma_{int}(T) = n\langle\sigma v\rangle \simeq \frac{3\zeta(3)}{2\pi^2}T^3 \times \frac{N\alpha^2}{T^2}, \tag{15}$$

where N is the number of modes. On the other hand, since the right-handed neutrinos are singlet under the SM gauge group, the only interaction through which the right-handed neutrino can be in thermal equilibrium is the Yukawa coupling in Eq. (8). Consider a pair annihilation process of the lightest heavy neutrino through the Yukawa couplings. The thermal averaged annihilation rate is roughly estimated as

$$\Gamma_{int}(T) = n\langle\sigma v\rangle \simeq \frac{3\zeta(3)}{2\pi^2}T^3 \times \frac{Ng_Y^4}{100}T^{-2}, \tag{16}$$

where N is the number of annihilation channels, g_Y stands for dominant Yukawa couplings, and the factor $1/100$ denotes the kinematical phase factor. Then, we found that the condition for particles to be coupled, $\Gamma_{int} \lesssim H$, is satisfied for

$$T_{FI} \lesssim 10^{10}\text{GeV} \times \left(\frac{T_t}{10^7\text{GeV}}\right)^{2/3}. \tag{17}$$

The thermal leptogenesis in the brane cosmology era can take place if the lightest heavy neutrino mass is in the range

$$T_t < M_1 < T_{FI}. \tag{18}$$

Recall that $M_1 \simeq 10^{9-10}$ GeV is required in order to generate the sufficient baryon asymmetry. This implies 10^{9-10} GeV $> T_t \gtrsim 10^{6-7}$ GeV $(10^{12-13}$ GeV $> M_5 \gtrsim 10^{10-11}$ GeV) from Eq. (18).

For this scenario, one should notice that inflation need to have a very efficient reheating with $T_R \simeq 10^{9-10}$ GeV, because the potential energy during inflation V_{inf} must be less than $M_5^4 \simeq (10^{10-11}$ GeV $)^4$ in order for inflation models to be consistently treated in the context of the five-dimensional theory. In fact, such inflation models are possible but very limited [17]. The viability of brane world inflation at ρ^2 dominant era is crucial for this scenario.

4. CONCLUSION

We have studied the thermal leptogenesis in the brane world cosmology. The nonstandard expansion law of the brane world cosmology affects the condition of the out-of-equilibrium decay of the lightest heavy neutrino, and moderates the upper bound on the lightest light neutrino mass.

As a result, the degenerate mass spectrum for the light neutrinos can be consistent with the successful thermal leptogenesis scenario, if the transition temperature is lower than the lightest heavy neutrino mass. We have verified that the lightest heavy neutrino can be in thermal equilibrium through the Yukawa couplings among light neutrinos and Higgs bosons and found the region of the transition temperature consistent with the successful thermal leptogenesis in the brane world era. Then, we obtain the constraint on T_t (M_5).

On the other hand, for this scenario, the required reheating temperature after inflation T_R is not so small compared with M_5. Hence, inflation must be a high energy model such as a chaotic inflation.

ACKNOWLEDGMENTS

The work of N.O. is supported in part by the Grant-in-Aid for Scientific Research in Japan (#15740164). The work of O.S. is supported by PPARC.

REFERENCES

1. A. D. Sakharov, Pisma Zh. Eksp. Teor. Fiz. **5**, 32 (1967) [JETP Lett. **5** 24 (1967)].
2. M. Fukugita and T. Yanagida, Phys. Lett. B **174**, 45 (1986).
3. T. Yanagida, in *Proceedings of Workshop on the Unified Theory and the Baryon Number in the Universe*, Tsukuba, Japan, edited by A. Sawada and A. Sugamoto (KEK, Tsukuba, 1979), p 95; M. Gell-Mann, P. Ramond, and R. Slansky, in *Supergravity*, Proceedings of Workshop, Stony Brook, New York, 1979, edited by P. Van Nieuwenhuizen and D. Z. Freedman (North-Holland, Amsterdam, 1979), p 315; R. N. Mohapatra and G. Senjanovic, Phys. Rev. Lett. **44**, 912 (1980).
4. For a review, e.g. W. Buchmuller and M. Plumacher, Int. J. Mod. Phys. A **15**, 5047 (2000); G. F. Giudice, A. Notari, M. Raidal, A. Riotto and A. Strumia, Nucl. Phys. B **685**, 89 (2004).
5. V. A. Kuzmin, V. A. Rubakov and M. E. Shaposhnikov, Phys. Lett. B **155**, 36 (1985).
6. For a review, e.g., D. Langlois, Prog. Theor. Phys. Suppl. **148**, 181 (2003).
7. L. Randall and R. Sundrum, Phys. Rev. Lett. **83**, 4690 (1999).
8. N. Okada and O. Seto, Phys. Rev. D **70**, 083531 (2004); T. Nihei, N. Okada and O. Seto, Phys. Rev. D **71**, 063535 (2005); Phys. Rev. D **73**, 063518 (2006).
9. N. Okada and O. Seto, Phys. Rev. D **71**, 023517 (2005).
10. N. Okada and O. Seto, Phys. Rev. D **73**, 063505 (2006).
11. For example, Y. Ashie *et al.* [Super-Kamiokande Collaboration], Phys. Rev. D **71**, 112005 (2005).
12. W. Buchmuller, P. Di Bari and M. Plumacher, Nucl Phys B 643, 367 (2002).
13. S. Davidson and A. Ibarra, Phys. Lett. B **535**, 25 (2002)
14. M. Fujii, K. Hamaguchi and T. Yanagida, Phys. Rev. D **65**, 115012 (2002).
15. W. Fischler, G. F. Giudice, R. G. Leigh and S. Paban, Phys. Lett. B **258**, 45 (1991); W. Buchmuller and T. Yanagida, Phys. Lett. B **302**, 240 (1993).
16. M. C. Bento, R. Gonzalez Felipe and N. M. C. Santos, Phys. Rev. D **73**, 023506 (2006).
17. E. J. Copeland and O. Seto, Phys. Rev. D **72**, 023506 (2005).

Type II Leptogenesis

Stefan Antusch

Departamento de Física Teórica C-XI and Instituto de Física Teórica C-XVI,
Universidad Autónoma de Madrid, Cantoblanco, E-28049 Madrid, Spain

Abstract. We discuss leptogenesis via the out-of-equilibrium decay of the lightest right-handed neutrino in type II see-saw scenarios, where, in addition to the type I see-saw, an additional direct mass term for the light neutrinos is present. The results for the decay asymmetries are presented in theories where this additional contribution stems from the induced vev of a Higgs triplet, and furthermore in an effective, model independent approach. We emphasize that when the neutrino mass scale increases, type II leptogenesis can be more efficient than for hierarchical light neutrinos, in sharp contrast to the type I see-saw case where successful thermal leptogenesis even imposes a bound on the neutrino mass scale of about 0.1 eV.

Keywords: Neutrino physics; Early universe; Leptogenesis
PACS: 14.60.Pq,14.60.St,98.80.Cq

INTRODUCTION

Leptogenesis [1] is one of the most attractive mechanisms for explaining the observed baryon asymmetry of the universe $n_B/n_\gamma \approx (6.0965 \pm 0.2055) \times 10^{-10}$ [2]. The asymmetry is generated in the lepton sector via the out-of-equilibrium decay of the same heavy right-handed neutrinos which are responsible for generating small neutrino masses in the type I see-saw scenario [3]. It is then partially converted into a baryon asymmetry via sphaleron transitions. In the type I see-saw mechanism, successful thermal leptogenesis (assuming hierarchical right-handed neutrino masses) imposes strong constraints on the parameters of the see-saw mechanism, in particular a lower bound on the masses of the right-handed neutrinos of about 10^9 GeV [4] and an upper bound on mass scale of the light neutrinos of about 0.1 eV [5].

In models with a left-right symmetric particle content like minimal left-right symmetric models, Pati-Salam models or Grand Unified Theories (GUTs) based on SO(10), the type I see-saw mechanism is typically generalized to a type II see-saw [6], where an additional direct mass term m_{LL}^{II} for the light neutrinos is present. The effective mass matrix of the light neutrinos is then given by

$$m_{LL}^\nu = m_{LL}^I + m_{LL}^{II} \,, \tag{1}$$

where $m_{LL}^I = -v_u^2 Y_\nu M_{RR}^{-1} Y_\nu^T$ is the type I see-saw mass matrix with Y_ν being the neutrino Yukawa matrix in left-right convention, M_{RR} the mass matrix of the right-handed neutrinos and $v_u = \langle h_u^0 \rangle$ is the vacuum expectation value (vev) which leads to masses for the up-type quarks. From a model independent perspective, the type II mass term can be considered as an additional contribution to the lowest dimensional effective neutrino mass operator. In most explicit models, the type II contribution stems from see-saw suppressed induced vevs of SU(2)$_L$-triplet Higgs fields. One motivation

CP881, *Cairo International Conference on High Energy Physics (CICHEP II)*,
edited by S. Khalil

for considering the type II see-saw is that it allows to construct unified flavour models for partially degenerate neutrinos in a natural way, e.g. via a type II upgrade [7], which is otherwise difficult to achieve in type I models.

In this talk, we discuss the impact of this additional contribution to neutrino masses on leptogenesis, following Ref. [11], focusing on differences between leptogenesis within the type I and type II see-saw mechanism.

TYPE I AND TYPE II SEESAW MECHANISM

Motivated by left-right symmetric unified theories, let us consider two generic possibilities for explaining the smallness of neutrino masses via the see-saw mechanism. They can be viewed as generating the effective dimension five operator for Majorana neutrino masses,

$$\mathscr{L}_\kappa^{SM} = \frac{1}{4}\kappa_{gf}(\overline{L^{C^g}}\cdot H)(L^f\cdot H)+\text{h.c.}, \tag{2a}$$

$$\mathscr{L}_\kappa^{MSSM} = -\frac{1}{4}\kappa_{gf}(\hat{L}^g\cdot\hat{H}_u)(\hat{L}^f\cdot\hat{H}_u)\big|_{\theta\theta}+\text{h.c.}, \tag{2b}$$

in the SM or the MSSM, respectively, from integrating out heavy fields. The dot indicates the $SU(2)_L$-invariant product, $(\hat{L}^f\cdot\hat{H}_u) := \hat{L}_a^f(i\tau_2)^{ab}(\hat{H}_u)_b$, with τ_A $(A\in\{1,2,3\})$ being the Pauli matrices. Superfields are marked by hats.

After electroweak symmetry breaking, Majorana masses for the light neutrinos,

$$\mathscr{L}_\nu = -\tfrac{1}{2}m_{LL}^\nu\overline{\nu}_L\nu_L^{Cf}, \text{ with } m_{LL}^\nu = -\frac{v_u^2}{2}(\kappa)^*, \tag{3}$$

emerge from this operator.

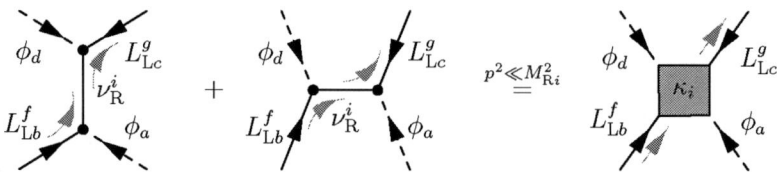

FIGURE 1. Generation of the dimension 5 neutrino mass operator in the type I see-saw mechanism.

In the type I see-saw mechanism, these heavy fields are the singlet (right-handed) neutrinos ν_{Ri} postulated in theories with a left-right symmetric particle spectrum. With neutrino Yukawa couplings being denoted by Y_ν and a mass matrix of the right-handed neutrinos M_{RR}, the low energy effective neutrino mass matrix is given by the type I see-saw formula

$$m_{LL}^I = -v_u^2Y_\nu M_{RR}^{-1}Y_\nu^T. \tag{4}$$

However, in left-right symmetric extensions of the SM, the mass matrix M_{RR} of the right-handed neutrinos can emerge either from renormalizable interactions, or from effective operators which are generated, for instance, at the Planck scale. In the first case, the theories predict the appearance of another heavy field capable of generating the neutrino mass operator, a SU(2)$_L$-triplet Δ_L which couples to two lepton doublets and thus caries two units of B-L charge. Its contribution to the neutrino mass operator is illustrated in figure 3. Equivalently, considering the left diagram in figure 3, one can understand the smallness of the type II see-saw contribution from the fact that the SU(2)$_L$-triplet Higgs obtains an induced vev after electroweak symmetry breaking, which is suppressed if the mass M_Δ of Δ_L is much larger than the electroweak vev v_u.

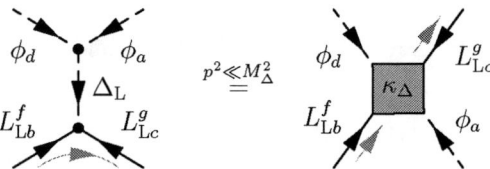

FIGURE 2. Extra diagram generating the dimension 5 neutrino mass operator in the type II see-saw mechanism from a SU(2)$_L$-triplet Higgs field.

In the presence of the SU(2)$_L$-triplet, there is an additional direct mass term for the neutrinos and the low energy effective neutrino mass matrix is given by

$$m_{LL}^\nu = m_{LL}^{II} + m_{LL}^I = m_{LL}^{II} - v_u^2 Y_\nu M_{RR}^{-1} Y_\nu^T . \tag{5}$$

As argued above, m_{LL}^{II} is naturally small if the triplets are heavy. In fact, in left-right symmetric unified theories, the generic size of both see-saw contributions m_{LL}^I and m_{LL}^{II} is v_u^2/v_{B-L} where v_{B-L} is the B-L breaking scale.

BARYOGENESIS VIA LEPTOGENESIS

In leptogenesis, the asymmetry is generated initially in the lepton sector by the out-of-equilibrium decay of the right-handed neutrinos.[1] It is then partially transformed into a baryon asymmetry via sphaleron conversion, which violates B and L, but conserves B-L. In the following, we will therefore write the negative of the lepton asymmetry as a B-L asymmetry. The ratio of the number density over the entropy density $Y_{B-L} = n_{B-L}/s$, can be written as[2]

$$Y_{B-L}^{SM} = -\eta \, \varepsilon_1 \, Y_{\nu_R^1}^{eq} , \tag{6a}$$

$$Y_{B-L}^{MSSM} = -\eta \left[\tfrac{1}{2}(\varepsilon_1 + \tilde{\varepsilon}_1) Y_{\nu_R^1}^{eq} + \tfrac{1}{2}(\varepsilon_{\tilde{1}} + \tilde{\varepsilon}_{\tilde{1}}) Y_{\tilde{\nu}_R^1}^{eq} \right] . \tag{6b}$$

ε_1 (and $\tilde{\varepsilon}_1$) are the decay asymmetries of the lightest right-handed neutrino into (s)lepton and Higgs(ino) and $\varepsilon_{\tilde{1}}$ (and $\tilde{\varepsilon}_{\tilde{1}}$) are the decay asymmetries of the lightest right-handed

---

[1] In the type II see-saw, the lepton asymmetry can also be generated via the decay of one or more SU(2)$_L$-triplets (see e.g. [8]), which is however beyond the scope of this talk.

[2] In this talk, we will not consider issues of charged lepton flavour in the Boltzmann equations and assume a hierarchical spectrum of the right-handed neutrinos (and $M_\Delta \gg M_{R1}$) such that, for instance, contributions to the decay asymmetry from the two heavier right-handed neutrinos can be ignored [9].

sneutrino,

$$\varepsilon_1 := \frac{\Gamma_{\nu_R^1 L} - \Gamma_{\nu_R^1 \bar{L}}}{\Gamma_{\nu_R^1 L} + \Gamma_{\nu_R^1 \bar{L}}}, \quad \tilde{\varepsilon}_1 := \frac{\Gamma_{\nu_R^1 \tilde{L}} - \Gamma_{\nu_R^1 \tilde{L}^*}}{\Gamma_{\nu_R^1 \tilde{L}} + \Gamma_{\nu_R^1 \tilde{L}^*}},$$

$$\varepsilon_{\tilde{1}} := \frac{\Gamma_{\tilde{\nu}_R^{1*} L} - \Gamma_{\tilde{\nu}_R^1 \bar{L}}}{\Gamma_{\tilde{\nu}_R^{1*} L} + \Gamma_{\tilde{\nu}_R^1 \bar{L}}}, \quad \tilde{\varepsilon}_{\tilde{1}} := \frac{\Gamma_{\tilde{\nu}_R^1 \tilde{L}} - \Gamma_{\tilde{\nu}_R^{1*} \tilde{L}^*}}{\Gamma_{\tilde{\nu}_R^1 \tilde{L}} + \Gamma_{\tilde{\nu}_R^{1*} \tilde{L}^*}}, \tag{7}$$

with the decay rate $\Gamma_{\nu_R^1 L} := \Sigma_{a,b,f} \Gamma(\nu_R^1 \to L_a^f H_u b)$.

$Y_{\nu_R^1}^{eq}$ and $Y_{\tilde{\nu}_R^1}^{eq}$ are the number densities of the neutrino and sneutrino at $T \gg M_{R1}$

$$Y_{\nu_R^1}^{eq} \approx \frac{45\,\zeta(3)}{\pi^4 g_* k}\frac{3}{4} \quad \text{and} \quad Y_{\tilde{\nu}_R^1}^{eq} \approx \frac{45\,\zeta(3)}{\pi^4 g_* k}, \tag{8}$$

if they were in thermal equilibrium, normalized with respect to the entropy density. g^* is the effective number of degrees of freedom, which amounts 106.75 in the SM and 228.75 in the MSSM, and k is the Boltzmann constant.

Furthermore, equation (6) provides the definition for the efficiency factor η for leptogenesis. In thermal leptogenesis, it can be computed from a set of coupled Boltzmann equations (for a recent review and references, see e.g. [10]). In non-thermal leptogenesis, e.g. via the decay of the inflaton, leptogenesis could be even more efficient, since ν_R^1 and $\tilde{\nu}_R^1$ could be almost completely out-of-equilibrium when they decay. However, such realizations of non-thermal leptogenesis depend on the specific inflation model.

The produced lepton asymmetry is then partially transformed into a baryon asymmetry via sphaleron conversion, resulting in the baryon asymmetry Y_B

$$Y_B = \alpha Y_{B-L}, \quad \text{with} \quad \alpha \approx \frac{24 + 4N_H}{66 + 13N_H} \tag{9}$$

and with N_H being the number of Higgs doublets. The produced baryon asymmetry in terms of the baryon to photon ratio (using $s/n_\gamma \approx 7.04k$) is then approximately given by

$$\frac{n_B^{SM}}{n_\gamma} \approx -0.97 \cdot 10^{-2} \varepsilon_1 \eta, \quad \frac{n_B^{MSSM}}{n_\gamma} \approx -1.04 \cdot 10^{-2} \varepsilon_1 \eta. \tag{10}$$

The important quantities for computing the produced baryon asymmetry are thus the decay asymmetry ε_1 and the efficiency factor η. While, to a good approximation, the latter is in general not affected by the type II see-saw contribution if the triplet is much heavier than M_{R1}, the decay asymmetry for type II leptogenesis is modified compared to the type I case, as we now discuss.

DECAY ASYMMETRIES

Right-Handed Neutrinos plus Triplets

The 1-loop diagrams which contribute to the decay $\nu_R^1 \to L_a^f H_u b$ in the MSSM in the type II see-saw mechanism where the direct mass term for the neutrinos stems from the

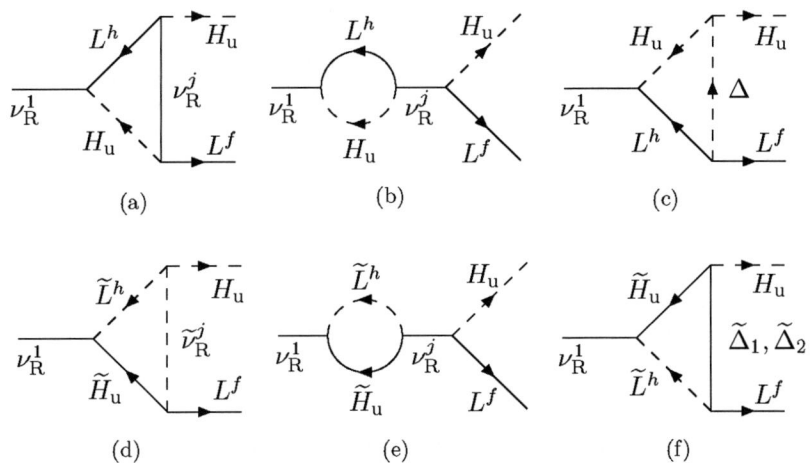

FIGURE 3. Loop diagrams in the MSSM which contribute to the decay $\nu_R^1 \to L_a^f H_{ub}$ for the case of a type II see-saw mechanism where the direct mass term for the neutrinos stems from the induced vev of a Higgs triplet. In diagram (f), $\tilde{\Delta}_1$ and $\tilde{\Delta}_2$ are the mass eigenstates corresponding to the superpartners of the $SU(2)_L$-triplet scalar fields Δ and $\bar{\Delta}$. The SM diagrams are the ones where no superpartners (marked by a tilde) are involved and where H_u is renamed to the SM Higgs.

induced vev of a Higgs triplet are shown in figure 3. The calculation of the corresponding decay asymmetries yields

$$\varepsilon_1^{(a)} = \frac{1}{8\pi} \frac{\sum_{j\neq 1} \text{Im}\left[(Y_\nu^\dagger Y_\nu)_{1j}^2\right]}{\sum_f |(Y_\nu)_{f1}|^2} \sqrt{x_j} \left[1 - (1+x_j) \ln\left(\frac{x_j+1}{x_j}\right)\right], \tag{11a}$$

$$\varepsilon_1^{(b)} = \frac{1}{8\pi} \frac{\sum_{j\neq 1} \text{Im}\left[(Y_\nu^\dagger Y_\nu)_{1j}^2\right]}{\sum_f |(Y_\nu)_{f1}|^2} \sqrt{x_j} \left[\frac{1}{1-x_j}\right], \tag{11b}$$

$$\varepsilon_1^{(c)} = -\frac{3}{8\pi} \frac{M_{R1}}{v_u^2} \frac{\sum_{fg} \text{Im}\left[(Y_\nu^*)_{f1}(Y_\nu^*)_{g1}(m_{LL}^{II})_{fg}\right]}{(Y_\nu^\dagger Y_\nu)_{11}} y \left[-1 + y \ln\left(\frac{y+1}{y}\right)\right], \tag{11c}$$

$$\varepsilon_1^{(d)} = \frac{1}{8\pi} \frac{\sum_{j\neq 1} \text{Im}\left[(Y_\nu^\dagger Y_\nu)_{1j}^2\right]}{\sum_f |(Y_\nu)_{f1}|^2} \sqrt{x_j} \left[-1 + x_j \ln\left(\frac{x_j+1}{x_j}\right)\right], \tag{11d}$$

$$\varepsilon_1^{(e)} = \frac{1}{8\pi} \frac{\sum_{j\neq 1} \text{Im}\left[(Y_\nu^\dagger Y_\nu)_{1j}^2\right]}{\sum_f |(Y_\nu)_{f1}|^2} \sqrt{x_j} \left[\frac{1}{1-x_j}\right], \tag{11e}$$

$$\varepsilon_1^{(f)} = -\frac{3}{8\pi} \frac{M_{R1}}{v_u^2} \frac{\sum_{fg} \text{Im}\left[(Y_\nu^*)_{f1}(Y_\nu^*)_{g1}(m_{LL}^{II})_{fg}\right]}{(Y_\nu^\dagger Y_\nu)_{11}} y \left[1 - (1+y) \ln\left(\frac{y+1}{y}\right)\right], \tag{11f}$$

where $y := M_\Delta^2/M_{R1}^2$ and $x_j := M_{Rj}^2/M_{R1}^2$ for $j \neq 1$ and where we assume hierarchical right-handed neutrino masses and $M_\Delta \gg M_{R1}$.

The new contributions in the type II see-saw mechanism are the diagrams (c) and (f) of figure 3, where virtual $SU(2)_L$-triplet scalar fields or their superpartners are exchanged in the loop. The MSSM result for the type II contributions has been derived in [11]. In the SM, the results in [11] correct the previous result of [12] by a factor of $-3/2$. The corrected result agrees in the limit $y \gg 1$ with the calculation in the effective approach [11] (which we will review below), where the particles much heavier than M_{R1} are integrated out.

The results for the contributions to the decay asymmetries from the triplet in the SM and from the triplet superfields in the MSSM are

$$\varepsilon_1^{\text{SM,II}} = \varepsilon_1^{(c)}, \tag{12a}$$

$$\varepsilon_1^{\text{MSSM,II}} = \varepsilon_1^{(c)} + \varepsilon_1^{(f)}. \tag{12b}$$

In the MSSM, we furthermore obtain

$$\varepsilon_1^{\text{MSSM,II}} = \widetilde{\varepsilon}_1^{\text{MSSM,II}} = \varepsilon_{\widetilde{1}}^{\text{MSSM,II}} = \widetilde{\varepsilon}_{\widetilde{1}}^{\text{MSSM,II}}. \tag{13}$$

The results corresponding to the diagrams (a), (b), (d) and (e) which contribute to ε_1^I in the type I see-saw in the SM and in the MSSM, have been presented first in [13]. The results for the type I contribution to the decay asymmetries in the SM and in the MSSM are

$$\varepsilon_1^{\text{SM,I}} = \varepsilon_1^{(a)} + \varepsilon_1^{(b)}, \tag{14a}$$

$$\varepsilon_1^{\text{MSSM,I}} = \varepsilon_1^{(a)} + \varepsilon_1^{(b)} + \varepsilon_1^{(d)} + \varepsilon_1^{(e)}. \tag{14b}$$

Again, in the MSSM, the remaining decay asymmetries are equal to $\varepsilon_1^{\text{MSSM,I}}$:

$$\varepsilon_1^{\text{MSSM,I}} = \widetilde{\varepsilon}_1^{\text{MSSM,I}} = \varepsilon_{\widetilde{1}}^{\text{MSSM,I}} = \widetilde{\varepsilon}_{\widetilde{1}}^{\text{MSSM,I}}. \tag{15}$$

Finally, the total decay asymmetries from the decay of ν_R^1 in the type II see-saw, where the direct mass term for the neutrinos stems from the induced vev of a Higgs triplet, are given by

$$\varepsilon_1^{\text{SM}} = \varepsilon_1^{\text{SM,I}} + \varepsilon_1^{\text{SM,II}}, \tag{16}$$

$$\varepsilon_1^{\text{MSSM}} = \varepsilon_1^{\text{MSSM,I}} + \varepsilon_1^{\text{MSSM,II}}. \tag{17}$$

It is interesting to note that the type I results can be brought to a form which contains the neutrino mass matrix using

$$\frac{1}{8\pi} \frac{\sum_{j \neq 1} \text{Im}\left[(Y_\nu^\dagger Y_\nu)_{1j}^2\right]}{\sum_f |(Y_\nu)_{f1}|^2} \frac{1}{\sqrt{x_j}} = -\frac{1}{8\pi} \frac{M_{R1}}{v_u^2} \frac{\sum_{fg} \text{Im}\left[(Y_\nu^*)_{f1}(Y_\nu^*)_{g1}(m_{LL}^I)_{fg}\right]}{(Y_\nu^\dagger Y_\nu)_{11}}. \tag{18}$$

In the limit $y \gg 1$ and $x_j \gg 1$ for all $j \neq 1$, which corresponds to a large gap between the mass M_{R1} and the masses M_{R2}, M_{R3} and M_Δ, we obtain the simple results for the decay asymmetries $\varepsilon_1^{\text{SM}}$ and $\varepsilon_1^{\text{MSSM}}$,

$$\varepsilon_1^{\text{SM}} = \frac{3}{16\pi} \frac{M_{R1}}{v_u^2} \frac{\sum_{fg} \text{Im}\left[(Y_\nu^*)_{f1}(Y_\nu^*)_{g1}(m_{LL}^I + m_{LL}^{II})_{fg}\right]}{\sum_h |(Y_\nu)_{h1}|^2}, \tag{19a}$$

$$\varepsilon_1^{\text{MSSM}} = \frac{3}{8\pi} \frac{M_{R1}}{v_u^2} \frac{\sum_{fg} \text{Im}\left[(Y_\nu^*)_{f1}(Y_\nu^*)_{g1}(m_{LL}^I + m_{LL}^{II})_{fg}\right]}{\sum_h |(Y_\nu)_{h1}|^2}. \tag{19b}$$

In the presence of such a mass gap, the calculation can also be performed in an effective approach after integrating out the two heavy right-handed neutrinos and the heavy triplet, generating contributions to the effective neutrino mass operator.

Effective Approach to Leptogenesis

Let us assume that the lepton asymmetry is generated via the decay of the lightest right-handed neutrino and that all other additional particles, in particular the ones which generate the type II contribution, are much heavier than M_{R1}. We furthermore assume that we can neglect their population in the early universe, e.g. that their masses are much larger than the reheating temperature T_R and that they are not produced non-thermally in a large amount. We also assume that they approximately do not contribute to washout processes.

For a minimal effective approach, it is convenient to isolate the type I contribution from the lightest right-handed neutrino as follows:

$$m_{LL}^{\nu} = -\frac{v_u^2}{2} \left[2(Y_\nu)_{f1} M_{R1}^{-1} (Y_\nu^T)_{1f} + \kappa'^* \right]. \tag{20}$$

κ' includes type I contributions from the heavier right-handed neutrinos, plus any additional (type II) contributions from heavier particles. Examples for realizations of the neutrino mass operator can be found, e.g., in [14]. At M_{R1}, the most minimal extension of the SM or the MSSM would then be to introduce the effective neutrino mass operator κ' plus one right-handed neutrino ν_R^1 with mass M_{R1} and Yukawa couplings $(Y_\nu)_{f1}$ to the lepton doublets L^f, defined as $-(Y_\nu)_{f1}(L^f \cdot H)\, \nu_R^1$ in the Lagrangian of the SM and, analogously, as $(Y_\nu)_{f1}(\hat{L}^f \cdot \hat{H}_u)\, \hat{\nu}^{C1}$ in the superpotential of the MSSM.

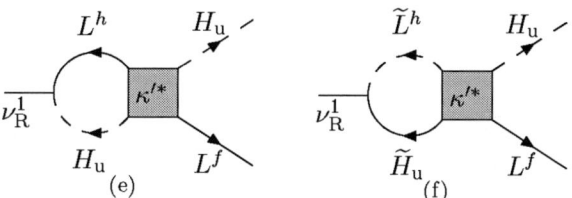

FIGURE 4. Loop diagrams contributing to the decay asymmetry via the decay $\nu_R^1 \rightarrow L_a^f H_{ub}$ in the MSSM with a (lightest) right-handed neutrino ν_R^1 and a neutrino mass matrix determined by κ'. Further contributions to the generated baryon asymmetry stem from the decay of ν_R^1 into slepton and Higgsino and from the decays of the sneutrino $\tilde{\nu}_R^1$. With H_u renamed to the SM Higgs, the first diagram contributes in the extended SM.

The contributions to the decay asymmetries in the effective approach stem from the interference of the diagram for the tree-level decay with the loop diagrams containing the effective operator, shown in figure 4. In the SM, the interference with diagram (a) of figure 4 gives the simple result [11]

$$\varepsilon_1^{SM} = \frac{3}{16\pi} \frac{M_{R1}}{v_u^2} \frac{\sum_{fg} \text{Im}\left[(Y_\nu^*)_{f1}(Y_\nu^*)_{g1}(m_{LL}^\nu)_{fg}\right]}{(Y_\nu^\dagger Y_\nu)_{11}}. \tag{21}$$

179

For the supersymmetric case, diagram (a) and diagram (b) contribute to ε_1 and we obtain [11]:

$$\varepsilon_1^{MSSM} = \frac{3}{8\pi} \frac{M_{R1}}{v_u^2} \frac{\sum_{fg} \text{Im}\left[(Y_v^*)_{f1}(Y_v^*)_{g1}(m_{LL}^v)_{fg}\right]}{(Y_v^\dagger Y_v)_{11}} . \tag{22}$$

Explicit calculation furthermore yields

$$\varepsilon_1^{MSSM} = \tilde{\varepsilon}_1^{MSSM} = \varepsilon_{\tilde{1}}^{MSSM} = \tilde{\varepsilon}_{\tilde{1}}^{MSSM} . \tag{23}$$

The results are independent of the details of the realization of the neutrino mass operator κ'. Note that, since the diagrams where the lightest right-handed neutrino runs in the loop do not contribute to leptogenesis, we have written $m_{LL}^v = -v_u^2(\kappa)^*/2$ instead of $m_{LL}'^v := -v_u^2(\kappa')^*/2$ in the formulae in equations (21) - (22). Having done this, the decay asymmetries are then seen to be directly related to the neutrino mass matrix m_{LL}^v.

For neutrino masses via the type I see-saw mechanism, they are in agreement with the known results [13] in the limit $M_{R2}, M_{R3} \gg M_{R1}$. In the limit $M_\Delta \gg M_{R1}$, the results obtained in the effective approach are also in agreement with our full theory calculation in the minimal type II scenarios with $SU(2)_L$-triplets in equation (11). In particular, we confirm the correction by the factor $-3/2$ in [11] compared to the previous result of [12] in the SM.

TYPE II BOUND ON DECAY ASYMMETRY AND ON M_{R1}

In the effective approach, we can calculate a model-independent upper bound for the decay asymmetries ε_1^{SM} and ε_1^{MSSM} from the requirement of successful thermal leptogenesis. With $m_{max}^v := \max(m_1, m_2, m_3)$ being the largest of the left-handed neutrino masses (at the energy scale M_{R1}), and using equations (21) and (22), the upper bounds [11]

$$|\varepsilon_1^{SM}| \leq \frac{3}{16\pi} \frac{M_{R1}}{v_u^2} m_{max}^v , \quad |\varepsilon_1^{MSSM}| \leq \frac{3}{8\pi} \frac{M_{R1}}{v_u^2} m_{max}^v \tag{24}$$

for the decay asymmetries can be derived. Thus, the upper bound increases with increasing mass scale of the light neutrinos. Note that compared to the low energy value, the neutrino masses at the scale M_{R1} are enlarged by renormalization group (RG) effects by $\approx +20\%$ in the MSSM and $\approx +30\%$ in the SM, which raises the bounds on the decay asymmetries by the same values (see e.g. figure 4 of [15]).

Using equation (10), for a given efficiency factor η and using an upper bound for m_{max}^v, it can be transformed into a lower type II bound for the mass of the lightest right-handed neutrino [11]:

$$M_{R1}^{SM} \geq \frac{16\pi}{3} \frac{v_u^2}{m_{max}^v} \frac{n_B/n_\gamma}{0.97 \cdot 10^{-2}\eta} , \quad M_{R1}^{MSSM} \geq \frac{8\pi}{3} \frac{v_u^2}{m_{max}^v} \frac{n_B/n_\gamma}{1.04 \cdot 10^{-2}\eta} . \tag{25}$$

The bound on M_{R1} is lower for a larger neutrino mass scale. It is shown in figure 5 as a function of the neutrino mass scale, i.e. of the mass of the lightest neutrino.

The situation in the type II framework is very different to the type I see-saw case: E.g., for a normal mass ordering, the type II bound on the decay asymmetry is proportional to m_3, whereas the type I bound is proportional to $\Delta m_{31}^2/m_3$ [4]. In addition, thermal type I leptogenesis gets less efficient for a larger neutrino mass scale since $\tilde{m}_1 \geq m_{min}^v$, with $m_{min}^v := \min(m_1, m_2, m_3)$. Together with an improved bound on the type I decay asymmetry [5], this strongly increases the type I bound on M_{R1} [4] for increasing m_{min}^v and finally leads to an upper bound for the absolute mass scale of the light neutrinos of $m_{min}^v \lesssim 0.1$ eV [5]. In the type II scenario where $\tilde{m}_1$ is in general independent of m_{max}^v, there is no bound on the neutrino mass scale from the requirement of successful leptogenesis. A neutrino mass scale $\lesssim 0.35$ eV on the contrary allows for a mass of the lightest right-handed neutrino M_{R1} of about an order of magnitude below the absolute bound in type I models. The type I lower bound on M_{R1} can be realized only for hierarchical light neutrino masses, and it dramatically increases with increasing neutrino mass scale m_{min}^v.

Lower bound on M_{R1} for various efficiencies η

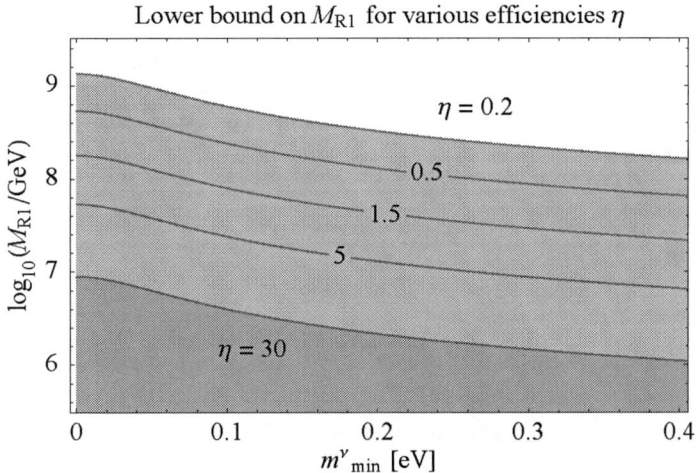

FIGURE 5. Lower bound on M_{R1} in type II leptogenesis in the MSSM as a function of the mass of the lightest neutrino $m_{min}^v := \min(m_1, m_2, m_3)$ for some values of the efficiency factor η and for a baryon to photon ratio $n_B = 6.5 \cdot 10^{-10}$. In the extreme cases for thermal leptogenesis, the maximal value of the efficiency factor is $\eta_{max}^{zero} \approx 0.2$ for a zero initial population of v_{R1} and $\eta_{max}^{dom.} \approx 30$ for a maximal initial population (approximate values taken from [16]).

SUMMARY, DISCUSSION AND CONCLUSIONS

We have discussed leptogenesis via the out-of-equilibrium decay of the lightest right-handed neutrino in type II see-saw scenarios, where, in addition to the type I see-saw, an additional direct mass term for the light neutrinos is present. The results for the decay asymmetries have been presented in theories where this additional contribution stems from the induced vev of a Higgs triplet, and furthermore in an effective, model independent approach. We have shown that when the neutrino mass scale increases and the light neutrino masses become quasi-degenerate, type II leptogenesis can be more efficient than for hierarchical light neutrinos, in sharp contrast to the type I see-saw case where the efficiency drops and successful thermal leptogenesis even imposes a bound on the neutrino mass scale of about 0.1 eV. It is interesting to note that the type II see-saw mechanism furthermore allows to explain such quasi-degenerate neutrino masses in an elegant way, by using SO(3) flavour symmetry or one of its discrete subgroups [7]. In unified flavour models, the additional type II contribution to neutrino masses can furthermore be used to improve consistency of classes of unified flavour models with respect to the requirements of thermal leptogenesis [17], which is an interesting application of type II leptogenesis. In summary, if the masses of the light neutrinos turn out to be in the quasi-degenerate regime, both, thermal leptogenesis as well as prospects of explaining the quasi-degenerate spectrum in unified flavour theories, would favor the type II see-saw mechanism over the type I see-saw.

ACKNOWLEDGMENTS

I would like to thank Steve King for his collaboration in the work presented here. I would also like to thank the organizers of CICHEP II and acknowledge supported by the EU 6th Framework Program MRTN-CT-2004-503369 "The Quest for Unification: Theory Confronts Experiment".

REFERENCES

1. M. Fukugita and T. Yanagida, Phys. Lett. **174B** (1986), 45.
2. D. N. Spergel *et al.*, arXiv:astro-ph/0603449.
3. P. Minkowski, Phys. Lett. B 67 421 (1977); M. Gell-Mann, P. Ramond and R. Slansky, in *Supergravity*, edited by P. van Nieuwenhuizen and D. Freedman, (North-Holland, 1979), p. 315; T. Yanagida, in *Proceedings of the Workshop on the Unified Theory and the Baryon Number in the Universe*, edited by O. Sawada and A. Sugamoto (KEK Report No. 79-18, Tsukuba, 1979), p. 95; R.N. Mohapatra and G. Senjanović, Phys. Rev. Lett. **44** (1980) 912.
4. S. Davidson and A. Ibarra, Phys. Lett. **B535** (2002), 25–32.
5. W. Buchmüller, P. Di Bari, and M. Plümacher, Nucl. Phys. **B665** (2003), 445–468.
6. M. Magg and C. Wetterich, Phys. Lett. B **94** (1980) 61; G. Lazarides, Q. Shafi and C. Wetterich, Nucl. Phys. B **181** (1981) 287; J. Schechter and J. W. F. Valle, Phys. Rev. D **22** (1980) 2227; R. N. Mohapatra and G. Senjanović, Phys. Rev. **D23** (1981), 165; C. Wetterich, Nucl. Phys. **B187** (1981), 343.
7. S. Antusch and S. F. King, Nucl. Phys. B **705** (2005) 239.
8. P. J. O'Donnell and U. Sarkar, Phys. Rev. **D49** (1994), 2118–2121; E. Ma and U. Sarkar, Phys. Rev. Lett. **80** (1998), 5716–5719; T. Hambye, E. Ma, and U. Sarkar, Nucl. Phys. **B602** (2001), 23–38; T. Hambye, M. Raidal and A. Strumia, Phys. Lett. B **632** (2006) 667.
9. A. Abada, S. Davidson, F. X. Josse-Michaux, M. Losada and A. Riotto, JCAP **0604** (2006) 004, E. Nardi, Y. Nir, E. Roulet and J. Racker, JHEP **0601** (2006) 16, A. Abada, S. Davidson, A. Ibarra, F. X. Josse-Michaux, M. Losada and A. Riotto, arXiv:hep-ph/0605281, S. Blanchet and P. Di Bari, arXiv:hep-ph/0607330; O. Vives, Phys. Rev. D **73** (2006) 073006.
10. W. Buchmuller, R. D. Peccei and T. Yanagida, Ann. Rev. Nucl. Part. Sci. **55** (2005) 311.
11. S. Antusch and S. F. King, Phys. Lett. B **597** (2004) 199.
12. T. Hambye and G. Senjanovic, Phys. Lett. B **582**, 73 (2004).
13. L. Covi, E. Roulet, and F. Vissani, Phys. Lett. **B384** (1996), 169–174.
14. E. Ma, Phys. Rev. Lett. **81** (1998), 1171–1174.
15. S. Antusch, J. Kersten, M. Lindner, and M. Ratz, Nucl. Phys. **B674** (2003), 401–433.
16. G. F. Giudice, A. Notari, M. Raidal, A. Riotto, and A. Strumia, Nucl. Phys. B **685**, 89 (2004).
17. S. Antusch and S. F. King, JHEP **0601** (2006) 117.

Neutralino dark matter from string scenarios

D. G. Cerdeño [1]

Institute for Particle Physics Phenomenology, University of Durham, DH1 3LE, U.K.

Abstract. The direct detection of neutralino dark matter is analysed within the context of orbifold scenarios from the heterotic superstring. In particular, the neutralino-proton cross section is computed and compared with the sensitivity of detectors, taking into account the most recent experimental and astrophysical constraints. In addition to the usual non-universalities of the soft terms in orbifold compactifications, due to their modular weight dependence, the contribution of a D-term, generated by the presence of an anomalous $U(1)$ is also investigated. The D-term contribution provides more flexibility in the non-universalities, and is crucial in avoiding dangerous charge and colour-breaking minima. Thanks to it, large neutralino detection cross sections can be obtained in regions of the parameter space fulfilling all experimental and astrophysical constraints.

Keywords: Dark matter, String theory
PACS: 95.35.+d, 11.25.Wx, 12.60.Jv

INTRODUCTION

The lightest neutralino, $\tilde{\chi}_1^0$, a particle predicted by supersymmetric extensions of the Standard Model, is the leading candidate for dark matter within the class of weakly interacting massive particles (WIMPs). The direct detection of WIMPs could take place through their elastic scattering with nuclei inside a detector, and many experiments around the world are currently looking for this signal. It is therefore necessary to study how large the detection cross section of neutralinos can be in order to determine the feasibility of such an observation. Exhaustive analyses of the neutralino-proton cross section have been carried out, mostly addressing the spin-independent contribution, $\sigma_{\tilde{\chi}_1^0 - p}$. The conditions under which large neutralino detection cross sections can be obtained in a general Supergravity (SUGRA) theory are now well understood and non-universal soft supersymmetry-breaking terms are known to play a crucial role.

In the Constrained MSSM (CMSSM) the soft terms of the minimal supersymmetric standard model (MSSM) - scalar masses, m_{ij}, gaugino masses, M_i, and trilinear couplings, A_{ijk} - are assumed to be universal at the unification scale, $M_{GUT} \approx 2 \times 10^{16}$ GeV, and are therefore given by only three parameters, (m, M, and A). Radiative electroweak symmetry breaking is imposed, which fixes the absolute value of the μ term, but leaves its sign undetermined. In addition, the ratio of the Higgs vacuum expectation values, $\tan\beta \equiv \langle H_u^0 \rangle / \langle H_d^0 \rangle$, is also a free parameter. In this scenario, and taking into account all kind of experimental and astrophysical constraints, the neutralino-proton cross section turns out to be constrained by $\sigma_{\tilde{\chi}_1^0 - p} \lesssim 3 \times 10^{-8}$ pb. Clearly, present experiments are not sufficient and only the planned 1 tonne Ge/Xe detectors would be able to test part of

---

[1] Work in progress, done in collaboration with T. Kobayashi and C. Muñoz.

the parameter space.

The non-universality of the soft parameters allows to increase $\sigma_{\tilde{\chi}_1^0-p}$ significantly. In particular, it is possible to enhance the scattering channels involving exchange of CP-even neutral Higgses by reducing the Higgs masses, and also by increasing the Higgsino components of the lightest neutralino. Non-universal values for the Higgs mass parameters, $m_{H_d}^2$ and $m_{H_u}^2$, can account for these effects. First, a decrease in the values of the Higgs masses can be obtained by increasing $m_{H_u}^2$ at the GUT scale (thus making it less negative at the EW scale) and/or decreasing $m_{H_d}^2$. More specifically, the value of the mass of the heaviest CP-even Higgs, H, can be very efficiently lowered under these circumstances. This can be understood by analysing the (tree-level) mass of the CP-odd Higgs A, which for $\tan^2\beta \gg 1$ can be approximated as $m_A^2 \approx m_{H_d}^2 - m_{H_u}^2 - M_Z^2$. Since the heaviest CP-even Higgs, H, is almost degenerate in mass with A, lowering m_A^2 we obtain a decrease in m_H which produces an increase in the scattering channels through Higgs exchange. Second, through the increase in the value of $m_{H_u}^2$ an increase in the Higgsino components of the lightest neutralino can also be achieved. Making $m_{H_u}^2$ less negative, its positive contribution to μ^2 in the minimisation of the Higgs potential would be smaller. Eventually $|\mu|$ will be of the order of the bino and wino masses and $\tilde{\chi}_1^0$ will then be a mixed Higgsino-gaugino state. Thus scattering channels through Higgs exchange become more important than in the CMSSM, where $|\mu|$ is large and $\tilde{\chi}_1^0$ is mainly bino. It is worth emphasizing that the effect of lowering the Higgs masses is typically more important, since it can provide large values for the neutralino-nucleon cross section even in the case of bino-like neutralinos.

Using the following parameterization for the non-universalities in the Higgs masses at the GUT scale,

$$m_{H_d}^2 = m^2(1 + \delta_{H_d}) , \quad m_{H_u}^2 = m^2(1 + \delta_{H_u}) , \tag{1}$$

the above mentioned effects can be achieved with $\delta_{H_u} > 0$ and $\delta_{H_d} < 0$. Large values for $\sigma_{\tilde{\chi}_1^0-p}$, within the reach of future experiments, are then possible while fulfilling all experimental and astrophysical constraints [1].

On the other hand, explicit examples of SUGRA scenarios are obtained at the low-energy limits of superstring theory. We will focus our attention on the $E_8 \times E_8$ heterotic superstring, for which a number of interesting four-dimensional vacua with particle content not far from that of the SUSY standard model were found (see e.g., the discussion in [2]). Interestingly, orbifold compactifications of the heterotic string generically lead to non-universal soft terms in the resulting SUGRA. We will investigate this possibility and its implications for neutralino dark matter detection computing the theoretical predictions for $\sigma_{\tilde{\chi}_1^0-p}$ and comparing them with the sensitivities of present and projected dark matter experiments. Furthermore, we will also analyse the D-term contribution, generated by the presence of an anomalous $U(1)$, which further contributes to the non-universality of the soft parameters.

HETEROTIC ORBIFOLD MODELS

These constructions have a natural hidden sector built-in: the complex dilaton field S arising from the gravitational sector of the theory, and the complex moduli fields T_i parameterizing the size and shape of the compactified space. The auxiliary fields of those gauge singlets can be the seed of SUSY breaking, solving the arbitrariness of SUGRA where the hidden sector is not constrained. In order to determine the pattern of soft terms it is important to know which field, either S or T_i, play the predominant role in the process of SUSY breaking. Thus, it is customary to introduce a parameterization for the vacuum expectation values (VEVs) of the dilaton and moduli

$$F^S = \sqrt{3}(S + S^*)m_{3/2}\sin\theta \quad , \quad F^i = \sqrt{3}(T_i + T_i^*)m_{3/2}\cos\theta\,\Theta_i \,, \tag{2}$$

where $i = 1, 2, 3$ labels the three complex compact dimensions, $m_{3/2}$ is the gravitino mass, and the angles θ and Θ_i (with $\sum_i |\Theta_i|^2 = 1$) parameterize the direction of the Goldstino in the S, T_i field space. Here we are neglecting phases and the cosmological constant vanishes by construction. Using this parameterization one obtains [3]

$$
\begin{aligned}
M_a &= \sqrt{3}m_{3/2}\sin\theta \,, \\
m_\alpha^2 &= m_{3/2}^2\left(1 + 3\cos^2\theta\sum_i n_\alpha^i\Theta_i^2\right), \\
A_{\alpha\beta\gamma} &= -\sqrt{3}m_{3/2}\left(\sin\theta + \cos\theta\sum_i\Theta_i\left[1 + n_\alpha^i + n_\beta^i + n_\gamma^i - (T_i + T_i^*)\partial_i\log\lambda_{\alpha\beta\gamma}\right]\right).
\end{aligned}
\tag{3}
$$

Notice that in general the soft terms (scalar masses and trilinear parameters) show a lack of universality due to the modular weight dependence. Here we will be more interested on the scalar masses. Assuming an overall modulus ($T = T_i$, $\Theta_i = 1/\sqrt{3}$) these read

$$m_\alpha^2 = m_{3/2}^2\left(1 + n_\alpha\cos^2\theta\right), \tag{4}$$

where the overall modular weights are defined as $n_\alpha = \sum_i n_\alpha^i$, which in the case of Z_n Abelian orbifolds can take the values $-1, -2, -3, -4, -5$. Fields belonging to the untwisted sector of the orbifold have $n_\alpha = -1$. Fields in the twisted sector but without oscillators have usually modular weight -2 and those with oscillators have $n_\alpha \leq -3$. If all modular weights of the standard model fields are equal, or if $\sin\theta = 1$ (the dilaton-dominated SUSY-breaking case), universality is restored.

On the other hand, the apparent success of the joining of gauge coupling constants at $\approx 2 \times 10^{16}$ GeV in the MSSM is not automatic in the heterotic superstring, where the natural unification scale is $M_{GUT} \simeq g_{GUT} \times 5.27 \cdot 10^{17}$ GeV with g_{GUT} the unified gauge coupling. Despite the sizable mismatch between both scales, this problem might be solved with the presence of large string threshold corrections [4, 5]. In a sense, what would happen is that the gauge coupling constants will cross at the MSSM unification scale and diverge towards different values at the heterotic string unification scale. These different values appear due to large one-loop stringy threshold corrections. It was found in [5] that these corrections can be obtained for restricted values of the modular weights of the fields. In fact, assuming generation independence for the n_α as well as $-3 \leq n_\alpha \leq -1$, the simplest possibility corresponds to taking the following values for the standard model fields:

$$n_{Q_L} = n_{d_R} = -1, \quad n_{u_R} = -2, \quad n_{L_L} = n_{e_R} = -3, \quad n_{H_u} + n_{H_d} = -5, -4 \,. \quad (5)$$

The associated soft sfermion masses are thus given by

$$
\begin{aligned}
m_{Q_L}^2, m_{d_R}^2 &= m_{3/2}^2 \left(1 - \cos^2 \theta\right) , \\
m_{u_R}^2 &= m_{3/2}^2 \left(1 - 2 \cos^2 \theta\right) , \\
m_{L_L}^2, m_{e_R}^2 &= m_{3/2}^2 \left(1 - 3 \cos^2 \theta\right) ,
\end{aligned}
\quad (6)
$$

whereas for the soft Higgs masses, choosing $n_{H_u} = -1$, $n_{H_d} = -3$, we have

$$m_{H_u}^2 = m_{3/2}^2 \left(1 - \cos^2 \theta\right) \quad , \quad m_{H_d}^2 = m_{3/2}^2 \left(1 - 3 \cos^2 \theta\right) . \quad (7)$$

These soft terms serve as an explicit model for the study of the dark matter cross section. Since they are completely determined in terms of just the gravitino mass and the Goldstino angle, we are left with just four free parameters, namely $m_{3/2}$, θ, $\tan\beta$ and the sign of μ.

Following the discussion of the Introduction, the Higgs modular weights giving rise to the soft masses (7), could induce an increase of $\sigma_{\tilde{\chi}_1^0-p}$ with respect to the universal case. In order to investigate this possibility we have performed a scan on the three-dimensional parameter space. We have allowed the gravitino mass to vary from 0 to 1 TeV, and values of $\tan\beta$ ranging from 10 to 50 have been analysed for the full range on the Goldstino angle, $0 \leq \theta < 2\pi$. The most recent experimental and astrophysical constraints have been included. In particular, the lower bounds on the masses of the supersymmetric particles and on the lightest Higgs have been implemented, as well as the experimental limits on the branching ratios of the $b \rightarrow s\gamma$ and $B_s^0 \rightarrow \mu^+\mu^-$ processes, and the constraint on the supersymmetric contribution to the muon anomalous magnetic moment. Also, the neutralino relic density is required to fulfil the WMAP bound on the amount of cold dark matter. Last, dangerous charge and colour breaking minima of the Higgs potential are avoided by excluding unbounded from below (UFB) directions [6]. Among these, the one labelled as UFB-3, where sleptons take non-vanishing VEVs (thus leading to a charge breaking minimum) is the most dangerous one, especially for light staus.

For a better understanding of all these constraints, we have represented in Fig. 1 their effect on the $(m_{3/2}, \theta)$ plane for $\tan\beta = 10$. Extensive (white) regions are excluded due to the occurrence of tachyonic masses for sleptons and the area in the parameter space where the lightest neutralino is the LSP are not very large. These regions occur for small values of $\cos\theta$, thus corresponding to a breaking of SUSY which is mainly due to the dilaton auxiliary term. The narrow (black) areas where the neutralino fulfils all experimental constraints and has the correct relic density occur due to a coannihilation

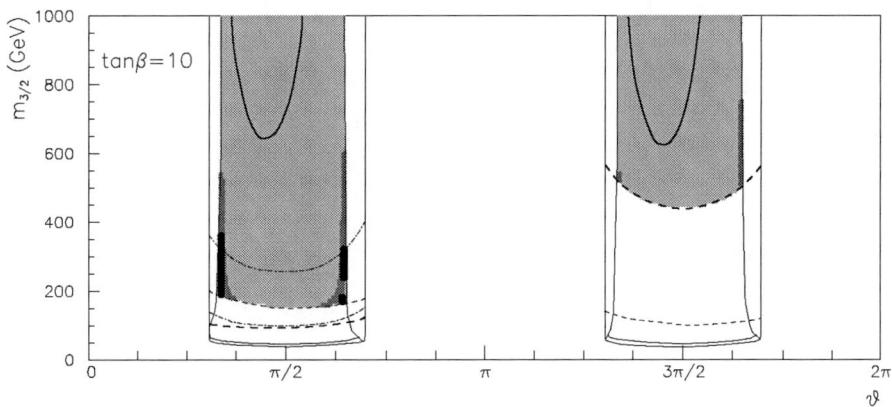

FIGURE 1. Effect of the different experimental constraints on the parameter space $(m_{3/2}, \theta)$ for $\tan\beta = 10$. Only the two areas centered around $\theta \approx \pi/2, 3\pi/2$ are free from tachyons in the slepton sector. In these regions, the narrow vertical white areas contained within solid lines correspond to those where the stau is the LSP, whereas in the thin horizontal region at very small gravitino masses, also bounded by solid lines, the LSP is the lightest sneutrino. The region below the thin dashed line is excluded by the lower bound on the Higgs mass. The region bounded by thin dot-dashed lines is favoured by a_μ^{SUSY} The area below the thick dashed line is excluded by $b \to s\gamma$. The light shaded area is favoured by all the experimental constraints, while the dark one fulfils in addition $0.1 \leq \Omega_{\tilde{\chi}_1^0} h^2 \leq 0.3$, and the black region on top of this indicates the WMAP range $0.094 < \Omega_{\tilde{\chi}_1^0} h^2 < 0.129$. Finally, the UFB constraints are only fulfilled in the area above the thick solid line.

effect with the stau (the next-to-lightest SUSY particle in this case). Finally, due to the smallness of the slepton masses, and more specifically, of the stau mass (6), the potential along the UFB-3 direction becomes very negative and easily gives rise to a minimum much deeper than the realistic one. In fact, no areas of the parameter space are found where the UFB-3 constraint is satisfied at the same time than the experimental and astrophysical ones. Interestingly, this happens for all values of $\tan\beta$ and trilinear parameter.

The theoretical predictions for the neutralino-nucleon cross section have been calculated in the experimentally accepted regions, and are represented versus the neutralino mass in Fig. 2 for $\tan\beta = 10, 20$ and 35. Despite the non-universality of the soft terms, these results resemble those of the CMSSM, as no high values are obtained. The μ parameter and the heavy Higgs masses are sizable, thus implying Bino-like neutralinos and a suppressed contribution to $\sigma_{\tilde{\chi}_1^0 - p}$ from Higgs-exchanging processes. After analysing the range $\tan\beta = 10$ to 50 we found that $\sigma_{\tilde{\chi}_1^0 - p} \lesssim 10^{-8}$ pb, the maximum values corresponding to $\tan\beta \approx 20$. These results are therefore beyond the present sensitivities of dark matter detectors and would only be partly within the reach of the projected 1 tonne detectors. In any case, as we have already emphasized above, all the points in Fig. 2 would be disfavoured due to the UFB-3 constraints.

Let us finally remark that other examples with different choices of modular weights

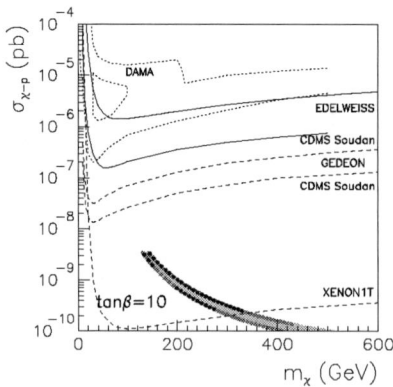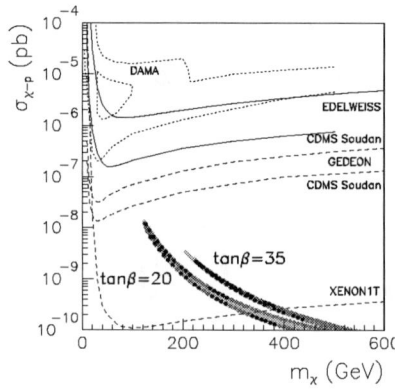

FIGURE 2. Theoretical predictions for $\sigma_{\tilde{\chi}_1^0-p}$, as a function of $m_{\tilde{\chi}_1^0}$, for the case with soft terms given by (6) and (7), with $\tan\beta = 10, 20$ and 35. The light grey dots represent points fulfilling all the experimental constraints The dark grey dots stand for those which satisfy in addition $0.1 \leq \Omega_{\tilde{\chi}_1^0}h^2 \leq 0.3$ and the black dots on top of these indicate those in agreement with the WMAP range. The sensitivities of present and projected experiments are also depicted with solid and dashed lines, respectively, in the case of an isothermal spherical halo model. The large (small) area bounded by dotted lines is allowed by the DAMA experiment when astrophysical uncertainties to this simple model are (are not) taken into account. None of these points fulfil the UFB constraints.

for the Higgs parameters satisfying (5) have been investigated, as well as the rest of the scenarios which give rise to gauge coupling unification with an overall modulus in this class of orbifold models [7]. Remarkably, none of these presents any point in the parameter space where the experimental and astrophysical bounds and the UFB constraints are satisfied simultaneously. In the following section we will see how this situation changes when the D-term contribution from an anomalous $U(1)$ is taken into account.

D-term contribution from an anomalous $U(1)$

An anomalous $U(1)$ is usually present after compactification, which is crucial for model building. It generates a FI contribution to the D-term [8], breaking extra $U(1)$ symmetries, and allowing the construction of realistic standard-like models. In order to cancel the FI term some scalar fields (with vanishing hypercharges), C_β, develop large VEVs along the D-flat direction, inducing the D-term contribution [9, 10, 11] to the soft scalar masses of the observable fields.[2] Totally, the soft scalar mass squared is given by

---

[2] There is no additional contributions to gaugino masses and A-terms when Higgs fields relevant to such symmetry breaking have less F-term than those of dilaton and moduli fields.

$$m_\alpha^2 = m_{3/2}^2 \left\{ 1 + n_\alpha \cos^2\theta + q_\alpha^A \frac{\sum_\beta (T+T^*)^{n_\beta} q_\beta^A |C_\beta|^2 \left[(6-n_\beta)\cos^2\theta - 5 \right]}{\sum_\beta (T+T^*)^{n_\beta} (q_\beta^A)^2 |C_\beta|^2} \right\}, \qquad (8)$$

where, q_β^A are the $U(1)$ charges of the fields C_β. The first two terms are the usual contributions (compare with (4)), and the third term is the D-term contribution. Obviously, if the observable fields have vanishing $U(1)$ charges, q_α^A, this contribution is also vanishing, and we recover (4). However, the observable fields have usually non-vanishing charges in explicit models [12], and the effect of this contribution must be taken into account in the analysis.

As we can see, the D-term contribution generates in general an additional non-universality among soft scalar masses, depending on q_α^A. Let us simplify the analysis considering the case where only a single field C develops a VEV. Thus the above result reduces to the following form:

$$m_\alpha^2 = m_{3/2}^2 \left\{ 1 + n_\alpha \cos^2\theta + \frac{q_\alpha^A}{q_C^A} \left[(6-n_C)\cos^2\theta - 5 \right] \right\}, \qquad (9)$$

where q_C^A and n_C are the $U(1)$ charge and modular weight of the field C, respectively. It is worth emphasizing here that even in the dilaton-dominated case ($\sin\theta = 1$) the soft scalar masses are non-universal. Hence, we can expect that for appropriate values of the $U(1)$ charges the above additional terms lead to interesting values for the dark matter cross section. Contrary to the cases without an anomalous U(1) analysed in the previous section, positive values for the non-universalities are now possible, by choosing $q_\alpha^A/q_C^A < 0$. This is welcome in order to enhance the stau masses and thus avoid the UFB constraints. In addition, the value of $m_{H_u}^2$ can be increased, and thus larger predictions for $\sigma_{\tilde{\chi}_1^0-p}$ may be obtained.

For instance, assuming that C is a twisted field with modular weight $n_C = -2$, and the ratios $q_{L_L,e_R}^A/q_C^A = -2$ with the slepton $U(1)$ charges (which are typically obtained in explicit models [12]), one has

$$m_{L_L,e_R}^2 = m_{3/2}^2 \left(11 - 19\cos^2\theta \right), \qquad (10)$$

for the slepton soft masses. In analogy to (1), the degree of non-universality can be defined as $\delta_{L_L,e_R} = 10 - 19\cos^2\theta$. Notice that this can be positive and sizable (especially in the dilaton limit). Due to such an increase in the stau mass, the UFB constraints are now easily fulfilled. Also, the allowed parameter space where the lightest neutralino is the LSP is more extensive and larger values of $\tan\beta$ are available. This allows the pseudoscalar and heavy scalar Higgs masses to decrease more efficiently, and entails an enhancement in $\sigma_{\tilde{\chi}_1^0-p}$, as discussed in the Introduction. However, the decrease in the pseudoscalar mass leads to large contributions to the branching ratios of $b \to s\gamma$ and $B_s^0 \to \mu^+\mu^-$, which therefore put stringent upper bounds on $\sigma_{\tilde{\chi}_1^0-p}$. This can be seen on the left-hand side of Fig. 3, where the theoretical predictions for $\sigma_{\tilde{\chi}_1^0-p}$ are represented for $\tan\beta = 45$ (the most favourable case). Although very large values could be obtained,

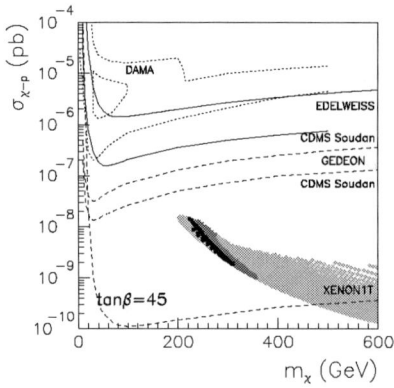 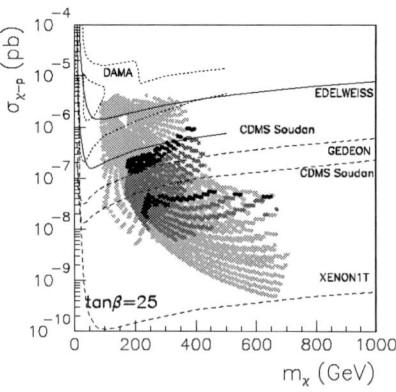

FIGURE 3. (Left) The same as Fig. 2 but with slepton masses given by (10) and with $\tan\beta = 45$, and (Right) when, in addition, Higgs masses are given by (11) and with $\tan\beta = 25$. In both cases all the points fulfil the UFB constraints.

all of which are in agreement with UFB and astrophysical constraints, the experimental bound on $B(B_s^0 \to \mu^+\mu^-)$, imposes $\sigma_{\tilde{\chi}_1^0-p} \lesssim 2 \times 10^{-8}$ pb.

Concerning the dark matter cross section, the situation can be further improved if we assume that, in addition to sleptons (10), also Higgses have non-vanishing anomalous $U(1)$ charge. For example, setting $q_{H_u}^A/q_C^A = -2$, using result (9) equation (7) is modified as

$$m_{H_u}^2 = m_{3/2}^2 \left(11 - 17\cos^2\theta\right) \quad , \quad m_{H_d}^2 = m_{3/2}^2 \left(1 - 3\cos^2\theta\right) . \tag{11}$$

The degree of non-universality, using notation (1), is now given by $\delta_{H_u} = 10 - 17\cos^2\theta$, and $\delta_{H_d} = -3\cos^2\theta$. In the dilaton-dominated case this turns out to be very large, $\delta_{H_u} = 10$.

Because of such large non-universalities the μ parameter can be significantly reduced, thus enhancing the Higgsino components of the lightest neutralino. This contributes to the increase of $\sigma_{\tilde{\chi}_1^0-p}$, even for moderate values of $\tan\beta$ (thus keeping the $B(B_s^0 \to \mu^+\mu^-)$ constraint under control). The neutralino-nucleon cross section is depicted on the right hand-side of Fig. 3 for an example with $\tan\beta = 25$, showing neutralinos with $\sigma_{\tilde{\chi}_1^0-p} \gtrsim 10^{-7}$ pb with a mass in the range $m_{\tilde{\chi}_1^0} \approx 200 - 400$ GeV that would be within the reach of the CDMS Soudan experiment. Once more, due to the large slepton masses (10), the UFB constraints are fulfilled by all the points.

The corresponding allowed area in the $(m_{3/2}, \theta)$ plane for this last example is shown in Fig. 4. The allowed regions in the parameter space are very narrow, since either the pseudoscalar mass-squared, m_A^2, or μ^2 easily become negative. For instance, the area around the dilaton-dominated case, where the non-universality is maximal, is completely ruled out for this reason. Also, unlike the cases without the $U(1)$ contribution, the areas where the neutralino has the correct relic density are due to an enhancement of the neutralino annihilation rate when $m_A \approx 2m_{\tilde{\chi}_1^0}$, which is now possible because

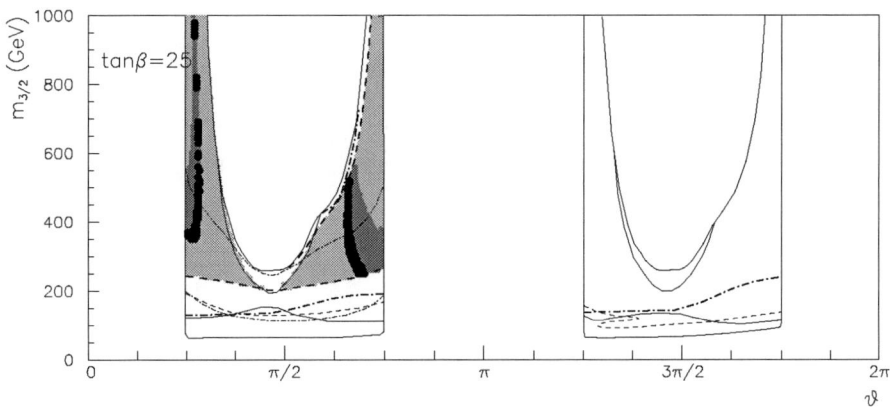

FIGURE 4. The same as Fig. 1 but with slepton masses given by (10), Higgs masses given by (11) and with $\tan\beta = 25$. Unlike Fig. 1, here all the points represented satisfy the UFB constraints.

of the decrease of the pseudoscalar mass. In addition, all these areas satisfy the UFB constraints.

The Higgs non-universality can be further increased if we also consider H_d to be charged under the anomalous $U(1)$. For example, taking $q_{H_d}^A/q_C^A = 1/2$, the soft masses for the Higgses become

$$m_{H_u}^2 = m_{3/2}^2\left(11 - 17\cos^2\theta\right) \quad , \quad m_{H_d}^2 = m_{3/2}^2\left(-\frac{3}{2} + \cos^2\theta\right) . \tag{12}$$

In this case the non-universalities are given by $\delta_{H_u} = 10 - 17\cos^2\theta$, and $\delta_{H_d} = -\frac{5}{2} + \cos^2\theta$, which implies that in the dilaton limit they become $\delta_{H_u} = 10$ and $\delta_{H_d} = -\frac{5}{2}$. The theoretical predictions for $\sigma_{\tilde{\chi}_1^0-p}$ are depicted in Fig. 5 for $\tan\beta = 20$. Values of $\sigma_{\tilde{\chi}_1^0-p}$ within the reach of present experiments can now be found for even smaller values of $\tan\beta$. The allowed regions in the parameter space are, nevertheless, narrower, since either m_A^2 or μ^2 easily become negative.

CONCLUSIONS

Large theoretical predictions for the direct detection of neutralino dark matter, within the reach of future dark matter experiments, are attainable in Supergravity theories which result from orbifold compactifications of the $E_8 \times E_8$ Heterotic superstring. This is possible thanks to the D-term contribution to the soft terms, which enhances and gives more flexibility to their non-universality. Neutralinos fulfilling all the experimental and astrophysical bounds, and with $\sigma_{\tilde{\chi}_1^0-p} \gtrsim 10^{-7}$ pb can be obtained in regions of the parameter space where also the UFB constraints are satisfied.

191

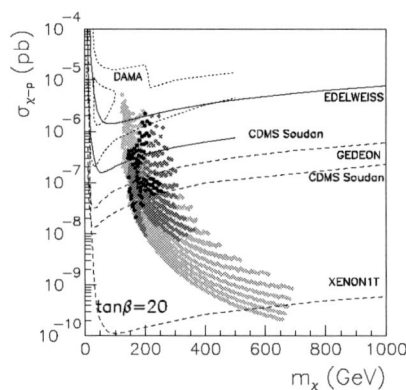

FIGURE 5. The same as Fig. 2 but with slepton masses given by (10), Higgs masses given by (12) and with $\tan\beta = 20$. All the points represented satisfy the UFB constraints.

REFERENCES

1. See e.g., S.Baek, D.G.Cerdeño, Y.G. Kim, P.Ko and C.Muñoz, *J.High Energy Phys.* **06** (2005) 017, and references therein.
2. C. Muñoz, *J. High Energy Phys.* **12** (2001) 015.
3. A. Brignole, L.E. Ibáñez and C. Muñoz, *Nucl.Phys.* **B422** (1994) 125; **B436** (1995) 747; T. Kobayashi, D. Suematsu, K. Yamada and Y. Yamagishi, *Phys. Lett.* **B348** (1995) 402;
4. K. Choi, *Phys. Rev.* **D37** (1988) 1564.
5. L.E. Ibáñez, D. Lüst and G.G. Ross, *Phys. Lett.* **B272** (1991) 251.
6. J.A. Casas, A. Lleyda and C. Muñoz, *Nucl. Phys.* **B471** (1996) 3.
7. L.E. Ibánez and D. Lüst, *Nucl. Phys.* **B382** (1992) 305.
8. E. Witten, *Phys. Lett.* **B149** (1984) 351; M. Dine, N. Seiberg and E. Witten, *Nucl. Phys.* **B289** (1987) 317; J.J. Atick, L.J. Dixon and A. Sen, *Nucl. Phys.* **B292** (1987) 109; M. Dine, I. Ichinose and N. Seiberg, *Nucl. Phys.* **B293** (1987) 253.
9. H. Nakano, *Prog. Theor. Phys. Suppl.* **123** (1996) 387; Y. Kawamura, *Phys. Rev.* **D53** (1996) 3779; *Prog. Theor. Phys. Suppl.* **123** (1996) 421.
10. Y. Kawamura and T. Kobayashi, *Phys. Lett.* **B375** (1996) 141 [arXiv:hep-ph/9601365], Erratum: **B388** 867 (1996); *Phys. Rev.* **D56** (1997) 3844 [arXiv:hep-ph/9608233].
11. Y. Kawamura, T. Kobayashi and T. Komatsu, *Phys. Lett.* **B400** (1997) 284; Y. Kawamura, *Phys. Lett.* **B446** (1999) 228.
12. J.A. Casas and C. Muñoz, *Phys. Lett.* **B209** (1988) 214; *Phys. Lett.* **B214** (1988) 63; A. Font, L.E. Ibáñez, H.P. Nilles and F. Quevedo, *Phys. Lett.* **B210** (1988) 101.
13. M. P. Brown, and K. Austin, *The New Physique*, Publisher Name, Publisher City, 2000, pp. 212–213.
14. M. P. Brown, and K. Austin, *Appl. Phys. Letters* **85**, 2503–2504 (2000).
15. R. Wang, "Title of Chapter," in *Classic Physiques*, edited by R. B. Hamil, Publisher Name, Publisher City, 2000, pp. 212–213.
16. C. D. Smith and E. F. Jones, "Load-Cycling in Cubic Press," in *Shock Compression of Condensed Matter-1999*, edited by M. D. F. et al., AIP Conference Proceedings 505, American Institute of Physics, New York, 1999, pp. 651–654.

Neutrino telescopes in the World

ERNENWEIN, J.-P.

GRPHE, Université de Haute Alsace, 61 rue Albert Camus, 68093 Mulhouse cedex, France.

Abstract. Neutrino astronomy has rapidly developed these last years, being the only way to get specific and reliable information about astrophysical objects still poorly understood.

Currently two neutrino telescopes are operational in the World: BAIKAL, in the lake of the same name in Siberia, and AMANDA, in the ices of the South Pole. Two telescopes of the same type are under construction in the Mediterranean Sea: ANTARES and NESTOR. All these telescopes belong to a first generation, with an instrumented volume smaller or equal to 0.02 km^3. Also in the Mediterranean Sea, the NEMO project is just in its starting phase, within the framework of a cubic kilometer size neutrino telescope study. Lastly, the ICECUBE detector, with a volume reaching about 1 km^3, is under construction on the site of AMANDA experiment, while an extension of the BAIKAL detector toward km^3 is under study. We will present here the characteristics of these experiments, as well as the results of their observations.

Keywords: Astrophysics, Telescope, Neutrino, Muon, Cherenkov
PACS: 95.55.Vj

THE NEUTRINO, A PROMISING NEW MESSENGER

The photon, from gamma rays to radio waves, is the most used probe in astrophysics: astronomy in visible and infra-red light has been developed on the ground, thanks to large telescopes located at high altitude (VLT in Chile, CFHT in Hawaii), while the microwave radiation is measured by means of satellites (COBE, WMAP, PLANCK SURVEYOR). The detection of radio waves is ensured by networks of ground based antennas (VLA). The low energy gamma rays (<30 GeV), X-rays and UV radiations are studied by experiments carried by satellites (EGRET and BATSE on CGRO, BEPPOSAX, CHANDRA, INTEGRAL, SWIFT), whereas the gamma rays of higher energy (>100 GeV) are accessible by Cherenkov detectors on the ground such as WHIPPLE, MAGIC, or HESS.

However, beyond 100 TeV, the interaction of the photon with the diffuse cosmological and infra-red backgrounds limits the depth of observation to 10 Mpc, which is the typical size of our local galaxy cluster.

Other probes used in experiments as HIRES, AGASA, and AUGER, are the proton and other nuclei constituting the Cosmic Rays. But at low energy, the trajectories of charged particles is bended by magnetic fields, making impossible the search for point sources. Moreover, beyond 100 EeV, the interaction of the proton with the diffuse photon backgrounds limits its path to 50 Mpc.

Therefore, the neutrino appears as an ideal messenger. It will not deviate because it does not carry any electrical charge, and is only sensitive to the weak interaction. It can thus traverse cosmological distances, from the heart of the sources to the Earth. The current neutrino telescopes have a sensitivity optimized beyond the TeV scale, but are operational at a hundred GeV.

CP881, *Cairo International Conference on High Energy Physics (CICHEP II)*,
edited by S. Khalil
© 2007 American Institute of Physics 978-0-7354-0382-6/07/$23.00

PRODUCTION OF NEUTRINOS IN THE UNIVERSE

The construction of neutrino telescopes is relevant only if this particle is produced within the various galactic and extragalactic objects, in sufficient flux to be detectable on Earth, and has an energy accessible to the telescopes.

In Supernova Remnants (SNRs) located in our galaxy, proton acceleration may lead to the production of detectable neutrinos. Current observations of SNRs in X and gamma rays do not make it yet possible to know precisely the production mechanisms of these radiations. Two models are possible:

- an electronic model, in which X-rays and gamma rays are produced by synchrotron radiation of electrons and by Inverse Compton effect, respectively.
- a hadronic model, in which protons are accelerated, generating pions by collision. The neutral pions decay in gamma photons, while decay cascades of charged pions provide neutrinos in the proportion $(v_e, v_\mu, v_\tau) = (1,2,0)$. On Earth, because of neutrino oscillations, this proportion becomes $(1,1,1)$.

In the last observations from the HESS experiment [1], the simplest electronic models are disfavored by the shape of the observed gamma spectra. A direct observation of neutrinos, even with a very low statistic, would then be a mean to clarify without ambiguity the processes involved in the SNRs. The lack of information on the production mechanisms of gamma rays within the SNRs make it difficult to estimate the neutrino flux and its energy spectrum according to the observed gamma ray flux.

Micro-quasars also constitute a promising source of high energy neutrinos ($\mathcal{O}(1 \text{ TeV})$). They consist in a central black hole ($M \sim 10 M_\odot$) whose accretion disc is fed by a companion star. Two jets, of typical length one light year, are emitted perpendicularly to the disc, and finish sometimes in lobes emitting radio waves. Like in the case of SNRs, the types of interactions within the jets are decisive for the existence of neutrino production [2][3]. The micro-quasars being gamma ray emitters, an evaluation of the neutrino flux can be carried out on the basis of photon flux [4]: a neutrino telescope like ANTARES or AMANDA requires from one to several years of observation to detect a given micro-quasar.

Outside our galaxy, Active Galactic Nuclei (AGN), made up of a super massive black hole (typically $10^8 M_\odot$) and jets perpendicular to the plan of the galaxy, may generate neutrinos if hadronic interactions within the jets are favored.

The Gamma Ray Bursts (GRBs) are observed by satellites since 1967. Most of the time, they are followed by an afterglow in X radiation, optical light, and sometimes radio waves. The isotropy of GRBs sources as well as measurements of their redshifts prove their mainly extragalactic origin. The production processes of these radiations are still poorly understood. They are related to the coalescence of two compact stars or to the collapse of a massive star in black hole. According to the most popular models (Cannonball [5][6], Fireball [7][8]), the energy of emitted neutrinos could be around 100 GeV (Cannonball) or 100 TeV (Fireball). If the fluxes are large enough, the observation by neutrino telescopes will be possible in a time window framing detection by the satellites and ground based observatories.

In the case of point sources, the angular resolution of neutrino telescopes plays an important part for the location, and therefore also for the identification of the objects at the origin of the detected neutrinos.

Diffuse neutrino fluxes: the sum of the whole galactic and extra-galactic sources leads to a flux of possibly detectable diffuse neutrinos. The theoretical estimate of this flux, strongly depending on the models, is adjusted by using normalization on the Cosmic Rays. The so-called Waxman Bahcall limit [9], for example, gives a muon neutrino flux $E^2 dN/dE < 2.2 \cdot 10^{-8}$ GeV cm^{-2}s^{-1}sr^{-1} at the Earth level, if neutrino oscillations are included.

Dark Matter: the indirect search for WIMPs (Weakly Interacting Massive Particles) is also a field of application for the neutrino telescopes. Indeed, these particles, which are neutralinos within the framework of some supersymmetric models, are likely to accumulate and annihilate in the center of massive objects, thus producing a cascade of particles among which neutrinos are produced, with a typical energy lower than the TeV scale. The studied massive objects are the Earth, the Sun, and the Galactic Center.

Atmospheric background noise: the Cosmic Rays, made up of protons, helium nuclei, and heavier nuclei such as carbon or iron, initiate cascades in the Earth's atmosphere, generating a great number of neutrinos and muons superimposed on the astrophysical signal:

- atmospheric neutrinos constitute an irreducible background noise, up-going as well as down-going, the Earth being transparent at these neutrino energies. The main difference between these neutrinos and those of astrophysical origin resides in their energy spectrum: the atmospheric neutrino flux decreases in $E^{-2.7 \to -3.7}$ whereas astrophysical fluxes decrease in E^{-2}. Besides, in the case of point sources, the angular resolution of telescopes allows to distinguish a possible neutrino excess in the observed object direction, compared to the diffuse atmospheric background.

- atmospheric muons being stopped by the Earth, they are downward only, but their number is approximately 10^6 times higher than the number of muons generated by the atmospheric neutrinos, at 2400 meters depth.

To reduce this background noise, the neutrino telescopes are optimized to observe upward muons, i.e. muons produced by neutrinos having crossed the Earth. Moreover, in order to minimize the number of downward atmospheric muons, the neutrino telescopes are located in-depth, using water or the ice as shielding. Indeed, the flux of vertical atmospheric muons is multiplied by a factor 20 between -3500 m and -1500 m.

NEUTRINO DETECTION WITH UNDERWATER TELESCOPES

Guiding detection principles

All the neutrino telescopes are based on the same basic principle: the detection of the Cherenkov light emitted by a transparent medium during the passage of a muon or more generally of one or more charged particles, and the reconstruction of the trajectory. The primary particle that one aims to observe being a neutrino, its weak interaction requires great detection volumes, realizable in ice or deep water. The interaction of the neutrino with a nucleus in the vicinity of the instrumented area produces one lepton of

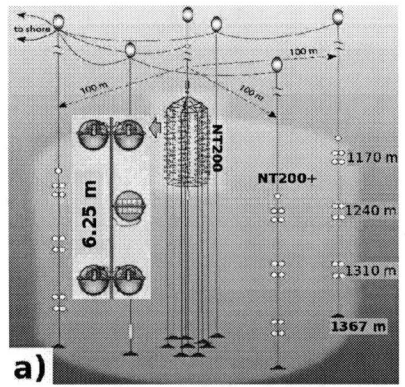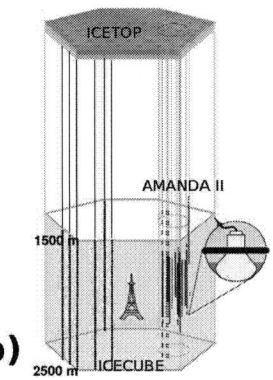

FIGURE 1. a) The BAIKAL detector [11], b) AMANDA and ICECUBE schemes [12][13]

the same flavor, accompanied by a shower at the vertex (charged current interaction: $W^{\pm}$), or only one shower at the vertex, the neutrino itself being only scattered (neutral current interaction: Z^{0}). The Cherenkov light produced by charged particles is detected by a set of photomultipliers tubes (PMTs) laid out regularly in a large volume. These PMTs, whose diameter ranges between 20 cm and 37 cm, depending on detectors, are housed in pressure improved glass spheres, the sphere unit + PMT being called optical module (OM). The PMT signals are sent to the surface by an electro-optical cable. The essential parameters are the time resolution ($\mathscr{O}(ns)$), the PMT sensitivity (single photo-electron), their dynamic range (up to 1000 photo-electrons), the optical module spatial distribution and numbers, and the optical quality of the medium in which the Cherenkov light propagates.

The BAIKAL experiment [10][11] is located in the Baikal lake at $54°50'$ N, $104°20'$ E, at 1367 m depth, 4 km from shore. A first prototype (NT36) was immersed in 1993, and in 1995 the BAIKAL collaboration announced the observation of the first upward atmospheric neutrinos. Since 1998 detector NT200 is operating. It consists in 8 lines of 12 floors, each floor gathering two optical modules, for a total of 192 PMTs whose diameter is 37 cm (figure 1a). Some photomultipliers are directed to the top, thus allowing a better rejection of the atmospheric muons. The installation and the mainte-nance of the detector are performed in Winter, when the ice covering the lake can be used as a platform. In 2005, 3 peripheral lines of 6 floors were added at a distance of 100 m around the core NT200, improving the sensitivity at very high energies. During the next decade, the BAIKAL collaboration aims to extend the telescope to 91 lines of 12 or 16 optical modules (1308 OMs), thus rising the effective volume to approximately 1 km^3 for the cascades of more than 100 TeV, but resulting in a threshold of 30 TeV for the muons.

The AMANDA experiment (Antarctic Muon and Neutrino Detector Array) [12][13] is located in the ices of the South Pole, at a depth ranging from 900 to 2350 m. The first four lines were operational in 1996. Then, in 1997, AMANDA-B10 (10 lines, 302 optical modules) started physics data taking. The detector in operation since 2000, AMANDA II, consists of 19 lines carrying 677 optical modules. The floors, each one containing one PMT of 20 cm size directed to the bottom, are spaced by 15 m (figure 1b).

In 2011, **the ICECUBE detector** [13] will be completed, gathering 80 lines 125 m apart, each line carrying 60 floors separated by 17 m. The 4800 PMTs of 25 cm size thus assembled at 2450 m depth will be supplemented by a detector located at the surface (ICETOP), made up of 160 Cherenkov tanks. ICETOP will mainly allow to improve the angular resolution and to identify the downward atmospheric muons, improving the rejection of this background noise. Currently, in 2006, 9 lines of ICECUBE and 16 tanks of ICETOP are operational.

The ANTARES experiment (Astronomy with a Neutrino Telescope and Abyss environmental **RES**earch) [14][15] is located at 42°50′ N, 6°10′ E, at 2475 m depth in the Mediterranean Sea, 40 km off the French coast. At the end of 2007, the detector will consist in 12 lines 70 m apart, each line carrying 25 floors separated by 14.5 m (figure 2). A floor contains 3 optical modules oriented at 45° towards the sea bed. ANTARES will thus gather 900 PMTs of 25 cm diameter. After the immersion in 2003 of a prototype of 5 storeys, the collaboration immersed in March 2005 an instrumented line equipped with three optical modules. In March 2006, the first line of 25 floors was connected to the junction box achieving communication with the coast, and is currently taking data. In July 2006, the second line was immersed and will be connected in September 2006. The immersion of the ANTARES lines will continue regularly until the end of 2007.

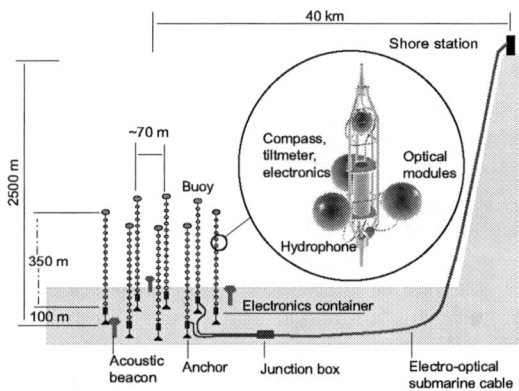

FIGURE 2. Schematic view of the ANTARES detector

The NESTOR experiment (Neutrino Extended Submarine Telescope with Oceanographic Research) [16] is located at 36°36′ N, 21°30′ E in the Mediterranean Sea, at a depth of 3800 m, 14 km from shore. The detector will consist of towers joining together 12 hexagonal floors equipped at each top with two optical modules, one directed downwards and the other looking upwards (figure 3a), resulting in a set of 144 PMTs of 37 cm diameter per tower. In 2003, a floor of reduced size was immersed, allowing the detection of upward muons and an estimate of the downward muon flux [17][18]. In 2007, the NESTOR collaboration intends to immerse 4 floors.

NEMO (NEutrino Mediterranean Observatory) [19] is a project which envisages the construction of a neutrino telescope offshore Capo Passero (Sicily), at the coordinates 36°20′ N, 16° E, at 3500 m depth. The detector will consist of towers, each one carrying

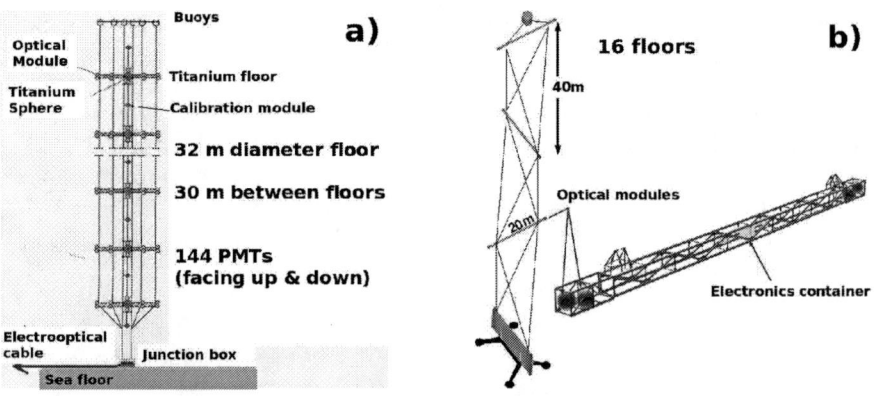

FIGURE 3. a) View of one NESTOR tower [16], b) View of one NEMO tower and floor [19]

16 floors spaced by 40 m. A floor will contain 4 optical modules: 2 directed horizontally and 2 directed to the bottom (figure 3b). A tower will thus gather 64 PMTs of 25 cm diameter. In 2006, a first phase will be the test of a mini tower of 4 reduced size floors (12 m) at 2000 m depth, offshore Catania.

Observed sky

The particular location of AMANDA/ICECUBE limits the observable sky to the Northern hemisphere, resulting in an angular coverage of 2π sr: AMANDA and ICECUBE do not see the Galactic Center. The other neutrino telescopes, laying at latitudes closer to the equator, and profiting from the Earth rotation, have a wider coverage of the sky, reaching 3.5π sr for the Mediterranean telescopes, and a little less for BAIKAL, but for a partial time of observation. For example, ANTARES can observe the Galactic Center during 2/3 of its operating time.

Site properties

The quality of the transparent medium is essential for a neutrino telescope, the absorption reducing photon statistics, and scattering introducing a delay and a dispersion in their time of arrival. These parameters substantially influence the energy threshold and the angular resolution of the telescopes. The absorption and scattering lengths in deep sea are respectively 60 m and 300 m at λ=470 nm, leading to a global attenuation length of about 50 m. The effective scattering length of Baikal lake deep water is similar, but the telescope suffers from a strong absorption length, of approximately 20 m at λ=470 nm, whereas AMANDA and ICECUBE experiments suffer from a short effective scattering length of 25 m, the absorption length of ice being 70 m (λ=470 nm). The resulting attenuation length is 20 m for ice as well as for Baikal deep water.

On the other hand, AMANDA/ICECUBE and BAIKAL are favored when we consider the optical background. Indeed, the optical modules are sensitive to the light of the bioluminescent organisms possibly present in the medium, and to the Cherenkov photons generated by ^{40}K beta radioactivity (1.12 MeV electron). As one can see in table 1, AMANDA and BAIKAL OMs are not exposed to ^{40}K radioactivity, which is absent in the ice and in fresh water, nor to bioluminescence light. The accidental hits only come from

TABLE 1. Counting rates induced by environnemental and instrumental backgrounds

Detector	PMT	counting rate	comment
BAIKAL	37 cm	0.1-0.3 kHz/2 OMs in 15 ns	few chemiluminescence, ^{40}K in sphere glass
AMANDA	20 cm	1 kHz/OM	photocathode noise, ^{40}K in sphere glass
ICECUBE	25 cm	0.7 kHz/OM	same as AMANDA, but digital OM
ANTARES	25 cm	60-100 kHz/OM	^{40}K + bioluminescence (-2500 m)
NESTOR	37 cm	50 kHz/OM	^{40}K + few bioluminescence (-3800 m)
NEMO	25 cm	30 kHz/OM	^{40}K + few bioluminescence (-3500 m)

^{40}K contained in the glass of the spheres, and, for BAIKAL, from some luminescence observed in the lake. This luminescence, most likely chemiluminescence, produces a seasonal background light of level similar to the level of light produced by ^{40}K in the sea. However, a coincidence window of 15 ns, required for the 2 photomultipliers of a same floor in BAIKAL detector, substantially reduces the number of accidental hits.

The counting rate induced by the environmental sources of light depends on the photocathode area and on the trigger threshold, taken at approximately 0.5 photo-electrons in table 1: in the Sea, the contribution of the ^{40}K is about 30 kHz for a 25 cm diameter PMT, the residuals coming from bioluminescent bacterias and organisms. NESTOR and NEMO, thanks to the large depth at which they are immersed, are not really affected by bioluminescence, contrarily to ANTARES, which is more subject to this seasonal background noise. Moreover, counting bursts, strongly correlated to the water current, are added to the baseline.

These sources of background noise produce mainly single photo-electron signals. The thermal noise of the photocathode is also of this level, and is relevant for all detectors, with however a weaker contribution in the case of AMANDA/ICECUBE.

Angular and energy resolution

The golden channel of neutrino telescopes is muon neutrino observation. Indeed, at the considered energies, the muon produced at the interaction point can travel several kilometers in water or through ice, thus making it possible to profit from an effective volume larger than the instrumented volume. In addition, at high energy (>10 TeV), the muon is essentially colinear to the neutrino, giving a precise information on the neutrino direction. The trajectory of the muon, which generates a Cherenkov cone of 41° half opening angle, is reconstructed by exploiting the arrival time and the number of Cherenkov photons collected by the PMTs, as well as the position of the fired optical modules. The time resolution depends on the photo-electron transit time spread (about 1.3 ns for 25 cm diameter PMTs), on the Cherenkov light scattering, on the readout electronics, and on the OM positioning precision in the case of moving detectors, like ANTARES ($\mathcal{O}$(10 cm), equivalent to ~0.5 ns). The time resolution is measured using laser and LED flashers deployed within detectors areas, and reaches about 2 ns. The muon track reconstruction method allows a good angular resolution, whereas the energy resolution is intrinsically bad, because of the probabilistic nature of the muon energy loss: above 1 TeV, the error on the energy reaches a factor of 2 or 3. The angular resolution presented in table 2 is the average angle between the incident muon neutrino and the reconstructed track of the muon. At low energy, it is dominated by the kinematics

TABLE 2. Angular resolution and effective area for 10 TeV muons

Detector	Angular resolution for muons	Effective area for muons
BAIKAL NT200	4°	2000 m^2 (1 TeV)
AMANDA 19 lines	2°-2.5°	40 000 m^2
ICECUBE	0.7°	0.85 km^2
ANTARES 12 lines	0.3°	35 000 m^2
NESTOR	<1° (6 towers)	20 000 m^2 (1 tower)
NEMO 9×9 towers, 140m spaced	0.1°	1 km^2

of the interaction, whereas at high energies (>10 TeV), it only depends on the quality of the reconstruction, which is better in sea water, due to its good optical properties.

The muon neutrinos interacting by charged current, although constituting the starting idea, are not the only detectable ones. The showers produced by other flavors (ν_e, ν_τ) are also of great interest. The reconstructed entity is not the track of only one particle (the muon), but the shower itself. The angular resolution is faded in favor of the energy resolution: the detector is used as a calorimeter more than as a telescope. Indeed, the isotropic Cherenkov light emission along the shower, and the small length of shower compared to muon track, degrade the angular resolution (10-30°), whereas the counting of photons constitutes a reliable estimator of shower energy, leading to a resolution of about 25-50%.

Effective area for μ and ν_μ

The effective area of a neutrino telescope is defined by the number of events detected per unit of time, divided by the incident particle flux. It is the equivalent area of a perfect detector in which any particle of the incident flux would be detected. Thus it accounts for the acceptance of the detector, for the efficiency of reconstruction, and in the case of neutrinos, for their interaction probability, which increases with energy. Table 2 shows the effective areas for high energy muons.

For up-going neutrinos, the absorption due to the crossing of the Earth, increasing with energy, has to be taken into account. Thus, in the case of telescopes of the size of ANTARES or AMANDA, the effective area for upward muon neutrinos grows with energy, reaches approximately 1 m^2 at 100 TeV, due to the increases of the neutrino cross section and of the muon path length, then falls beyond that, because of the Earth absorption (figure 4). Cubic kilometer size telescopes (ICECUBE, NEMO 81 towers) make it possible to improve the neutrino effective area by a factor 10 [20].

RESULTS OF THE OBSERVATIONS OF AMANDA AND BAIKAL

The first phase of NEMO will start in 2006. Since March 2006, ANTARES has a complete line in operation and, during the previous years, has performed an extensive study of the bioluminescence thanks to several prototypes [21]. In 2003, NESTOR has immersed one floor and has measured the atmospheric muon flux, a second step being the immersion of 4 floors in 2007. Currently, only BAIKAL and AMANDA are able to operate for neutrino astrophysics. Indeed, in addition to the measurement of atmospheric neutrinos flux, they

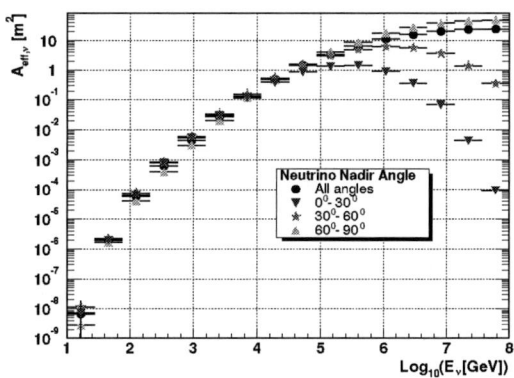

FIGURE 4. Effective area for muon neutrino as a function of energy (ANTARES experiment)

have established the first limits (90% C.L.) on several astrophysical neutrino fluxes: point sources (AMANDA) and diffuse fluxes (AMANDA and BAIKAL). The observation by AMANDA (analyses from 2000 to 2004) of known point sources (SNRs, Microquasars, Blazars), has not highlighted any excess of neutrinos compared to the atmospheric background, in the direction of the studied sources. In the same way, no signal was observed coming from GRBs, as well as in the direction of the galactic plane. Lastly, no evidence of diffuse flux was observed.

In accordance with AMANDA, no diffuse flux of neutrinos has been observed by BAIKAL (analyses from 1998 to 2002). The limits on diffuse fluxes established by these two experiments are deferred on the figure (figure 5), and are compared to the sensitivities announced by the telescopes in construction. The cascade analysis gives the limits $E^2\Phi_\nu < 8.1 \cdot 10^{-7}$ GeV cm^{-2}s^{-1}sr^{-1} for BAIKAL, and $E^2\Phi_\nu < 8.6 \cdot 10^{-7}$ GeV cm^{-2}s^{-1}sr^{-1} for AMANDA, while the pure muon neutrino analysis leads to the limit $E^2\Phi_{\nu_\mu} < 2.1 \cdot 10^{-7}$ GeV cm^{-2}s^{-1}sr^{-1} in [50 TeV, 1 PeV] for AMANDA. Neutrino telescopes of the size of AMANDA, BAIKAL, ANTARES or NESTOR having a sensitivity lower than Waxman Bahcall limit, it is necessary to envisage larger volumes of detection, as ICECUBE.

CONCLUSION

BAIKAL and AMANDA are the first underwater/ice neutrino telescopes ever built. They have already proved the feasability and interest of large Cherenkov detectors for high energy neutrino astrophysics. Currently, the first reliable results about atmospheric neutrinos are available, and preliminary astrophysics results are produced.

There is no evidence yet for astrophysics high energy neutrinos, neither coming from point sources, nor belonging to diffuse flux. However, it is necessary to get more statistics to be conclusive, the simplest way being to operate for a longer time, the most efficient being to increase the instrumented volume. AMANDA is growing towards

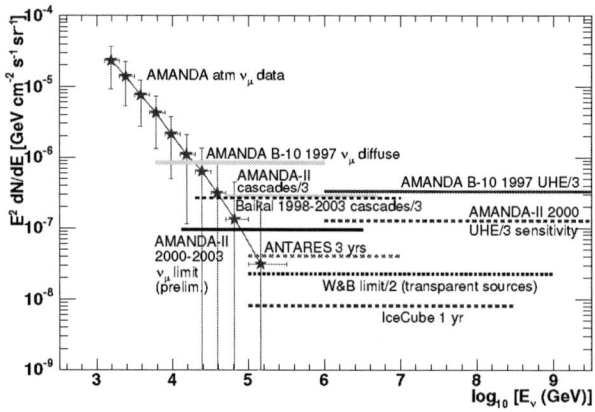

FIGURE 5.   Neutrino telescope limits or sensitivities on diffuse fluxes [11][13][22]

cubic kilometer size, thanks to ICECUBE construction. BAIKAL also intends to extend its volume, but specializing in high energies. In the Mediterranean Sea, in parallel to their own construction, the 3 current experiments collaborate in order to study the design of a future cubic kilometer neutrino telescope. The reached sensitivities will allow to test the theoritical limits on diffuse fluxes, not accessible with current neutrino telescopes, and to be more powerful for point source searches.

REFERENCES

1. F. Aharonian et al, *Astron. Astrophys.* **449** (2006), 223-242
2. C. Distefano, D. Guetta, E. Waxman and A. Levinson, *Astrophys. J.* **575** (2002), 378-383
3. A. Levinson and E. Waxman, *Phys. Rev. Lett.* **87** (2001), 171101, 5pp
4. F. Aharonian, L. Anchordoqui, D. Khangulyan and T. Montaruli, *J. Phys. Conf. Ser.* **39** (2006), 408-415
5. A. Dar and A. De Rùjula, *Phys. Rept.* **405** (2004), 203-278, and *astro-ph/0105094* (2001)
6. S. Ferry, *PhD Thesis* (Université Louis Pasteur, Strasbourg, 2004)
7. E. Waxman and J. Bahcall, *Phys.Rev.Lett.* **78** (1997), 2292-2295
8. A. Kouchner, *PhD Thesis* (Université Paris 7, 2001)
9. E. Waxman and J. Bahcall, *Phys. Rev. D* **59** (1999), 023002 and *Phys. Rev. D* **64** (2001), 023002
10. BAIKAL collaboration, *Astropart. Phys.* **25** (2006), 140-150
11. Zh. Dzhikibaev for the BAIKAL collaboration, *Neutrino 2006 conference*, Santa Fe (June 2006)
12. E. Bernardini for the AMANDA collaboration, *VLVnT2 conference*, Catania (November 2005)
13. G.C. Hill for the ICECUBE collaboration, *Neutrino 2006 conference*, Santa Fe (June 2006)
14. V. Bertin for the ANTARES collaboration, *Neutrino 2006 conference*, Santa Fe (June 2006)
15. ANTARES collaboration, *Nucl. Instrum. Meth.* **A555** (2005), 132-141
16. L. Resvanis for the NESTOR collaboration, *VLVnT2 conference*, Catania (November 2005)
17. NESTOR collaboration, *Astropart. Phys.* **23** (2005), 377-392
18. NESTOR collaboration, *Nucl. Instrum. Meth.* **A552** (2005), 420-439
19. E. Migneco for the NEMO collaboration, *VLVnT2 conference*, Catania (November 2005)
20. M. Circella for the NEMO collaboration, *29th International Cosmic Ray Conference*, Pune (2005)
21. ANTARES collaboration, *astro-ph/0606229* (2006), 18pp
22. T. Montaruli for the ANTARES collab., *TeV Particle Astrophysics Workshop*, Fermilab (August 2005)

Status and perspectives in B-Physics

Fernando Ferroni

Università di Roma La Sapienza & INFN Roma

Abstract. The two competing B-factories (KEKB and PEPII) are performing spectacularly well and the experiments they host (Belle and BaBar), have harvested an impressive amount of data. CP violation in B-physics has been established and it has become a precision measurement. It is time to look at the future development of the field with special attention to the e^+e^- option at very high luminosity.

Keywords: CP Violation
PACS: 12.15Hh

INTRODUCTION

The mystery of our existence is based on the premise that although our theory of fundamental sub-atomic particle interactions (the Standard Model) places matter and anti-matter at nearly equal footing, our Universe appears to be composed of matter only. Indeed the Standard Model accounts for small differences in the interactions of matter and anti-matter through a phenomenon known as CP violation. This difference however falls short, by orders of magnitude, in accounting for the observed matter asymmetry in the Universe. Therefore investigations of the matter/anti-matter asymmetries are of vital importance for the possible understanding of what could exist beyond the Standard Model. Two powerful particle accelerators, KEKB [1] and PEP-II [2] were built to study CP violation in the B meson decays. B-physics has some distinct advantage over K-physics for studying CP violation. Indeed the asymmetries expected in some decay channel of the B-mesons can be very large. The amount of CP violation in the theoretical framework of the Standard Model is fixed by the area of the Unitary Triangle $(O(10^{-6}))$ and therefore, with the caveat of choosing rather rare decays, the ratio of the difference of the decay width with their sum can easily be of $(O(1))$. Better that any word the following plot (Fig. 1), showing the direct CP asimmetry observed in the decay of $B^0 \rightarrow K^+\pi^-$ makes the case.

THE UNITARY TRIANGLE

At the beginning of the story, looking back into the proposal justifying the investment needed for the construction of the two competing B-factories and their relative detectors, only one goal was clearly outlined. The measurement of the angle β of the Unitary Triangle was deemed to be achievable under (almost) any condition and in particular if it were the one foreseen in the framework of Standard Model. A luminosity of 30 fb^{-1}/year was given as a measure of the success of the machine and a couple of years

CP881, *Cairo International Conference on High Energy Physics (CICHEP II)*,
edited by S. Khalil
© 2007 American Institute of Physics 978-0-7354-0382-6/07/$23.00

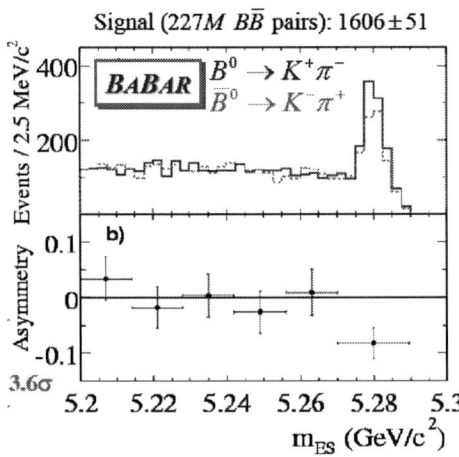

FIGURE 1. Difference (above) and asymmetry (below) in the decay of B^0 and its antiparticle in $K\pi$.

were expected for establishing the existence of CP violation in the B sector.

Very little words were spent on the possibility of measuring the angle α. In particular the necessity of performing the isospin analysis for disentangling the 1/2 and the 3/2 contribution to $B \to \pi\pi$ was clearly advocated and no reliable estimate of the error was possible. Concerning γ there was not even a simulation of the usable channels. It was said that somehow an error in the region of $20°$ could have been achievable by exploiting several (yet obscure) decay modes.

It is somewhat amazing to revisit these statements[3] in the light of what has been indeed achieved.

β: The value averaged out of the two determinations is $sin2\beta = (0.685 \pm 0.032)$. It corresponds to an integrated luminosity of about 0.5 ab^{-1}. What you learn from these numbers are two things. The first is the precision of the measurement (about 5%) and the second is the unbelievable statistics collected by BaBar[4] and Belle[5].

α: Here the channel that at the time of the letter of intent was considered to be the most promising ($B \to \pi\pi$) has shown the fundamental problem of the practical impossibility of performing the isospin analysis since all the branching fractions are small and in particular the $B \to \pi^0\pi^0$ (whose B.R. is around 1.5×10^{-6}) is low enough to yield too few events for the purpose but too big to be neglected and allow to utilize the Quinn-Grossman bound. The channel that has been discovered to serve the purpose better is $B \to \rho\rho$. Put together with $B \to \rho\pi$ that helps in reducing the ambiguities, it allows a determination of α rather precise: $\alpha = (99^{+13}_{-8})°$.

γ: A complete change of paradigm happened in this case. The favoured approach is the one of the analysis of the charged B decays like $B^- \to D^{(*)0}K^-$ followed by the decay $D^0 \to K_S\pi\pi$. The latter is a CP eigenstate and it is common both to D^0 and $\bar{D}^0$. Here the interference appears and a Dalitz analysis of the decay allows to optimize the sensitivity to the angle γ. The limitation so far is the statistics. One has to fight the Cabibbo suppression of one channel with respect to the other. The actual result is $\gamma = (68 \pm 17)°$. It is extraordinary that something that was not even thought realistically measurable by the experiments at the B-factories is now routinely addressed and looks a mere matter of statistics. Fig. 2 shows the status [6] of the Unitary Triangle.

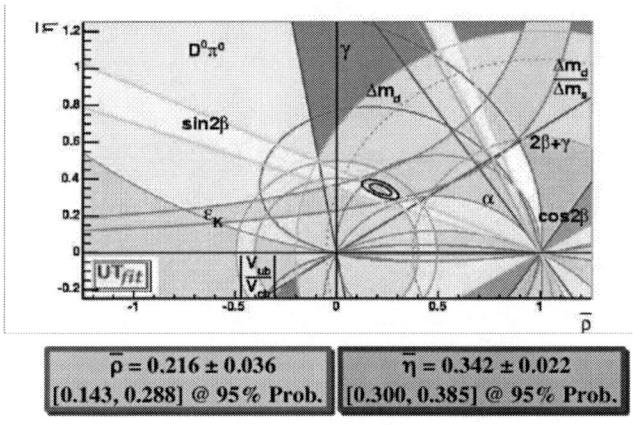

$$\bar{\rho} = 0.216 \pm 0.036$$
$$[0.143, 0.288] @ 95\% \text{ Prob.}$$

$$\bar{\eta} = 0.342 \pm 0.022$$
$$[0.300, 0.385] @ 95\% \text{ Prob.}$$

FIGURE 2. Unitary Triangle as at the time of the conference.

The conclusion of this review of the status of the Unitary Triangle is rather simple. The B-factories have shown a superior ability in dealing with the CP violation in the B sector. They looks a terrific tool for investigating this phenomenon in depth. The only limitation is in the statistics. By the time of the end of their adventure at the end of the decade they will have at most collected four times more data. A significant, but not resolutive, factor two in the errors.

THE CHALLENGE FOR THE FUTURE

As shown in the previous section, the amount of CP violation so far found in the B sector is compatible with the assumption that the Standard Model provides the only source of such a phenomenon. However the level of precision so far reached, and as we discussed even reachable in a near future with the exploitation of the existing facilities, is inadequate for the challenge posed by the measurement of the CP violation in the so-called penguin modes. The most famous of them is $B \to \Phi K$. This decay, due to a loop-mediated $b \to s$ transition is purely due to a penguin diagram, therefore pretty rare. The weak phase of the loop is the same as in the golden channel ($B \to J\Psi K$), namely β. However if any new, yet unknown, particle would replace the Standard Model loop possibly with a different weak phase, one should measure an angle different from the expected one. Fig. 3 shows the status of the art of the measurements in these channels. There is no strong conclusion possible. Some deviation is observed both in the main channel and in the average of all of the likes. However the deviation although suggestive is not statistically significant and most likely it will not become such by reducing the error by only a factor 2 as expected at the end of the day.

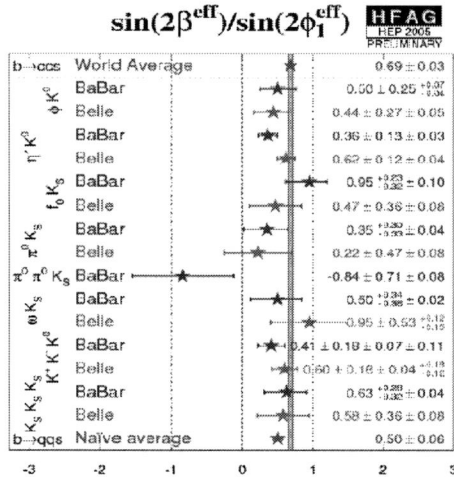

FIGURE 3. The compilation of the measurement of 'sin2β' in the $b \to s$, penguin-dominated channels.

The future of B-Physics lies therefore in the approach to New Physics that one could call 'shake the box' as opposite to 'open the box'. Although the main path to New Physics is definitely the observation of new particles through a peak in a mass spectrum which should be possible at the experiment at the LHC, one shall not dismiss the road of the radiative corrections. As an example to illustrate the power of such a method we will remind the prediction of the top mass through the electro-weak fit at the LEP data. This method could allow to assert the existence of New Physics even in the unlucky case that its scale exceeds the reach of the LHC. In a less extreme scenario, assuming the observation of SUSY at LHC, the studies performed in the sector of the Flavor Physics will allow to explore the flavor structure of this new particles and this could possibly be the only way to do it before (and if) a Linear Collider is built.

THE SUPER e^+e^- OPTION

The terrific power of the e^+e^- option has been demonstrated beyond any doubt by the success of the two existing B-factories. However, for how good they are, they will fail for pure statistical reasons to cope with the challenge posed by the unreasonable resistance of the Standard Model to show any sign of weakness. The name of the future game is no longer to establish the existence of CP violation in B-physics but rather to exploit it to probe New Physics through tiny deviations from expectation. To observe, in other words, the effect of the virtual loops of particles whose mass scale is much higher. This task calls for a a machine with much higher luminosity (to improve errors by a factor 10 you need to increase luminosity by 100) and an experiment able to take the challenge in a manner qualitatively like to the existing ones. There is the possibility, studied at KEK, to stress at its limit the concept of the existing machine. More power, more current, crab crossing will be the main ingredients. It promises a peak luminosity of a few 10^{35}cm^{-2}s^{-1} and a background high but possibly manageable. It is a progress of a factor 10-30 with respect to what exists now.

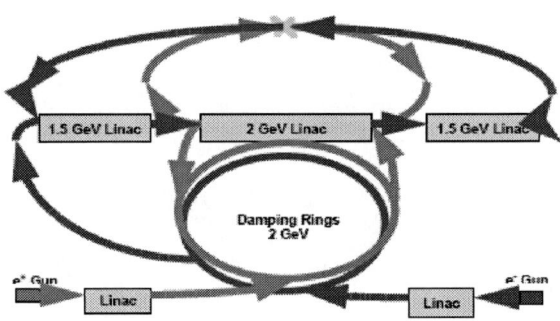

FIGURE 4.   One of the option under discussion for an innovative super-B factory.

Recently a new concept [7] has been put forward. A sort of 'linear' B-factory where the electron and positron will circulate into rings that are very similar to the damping rings of the future International Linear Collider (ILC)[8] and the interaction region will also resemble a lot to the one of the ILC making use of the final focus technique. In such a way a limited current might produce a very high luminosity, possibly in excess of $1 \times 10^{36} \mathrm{cm}^{-2} \mathrm{s}^{-1}$. This possibility is very appealing since it might allow a gain in luminosity of two orders of magnitude with respect to the existing machines while keeping the background low. One of the options under discussion at the time of this workshop is shown in Fig. 4.

As an illustration of the physics potential of a machine capable of supplying an experiment with several dozens of inverse femtobarn we present the precision reachable in some of the channels belonging to the crucial $b \to s$ promised land (Fig. 5).

CPV in Rare Decays		e^+e^- Precision		
Measurement	Goal	3/ab	10/ab	50/ab
$S(B^0 \to \phi K_s^0)$	$\approx 5\%$	16%	8.7%	3.9%
$S(B^0 \to \eta' K_s^0)$	$\approx 5\%$	5.7%	3%	1%
$S(B^0 \to K_s^0 \pi^0)$		8.2%	5%	4%
$S(B^0 \to K_s^0 \pi^0 \gamma)$	SM: $\approx 2\%$	11%	6%	4%
$A_{CP}(b \to s\gamma)$	SM: $\approx 0.5\%$	1.0%	0.5%	0.5%
$A_{CP}(B \to K^*\gamma)$	SM: $\approx 0.5\%$	0.6%	0.3%	0.3%

FIGURE 5. Physics reach for several $b \to s$ transitions as a function of the collected luminosity.

CONCLUSIONS

There certainly is a physics case for the construction of a super-B factory and for an experiment collecting $O(50 \text{ fb}^{-1})$. If this project could be realized together with the developments of the International Linear Collider this would be an additional bonus. The super-B factory whose role will be in principle to cast a light on the couplings of the Super particles through their effect in modifying the branching fractions and the CP asymmetries in B-physics, might lead even to the discovery of New Physics if this sits at a scale unaccessible to LHC.

We look forward for a time when the Unitary Triangle will be known with the precision shown in Fig. 6.

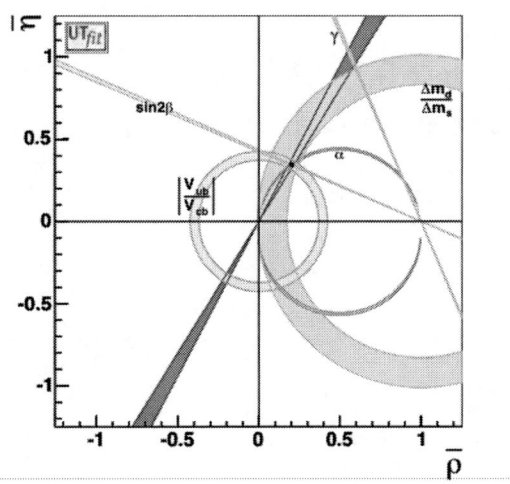

FIGURE 6.   Unitary Triangle as it could look like after the super-B era

REFERENCES

1. S. Kurokawa and E. Kikutani, *Nuclear Instrument and Methods in Physics Research*, **A499**, 1 (2003)
2. *PEPII- An asymmetric B Factory, Conceptual Design Report*, SLAC-418, LBL-5379, 1993.
3. *BaBar Physics Book*, SLAC-R-504, 1999.
4. B. Aubert et al., *Nuclear Instrument and Methods in Physics Research*, **A479**, 1 (2002)
5. A. Abashian et al., *Nuclear Instrument and Methods in Physics Research*, **A479**, 117 (2002)
6. http://utfit.roma1.infn.it/, http://ckmfitter.in2p3.fr
7. J. Albert et al., *SuperB: A linear high luminosity B Factory*, INFN-AE-05-08, 2005.
8. http://linearcollider.org/

Status of the Cabibbo-Kobayashi-Maskawa matrix and Unitarity Triangle fits

M. Bona*, M. Ciuchini†, E. Franco**, V. Lubicz‡, G. Martinelli**,
F. Parodi§, M. Pierini¶, P. Roudeau‖, C. Schiavi§, L. Silvestrini**,
V. Sordini‖, A. Stocchi‖ and V. Vagnoni††

*Lab. d'Annecy-le-Vieux de Physique des Particules LAPP, IN2P3/CNRS, Université de Savoie
†Dip. di Fisica, Università di Roma Tre and INFN, Sez. di Roma III, Roma, Italy
**Dip. di Fisica, Università di Roma "La Sapienza" and INFN, Sez. di Roma, Roma, Italy
‡Dip. di Fisica, Università di Roma Tre and INFN, Sez. di Roma III, I-00146 Roma, Italy
§Dip. di Fisica, Università di Genova and INFN, Genova, Italy
¶Department of Physics, University of Wisconsin, Madison, USA
‖Lab. de l'Accélérateur Linéaire, IN2P3-CNRS et Univ. de Paris-Sud, Orsay Cedex, France
††Corresponding author, INFN, Sez. di Bologna, Bologna, Italy

Abstract. The status of the Unitarity Triangle analysis realized by the UT*fit* Collaboration is presented. The most recent determinations of theoretical and experimental parameters are used in order to over-constrain the apex of the Unitarity Triangle in the Standard Model. In addition, we present the analysis of the Unitarity Triangle beyond the Standard Model, by parametrizing New Physics contributions in $\Delta F = 2$ processes. With the new measurements from the Tevatron, namely the mass difference Δm_s, the width difference $\Delta \Gamma_s$ and the di-muon asymmetry, it is possible to establish significant bounds on New Physics parameters also in the B_s sector. The results and the plots presented in this paper can be found at the URL http://www.utfit.org, where they are continuously kept up-to-date.

Keywords: Cabibbo-Kobayashi-Maskawa matrix, Unitarity Triangle, New Physics
PACS: 13.25.Hw, 12.15.Hh

INTRODUCTION

The analysis of the Unitarity Triangle (UT) and CP violation represents at the moment one of the key places where the Standard Model (SM) can be tested with outstanding precision, hence it provides a great opportunity to search and discover New Physics (NP) effects beyond the SM. During the recent years an unprecedented amount of new measurements, thanks to the successes of the B factories and the Tevatron, has been made available to the UT analysis, allowing for a precise determination of the parameters of the Cabibbo-Kobayashi-Maskawa (CKM) matrix.

The measurements used in the Standard Model UT analysis include the sides of the UT, namely $|V_{ub}|/|V_{cb}|$ from semileptonic decays, and the magnitudes of the mixing amplitudes of the neutral B mesons Δm_d and Δm_s, as well as ε_K, which parametrizes the indirect CP violation in the neutral Kaon system. Besides these quantities, which require the employment of non-perturbative hadronic parameters coming from lattice computations in order to be obtained, as in the case of $|V_{ub}|/|V_{cb}|$, or related to the CKM factors $\bar{\rho}$ and $\bar{\eta}$ in the other cases, several determinations of the UT angles α, β and γ

CP881, *Cairo International Conference on High Energy Physics (CICHEP II)*,
edited by S. Khalil

not requiring aid from the lattice are available.

Such an abundance of measurements, which can be basically unaffected from NP as in the case of $|V_{ub}|/|V_{cb}|$ and of the tree-level determination of γ from $B \to D^{(*)} K^{(*)}$, or alternatively affected by the presence of NP contributions in different ways, allows for a simultaneous determination of SM and NP parameters in the flavour sector. As it will be shown, the UT analysis in the presence of NP has reached nowadays an accuracy comparable to the SM analysis, thus providing very stringent constraints on NP contributions to $\Delta F = 2$ processes.

The **UT***fit* Collaboration has already published a series of papers, where the interested reader can find all the details not reported in this brief document [1, 2, 3, 4].

In the following we will show first the results of the SM analysis, i.e. assuming the validity of the SM. Then, upon the introduction of a generalized parametrization of $\Delta F = 2$ processes, we will discuss the UT fit in the presence of NP, thus showing the bounds so obtained on NP quantities. By restricting the NP scenario to Minimal Flavour Violation (MFV) models, both in the large and small $\tan \beta$ regimes, we will estimate the corresponding NP scales which are probed by means of the currently available experimental information.

STANDARD MODEL ANALYSIS

Before the advent of the B factories, UT fits were performed by only using measurements of the sides and indirect CP violation in neutral Kaon system. Several determinations of UT angles, thanks to the results coming from the B factories, allowed for a dramatical improvement of our knowledge of the UT.

Several measurements at present bound the range of the $\bar{\rho}$ and $\bar{\eta}$ parameters. Here below a brief description of the most relevant ones in use in our UT fit are described:

- The rates of charmed and charmless semileptonic B decays decays which allow to measure the ratio $|V_{ub}| / |V_{cb}|$.
- The $B_d^0 - \bar{B}_d^0$ mass difference between the light and heavy mass eigenstates of the $B_d^0 - \bar{B}_d^0$ system Δm_d.
- The mass difference of the $B_s^0 - \bar{B}_s^0$ system Δm_s, compared to Δm_d, $\Delta m_d / \Delta m_s$.
- The ε_K parameter, which measures CP violation in the neutral kaon system.
- $\sin 2\beta$ from the golden modes $B_d^0 \to c\bar{c} K^0$, that can be determined almost without hadronic uncertainties.
- The angle α, that can be obtained from the $B \to \pi\pi$ and $B \to \rho\rho$ decays, assuming the SU(2) flavour symmetry and neglecting the contributions of electroweak penguins. It can also be obtained using a time-dependent analysis of $B \to (\rho\pi)^0$ decays on the Dalitz plane. Fig. 1 shows the combination of the BaBar and Belle results including all the three methods, as well as and the bound on the $\bar{\rho} - \bar{\eta}$ plane. We obtain:

$$\alpha = ([7, 9]^\circ \cup [80, 107]^\circ \cup [159, 177]^\circ \text{ @95\%}) \tag{1}$$

The multimodal shape of the α p.d.f. is due to the intrinsic ambiguities of the methods used for its determination. The SM solution corresponds to the central

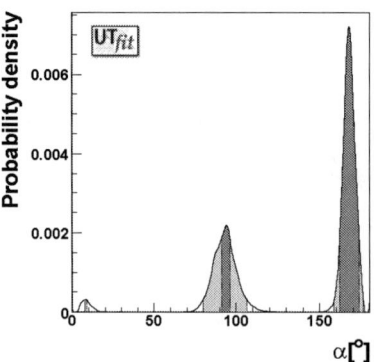
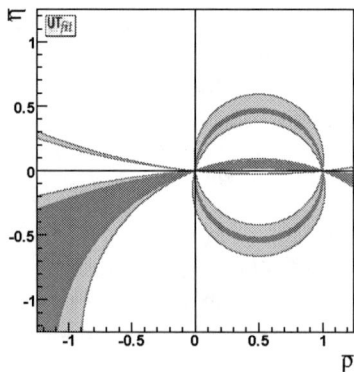

FIGURE 1. Left plot: p.d.f. for α resulting from the combination of the BaBar and Belle results. Right plot: corresponding bound on the $\bar{\rho} - \bar{\eta}$ plane, 68% and 95% confidence regions.

peak, and just considering it (e.g., using other constraints to discard the non-SM solutions) we get $\alpha = (92 \pm 7)^\circ$. The angle γ that can be extracted from the tree-level decays $B \to DK$, using the fact that a charged B can decay into a $D^0(\overline{D}^0)K$ final state via a $V_{cb}(V_{ub})$ mediated process. CP violation occurs if the D^0 and the $\overline{D}^0$ decay to the same final state. The same argument can be applied to $B \to D^*K$ and $B \to D^{(*)}K^*$ decays. Three methods have been proposed:

- The Gronau-London-Wyler method (GLW). It consists in reconstructing the neutral D meson in a CP eigenstate: $B^\pm \to D^0_{CP\pm}K^\pm$, where $D^0_{CP\pm}$ are the CP eigenstates of the D meson.
- The Atwood-Dunietz-Soni method (ADS). It consists in forcing the $\bar{D}^0$ (D^0) meson, coming from the Cabibbo-suppressed (Cabibbo-allowed) $b \to u$ ($b \to c$) transition to decay into the Cabibbo-allowed (Cabibbo-suppressed) $K\pi$ final state. In this way, one can look at the interference between two amplitudes.
- Dalitz method. It consists in studying the interference between the $b \to u$ and the $b \to c$ transitions using the Dalitz plot of D mesons reconstructed into three-body final states (such as $D^0 \to K_s\pi^-\pi^+$). The advantage of this method is that the full sub-resonance structure of the three-body decay is considered, including interferences such as those used for GLW and ADS methods plus additional interferences due to the overlap between broad resonances in some regions of the Dalitz plot.

Fig. 2 shows the combination of the BaBar and Belle results including all the three methods, as well as and the bound on the $\bar{\rho} - \bar{\eta}$ plane.
We obtain:

$$\gamma = (78 \pm 30)^\circ \cup (-102 \pm 30)^\circ \tag{2}$$

As in the case of the α angle, the p.d.f. for γ is multimodal due to a remaining two-fold ambiguity.

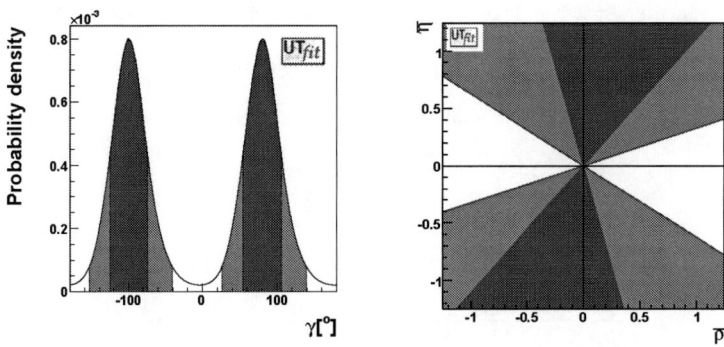

FIGURE 2. Left plot: p.d.f. for γ resulting from the combination of the BaBar and Belle results. Right plot: corresponding bound on the $\bar{\rho} - \bar{\eta}$ plane, 68% and 95% confidence regions.

TABLE 1. Values of the relevant input quantities used in the UT fit. The Gaussian and the flat contributions to the uncertainty are given in the third and fourth columns respectively (for details on the statistical treatment see [5]).

Parameter	Value	Gaussian (σ)	Uniform (half-width)
λ	0.2258	0.0014	-
$\|V_{cb}\|$(excl.)	41.3×10^{-3}	1.0×10^{-3}	1.8×10^{-3}
$\|V_{cb}\|$(incl.)	41.6×10^{-3}	0.7×10^{-3}	-
$\|V_{ub}\|$(excl.)	35.0×10^{-4}	4.0×10^{-4}	-
$\|V_{ub}\|$(incl.)	44.9×10^{-4}	3.3×10^{-4}	-
Δm_d	0.507 ps^{-1}	0.005 ps^{-1}	-
Δm_s	17.33 ps^{-1}	$^{+0.33}_{-0.18} \pm 0.07 \text{ ps}^{-1}$	-
$f_{B_s}\sqrt{\hat{B}_{B_s}}$	262 MeV	35 MeV	-
$\xi = \dfrac{f_{B_s}\sqrt{\hat{B}_{B_s}}}{f_{B_d}\sqrt{\hat{B}_{B_d}}}$	1.23	0.06	-
$\hat{B}_K$	0.79	0.04	0.08
ε_K	2.280×10^{-3}	0.013×10^{-3}	-
f_K	0.160 GeV	fixed	
Δm_K	$0.5301 \times 10^{-2} \text{ ps}^{-1}$	fixed	
$\sin 2\beta$	0.675	0.026	-
$\overline{m}_t$	163.8 GeV	3.2 GeV	-
$\overline{m}_b$	4.21 GeV	0.08 GeV	-
$\overline{m}_c$	1.3 GeV	0.1 GeV	-
$\alpha_s(M_Z)$	0.119	0.003	-

In Tab. 1 we summarize the values of the relevant input parameters used in the fit.

TABLE 2. Determination of UT parameters from the SM fit.

Parameter	Output	Parameter	Output
$\bar{\rho}$	0.162 ± 0.029	$\bar{\eta}$	0.342 ± 0.017
$\alpha[^\circ]$	92.9 ± 4.4	$\beta[^\circ]$	22.1 ± 0.9
$\gamma[^\circ]$	64.6 ± 4.4	Δm_s [ps^{-1}]	17.4 ± 0.3
$\sin 2\beta$	0.698 ± 0.023	Imλ_t [10^{-5}]	13.7 ± 0.6
$V_{ub}[10^{-3}]$	3.67 ± 0.15	$V_{cb}[10^{-2}]$	4.16 ± 0.06
$V_{td}[10^{-3}]$	8.51 ± 0.28	$|V_{td}/V_{ts}|$	0.208 ± 0.007
R_b	0.379 ± 0.015	R_t	0.904 ± 0.030

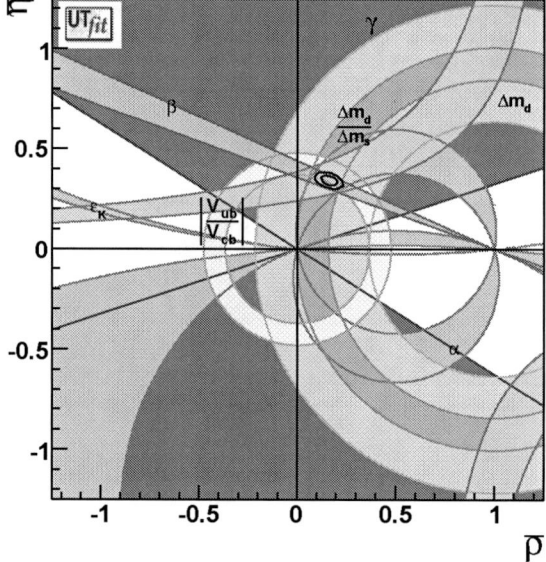

FIGURE 3. Determination of $\bar{\rho}$ and $\bar{\eta}$ from constraints on $|V_{ub}|/|V_{cb}|$, Δm_d, Δm_s, ε_K, β, γ, and α. 68% and 95% total probability contours are shown, together with 95% probability regions from the individual constraints.

The output of the SM fit including all the constraints is reported in Tab. 2 and in Fig. 3. By looking in more detail at Fig. 3, it is interesting to note that the 95% confidence regions depicted by the $\sin 2\beta$ and $|V_{ub}|/|V_{cb}|$ constraints, being them two of the most precise ones used in the fit, show just a bare agreement. In particular, in our analysis we find that while the experimental value of $\sin 2\beta$ is in good agreement with the rest of the fit, the same does not hold for $|V_{ub}|/|V_{cb}|$, which is rather on the higher side. It can be shown that this is due to a large value of the inclusive determination of $|V_{ub}|$. Unless this discrepancy should be considered as a hint of NP, it has to be explained by the uncertainties of the theoretical approaches. It is worth recalling that the value of $|V_{ub}|$ that is extracted from the experiments also relies on non perturbative hadronic quantities.

Nevertheless, it should be remarked that the UT analysis has shown so far an impressive success of the CKM picture in describing CP violation in the SM.

NEW PHYSICS ANALYSIS

Starting from a tree-level determination of $\bar{\rho}$ and $\bar{\eta}$, we perform the UT analysis in general extensions of the SM with arbitrary NP contributions to $|\Delta F| = 2$ processes. We will show that the recent measurements from B factories and the Tevatron allow us to determine with good precision the shape of the UT even in the presence of NP, and to draw significant bounds on NP contributions to $|\Delta F| = 2$ processes.

In particular, three results have dramatically improved the knowledge of the UT beyond the SM, extending our view in the B_s sector. First, the measurement of Δm_s from CDF [6], which reduces the uncertainty of the SM fit and has a strong impact on the determination of the Universal Unitarity Triangle (UUT) [7] in models with Minimal Flavour Violation (MFV) [8, 9, 10, 11, 12, 13, 14]. Moreover, it allows for the first time to put a bound on NP corrections to the magnitude of the B_s mixing amplitude. As far as the B_s mixing phase is concerned, useful information to put relevant bounds can be extracted from the measurement of the dimuon asymmetry in $p\bar{p}$ collisions by the D0 experiment [15] and of the B_s width difference.

Assuming that NP enters observables in the flavour sector only at the loop level. For this reason the constraints from V_{ub}/V_{cb} and γ from the interference between $b \to c$ and $b \to u$ transitions to DK final states are basically NP-free. Being the mixing processes described by a single amplitude, they can be parameterized without loss of generality in terms of two parameters quantifying the difference of the amplitude with respect to the SM one [16, 17, 18, 19, 20]. In the case of $B_q^0 - \bar{B}_q^0$ mixing we define

$$C_{B_q} e^{2i\phi_{B_q}} = \frac{\langle B_q^0 | H_{\text{eff}}^{\text{full}} | \bar{B}_q^0 \rangle}{\langle B_q^0 | H_{\text{eff}}^{\text{SM}} | \bar{B}_q^0 \rangle}, \qquad (q = d, s) \qquad (3)$$

where $H_{\text{eff}}^{\text{SM}}$ includes only the SM box diagrams, while $H_{\text{eff}}^{\text{full}}$ includes also the NP contributions. In the absence of NP, $C_{B_q} = 1$ and $\phi_{B_q} = 0$. The experimental quantities determined from the $B_q^0 - \bar{B}_q^0$ mixings are related to their SM counterparts and the NP parameters by the following relations:

$$\Delta m_q^{\text{exp}} = C_{B_q} \Delta m_q^{\text{SM}}, \quad \beta^{\text{exp}} = \beta^{\text{SM}} + \phi_{B_d}, \quad \alpha^{\text{exp}} = \alpha^{\text{SM}} - \phi_{B_d}, \quad \beta_s^{\text{exp}} = \beta_s^{\text{SM}} - \phi_{B_s} \quad (4)$$

For the $K^0 - \bar{K}^0$ mixing is instead convenient to introduce a single parameter:

$$C_{\varepsilon_K} = \frac{\text{Im}[\langle K^0 | H_{\text{eff}}^{\text{full}} | \bar{K}^0 \rangle]}{\text{Im}[\langle K^0 | H_{\text{eff}}^{\text{SM}} | \bar{K}^0 \rangle]}. \qquad (5)$$

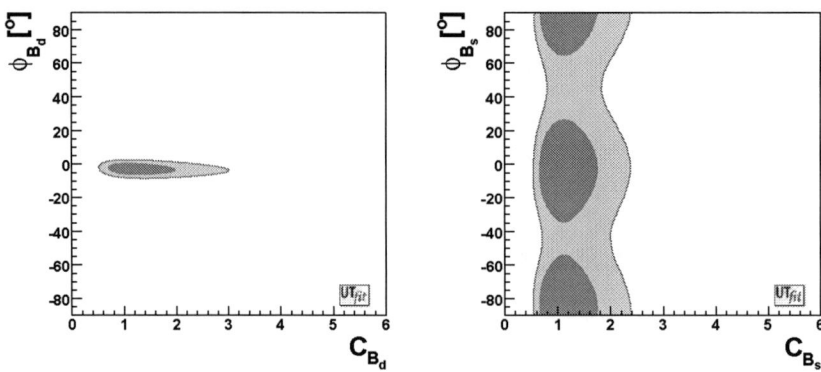

FIGURE 4. Constraints on the $C_{B_q} - \phi_{B_q}$ planes, from the NP generalized UT fit: 68% and 95% confidence regions.

TABLE 3. Determination of UT and NP parameters from the NP generalized fit.

Parameter	Output	Parameter	Output
C_{B_d}	1.25 ± 0.43	$\phi_{B_d}[°]$	-2.9 ± 2.0
C_{B_s}	1.13 ± 0.35	$\phi_{B_s}[°]$	$(-3 \pm 19) \cup (94 \pm 19)$
C_{ε_K}	0.92 ± 0.16		
$\bar{\rho}$	0.20 ± 0.06	$\bar{\eta}$	0.36 ± 0.04
$\alpha[°]$	93 ± 9	$\beta[°]$	24 ± 2
$\gamma[°]$	62 ± 9	$\mathrm{Im}\lambda_t[10^{-5}]$	14.6 ± 1.4
$V_{ub}[10^{-3}]$	4.01 ± 0.25	$V_{cb}[10^{-2}]$	4.15 ± 0.07
$V_{td}[10^{-3}]$	8.3 ± 0.6	$|V_{td}/V_{ts}|$	0.203 ± 0.015
R_b	0.42 ± 0.03	R_t	0.89 ± 0.06
$\sin 2\beta$	0.75 ± 0.04	$\sin 2\beta_s$	0.039 ± 0.004

which implies the following relation for the measured value of ε_K:

$$\varepsilon_K^{\mathrm{exp}} = C_{\varepsilon_K} \, \varepsilon_K^{\mathrm{SM}} \tag{6}$$

In fact, Δm_K is not considered because the long distance effects are not well under control. With these definitions, NP effects which enter the present analysis are parameterized in terms of 5 real quantities, C_{B_d}, ϕ_{B_d}, C_{B_s}, ϕ_{B_s} and C_{ε_K}.

The results of the fit are summarized in Tab. 3. The bounds on the two ϕ_B vs C_B planes are given in Fig. 4. The distributions for C_{B_q}, ϕ_{B_q} and C_{ε_K} are shown in Fig. 5, and in the same figure also the fit result in the $\bar{\rho} - \bar{\eta}$ plane is depicted. We see that the *non-standard* solution for the UT with its vertex in the third quadrant, which was present in

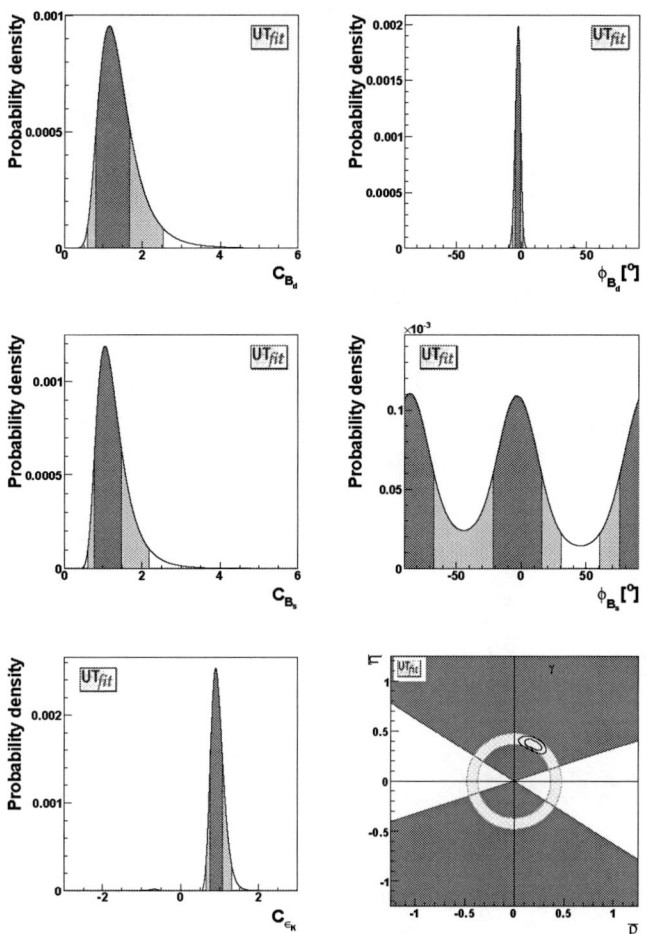

FIGURE 5. Constraints on ϕ_{B_q}, C_{B_q} and C_{ε_K} coming from the NP generalized analysis. The bottom-right plot shows instead the 68% and 95% confidence regions in the $\bar{\rho} - \bar{\eta}$ plane as resulting from the NP generalized fit, superimposed to the 95% confidence regions determined by the $|V_{ub}|/|V_{cb}|$ and γ constraints only.

a previous analysis [2], is now absent thanks to the improved value of $A_{\rm SL}$ by the BaBar Collaboration and to the measurement of $A_{\rm CH}$ by the D0 Collaboration. Furthermore, the measurement of Δm_s strongly constrains C_{B_s}, so that C_{B_s} is already known better than C_{B_d}. Finally, $A_{\rm CH}$ and $\Delta\Gamma_s$ provide the first relevant constraints on ϕ_{B_s}.

UNIVERSAL UNITARITY TRIANGLE ANALYSIS

In the context of MFV extensions of the SM, it is possible to use the so called UUT construction in order to determine the parameters of the CKM matrix independently of NP effects. To this end one has to use all the constraints from tree-level processes and from the angle measurements, as well as the $\Delta m_d/\Delta m_s$ ratio, which in MFV scenarios are NP-free. Instead, ε_K, Δm_d and Δm_s may receive NP contributions, because of the shifts δS_0^K and δS_0^B of the Inami-Lim functions in the K-$\bar{K}$ and $B_{d,s}$-$\bar{B}_{d,s}$ mixings. With only one Higgs doublet or at small $\tan\beta$ these two contributions are forced to be equal. Instead, for large $\tan\beta$, the two quantities are in general different. In both cases, one can use the output of the UUT given in Tab. 4 and in Fig. 6 to constrain $\delta S_0^{K,B}$. We get $\delta S_0 = \delta S_0^K = \delta S_0^B = -0.12 \pm 0.32$ for small $\tan\beta$, while for large $\tan\beta$ we obtain $\delta S_0^B = 0.26 \pm 0.72$ and $\delta S_0^K = -0.18 \pm 0.38$. The output distributions for δS_0^B and δS_0^K are also given in Fig. 6. Using the procedure detailed in [13], these bounds can be translated into lower bounds on the MFV scale Λ:

$$\Lambda \; > \; 5.9 \text{ TeV @95\% Prob. for small } \tan\beta$$
$$\Lambda \; > \; 5.4 \text{ TeV @95\% Prob. for large } \tan\beta \tag{7}$$

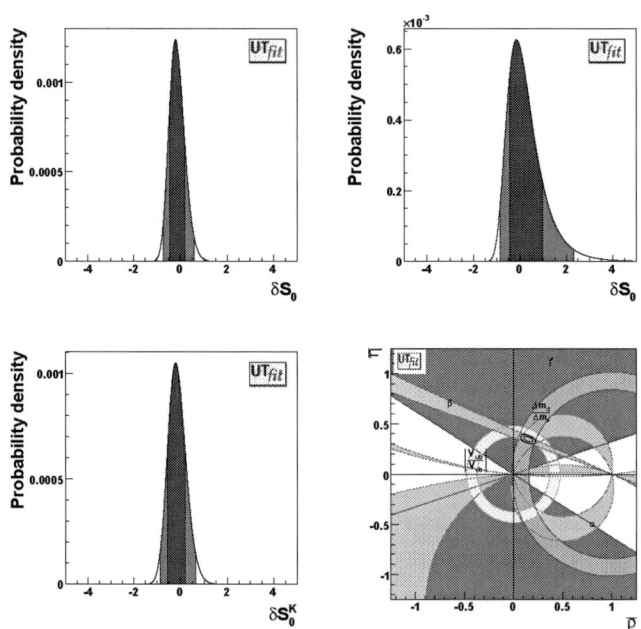

FIGURE 6. P.d.f. of δS_0, δS_0^B and δS_0^K (see the text for details). Bottom-right: Determination of $\bar{\rho}$ and $\bar{\eta}$ from the constraints on α, β, γ, $|V_{ub}/V_{cb}|$, and $\Delta m_d/\Delta m_s$ (UUT fit).

TABLE 4. Determination of UUT parameters from the constraints on α, β, γ, $|V_{ub}/V_{cb}|$, and $\Delta m_d/\Delta m_s$ (UUT fit).

Parameter	Output	Parameter	Output
$\overline{\rho}$	0.154 ± 0.032	$\overline{\eta}$	0.347 ± 0.018
$\alpha[°]$	91 ± 5	$\beta[°]$	22.2 ± 0.9
$\gamma[°]$	66 ± 5	$\sin 2\beta_s$	0.037 ± 0.002

CONCLUSIONS

In this paper we have presented an updated analysis of the UT in the SM, using all the relevant measurements available from the B factories and the Tevatron, in particular the recent measurement of δm_s from CDF.

Then, we have performed a model-independent analysis of the UT in general extensions of the SM with loop-mediated contributions to FCNC processes. We have shown how the great number of measurements nowadays available allow for a simultaneous determination of the CKM parameters, together with the NP contributions to $|\Delta F| = 2$ processes in the K^0, B^0 and B^0_s sectors.

Finally, we have analyzed in detail the UUT, showing that it is possible to constrain the additional NP parameters very accurately. In this way, we have been able to probe the NP scale in MFV scenarios in the large and small $\tan\beta$ limits up to about 5-6 TeV, to be compared with the SM reference scale of 2.4 TeV.

REFERENCES

1. M. Bona *et al.* [UTfit Collaboration], JHEP **0507**, 028 (2005) [arXiv:hep-ph/0501199].
2. M. Bona *et al.* [UTfit Collaboration], JHEP **0603**, 080 (2006) [arXiv:hep-ph/0509219].
3. M. Bona *et al.* [UTfit Collaboration], arXiv:hep-ph/0606167.
4. M. Bona *et al.* [UTfit Collaboration], arXiv:hep-ph/0605213.
5. M. Ciuchini *et al.*, JHEP **0107** (2001) 013 [arXiv:hep-ph/0012308].
6. [CDF - Run II Collaboration], Phys. Rev. Lett. **97** (2006) 062003 [arXiv:hep-ex/0606027].
7. A. J. Buras *et al.*, Phys. Lett. B **500**, 161 (2001) [arXiv:hep-ph/0007085].
8. E. Gabrielli and G. F. Giudice, Nucl. Phys. B **433**, 3 (1995) [Erratum-ibid. B **507**, 549 (1997)] [arXiv:hep-lat/9407029];
9. M. Misiak, S. Pokorski and J. Rosiek, Adv. Ser. Direct. High Energy Phys. **15**, 795 (1998) [arXiv:hep-ph/9703442];
10. M. Ciuchini *et al.*, Nucl. Phys. B **534**, 3 (1998) [arXiv:hep-ph/9806308];
11. C. Bobeth *et al.*, Nucl. Phys. B **726**, 252 (2005) [arXiv:hep-ph/0505110].
12. M. Blanke *et al.*, arXiv:hep-ph/0604057.
13. G. D'Ambrosio *et al.*, Nucl. Phys. B **645**, 155 (2002) [arXiv:hep-ph/0207036].
14. G. Isidori and P. Paradisi, arXiv:hep-ph/0605012.
15. http://www-d0.fnal.gov/Run2Physics/WWW/results/prelim/B/B29/B29.pdf
16. J. M. Soares and L. Wolfenstein, Phys. Rev. D **47**, 1021 (1993);
17. N. G. Deshpande, B. Dutta and S. Oh, Phys. Rev. Lett. **77**, 4499 (1996) [arXiv:hep-ph/9608231];
18. J. P. Silva and L. Wolfenstein, Phys. Rev. D **55**, 5331 (1997) [arXiv:hep-ph/9610208];
19. A. G. Cohen *et al.*, Phys. Rev. Lett. **78**, 2300 (1997) [arXiv:hep-ph/9610252].
20. Y. Grossman, Y. Nir and M. P. Worah, Phys. Lett. B **407**, 307 (1997) [arXiv:hep-ph/9704287].

Ultra High Energy Cosmic Rays and the Pierre Auger Observatory

Michael Unger for the Pierre Auger Collaboration

Forschungszentrum Karlsruhe, Institut fuer Kernphysik, Postfach 3640, D-76021 Karlsruhe

Abstract. We present first physics results from the ultra high energy cosmic ray data collected with the southern Pierre Auger Observatory: A search for anisotropies near the direction of the Galactic Centre, a limit on the photon fraction in cosmic rays with energies above 10^{19} eV and an estimate of the differential energy spectrum above $3 \cdot 10^{18}$ eV.

Keywords: Cosmic Rays, UHECR, Pierre Auger Observatory
PACS: 96.50.S, 13.85.Tp, 95.55.Vj , 98.70.Sa

INTRODUCTION

Over forty years after the first measurement of a cosmic ray shower with an energy of 10^{20} eV [1] the nature and origin of ultra high energy cosmic rays (UHECRs) is still unknown (see [2] for a recent review on this topic). The major open experimental questions are the following:

Is there an end to the cosmic ray spectrum?

While traversing the astronomical distances from their source to earth, UHECRs are expected to loose energy due to interactions with the photons of the cosmic microwave background radiation. For primary protons, the centre of mass energy exceeds the threshold for pion production above $5 \cdot 10^{19}$ eV and correspondingly a strong suppression of the flux from distant UHECR sources is expected above this so-called GZK-cutoff energy [3, 4]. The flux measurements at the highest energies from the two experiments with the so far largest exposures, AGASA [5] and HiRes [6], do however not agree with each other (cf. Fig. 1). Whereas the HiRes data seem to confirm the GZK effect, the AGASA results do not show a clear sign of a flux suppression even above 10^{20} eV.

Where do UHECRs come from?

General arguments about the size and magnetic field of the region where the UHECRs are accelerated [7] leave only few known objects like active galactic nuclei as candidates for the astrophysical origin of UHECRs. If the flux of cosmic rays indeed shows a GZK feature at the highest energies, this implies that their sources must be within our cosmological neighbourhood at distances smaller than about 50 Mpc. Does this proximity open a window for charged particle astronomy to identify the accelerators of UHECRs? Current data is, again, inconclusive: The AGASA collaboration claims a significant clustering of the arrival directions of the highest energy events [8], which is,

CP881, *Cairo International Conference on High Energy Physics (CICHEP II),*
edited by S. Khalil

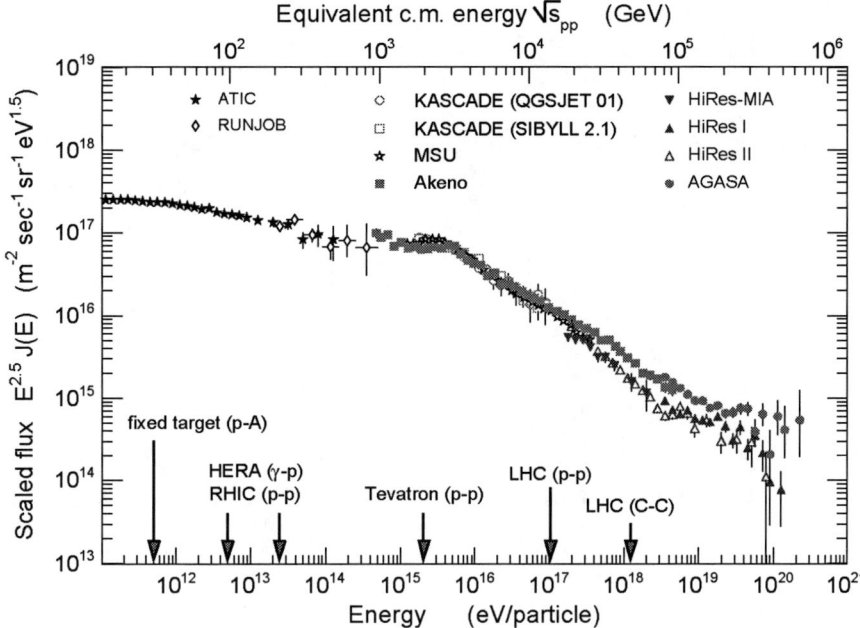

FIGURE 1. Scaled cosmic ray particle flux as a function of particle energy. The upper x-axis shows the equivalent centre of mass energy for collisions of cosmic rays with nuclei in the atmosphere of the earth. Arrows indicate centre of mass energies of man-made accelerators (taken from [14]).

however, not confirmed by a combined analysis of the HiRes and AGASA data [9].

What is the mass composition of UHECR particles?

Alternative models to the 'bottom-up' acceleration of UHECRs in astrophysical sources postulate a 'top-down' production in the decay of super-heavy relic particles or topological defects (see [10] for a recent review). A common prediction of the 'top-down' models is a substantial fraction of primary photons at the highest energies and therefore a good experimental sensitivity to the chemical composition of UHECRs is of particular importance to verify or refute these theories.

At lower energies, around the so-called 'ankle' between 10^{18} and 10^{19} eV (cf. Fig. 1), a measurement of the mass composition is vital to confirm the transition from galactic to extragalactic cosmic rays predicted by 'bottom-up' scenarios [11, 12, 13].

Do we understand air shower physics at ultra high energies?

As the flux of UHECRs is extremely low (one particle per century and square kilometre above 10^{20} eV) they can not be measured directly, but one relies on the indirect detection via extensive air showers induced by the collision of UHECRs with the nuclei of the atmosphere. The centre of mass energies that occur in these collisions are sketched in Fig. 1, and it can be seen that they exceed even the forthcoming LHC energies by far.

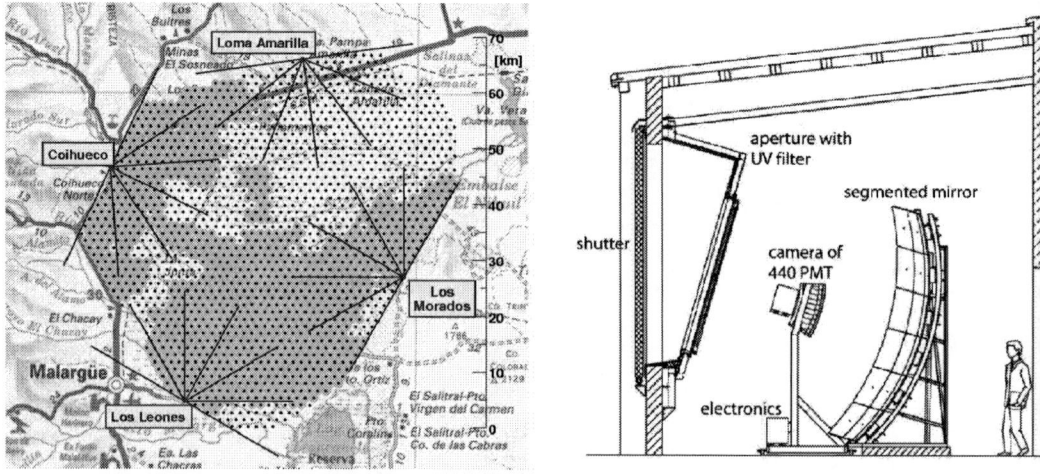

FIGURE 2. Left: Layout of the Pierre Auger Observatory. Black dots show the positions of surface detector tanks, the yellow areas denote tanks currently (July 2006) in operation. The field of view of the fluorescence telescopes are indicated as straight lines. Right: Schematic view of a fluorescence telescope.

While this opens the possibility to study hadronic processes at energies inaccessible to current colliders, it also leads to uncertainties in the estimated properties of the primary particle if these are inferred using non-perturbative hadronic interaction models that can not be checked with accelerator data [15].

THE PIERRE AUGER OBSERVATORY

The Pierre Auger Observatory is aimed to provide a large sample of high quality UHECR data to help to answer these questions. It will consist of two giant air shower detectors located in the northern and southern hemisphere.

The southern detector [16] is currently under construction in Malargüe, Argentina and due to be finished at the beginning of 2007. It is the first apparatus which allows for a combined measurement of the longitudinal energy deposit of a cosmic ray shower in the atmosphere and its lateral particle density distribution on the ground.

The layout of the experiment is sketched in Fig. 2. The surface array [17] will cover an area of 3000 km^2 on which 1600 water Cherenkov detectors will be deployed. These are cylindrical tanks filled with 12 tons of ultra-pure water in which the Cherenkov light produced by traversing secondary cosmic ray particles is detected with three photomultiplier tubes. Presently (July 2006) about 1000 tanks are in operation and taking data.

Four fluorescence detectors (FDs) [18] are placed at the sides of the array. Each houses six Schmidt telescopes with a 30° × 30° field of view (cf. Fig. 2). Currently three fluorescence detectors are taking data in clear new- to half-moon nights, with a corresponding duty cycle between 10 and 15%. The fourth telescope building is under construction and will be finished at the end of this year.

From this hybrid, i.e. SD and FD, approach [19] both sub-detectors gain substantially in precision: On the one hand, the energy scale of the SD can be determined from events measured simultaneously with the fluorescence detector. This has the advantage, that no shower simulations are needed to determine the SD energy calibration, and thus the energy scale is unaffected by systematic uncertainties inherent to hadronic interaction models at ultra high energies.

On the other hand, the geometrical reconstruction of the FD is much more precise, when even a single triggered surface detector tank can be used to constrain the intercept of the shower track with the ground plane.

Moreover, the measurement redundancy of the independent detection of the same shower in the surface detector and up to four fluorescence telescopes, can be used to infer the trigger efficiencies and measurement resolutions of the sub-detectors.

With the southern site nearing its completion the collaboration has selected southeast Colorado, USA, to host the northern detector and started to perform related R&D work.

FIRST RESULTS

Although the Pierre Auger Observatory is still in the phase of construction, it is routinely taking high quality data with the already installed detector components. In the following sections we will discuss physics results obtained from this very first data set.

Anisotropy Studies around the Galactic Centre

The galactic centre (GC) region provides an attractive candidate for anisotropy studies with the Pierre Auger Observatory. On the one hand, past observations from the AGASA [20] and SUGAR [21] experiments indicate an excess of cosmic rays from the GC at energies above 10^{18} eV. On the other hand, since the galactic centre harbours a very massive black hole, it provides a natural candidate for a cosmic ray accelerator to very high energies. Finally, its astronomically close distance to the earth could allow neutrons to reach us without a directional distortion due to the galactic magnetic field, since their mean decay length $\gamma c \tau$ happens to be almost exactly the earth-GC distance at 10^{18} eV.

The analysis [22, 23] was performed with both, surface and fluorescence detector data. The latter provides a better angular resolution (typically below than $1°$), whereas the SD resolution is $2.2°$ for events with only three tanks improving to below $1.4°$ for five or more tanks [24]. This somewhat worse resolution of the surface detector is however compensated by a much larger number of events due to its nearly 100% duty cycle, which is the reason we will concentrate on the SD results in the following.

A key ingredient to the anisotropy study is a precise determination of the background from isotropic cosmic rays. This is studied with two methods, where one employs a semi-analytical method to calculated the exposure of the sky cell around the GC, whereas the other method uses the shuffling technique, in which the background is obtained by the data itself via randomised arrival times and azimuth angles. Both methods

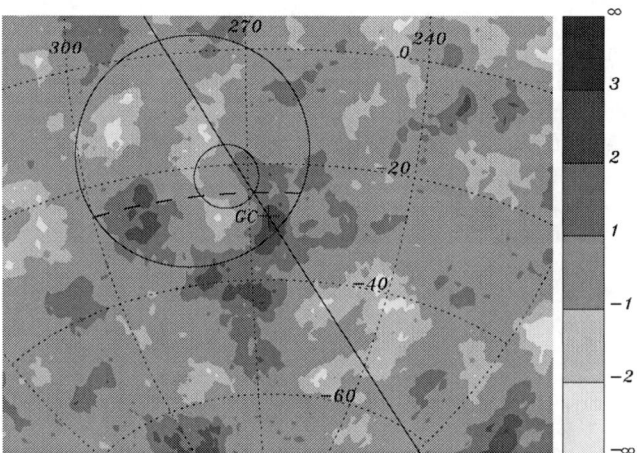

FIGURE 3. Map of overdensity significances near the galactic centre. The galactic plane is shown as a solid line, the galactic centre as a cross, the SUGAR signal region as a small circle and the AGASA signal region as a large circle, where the dashed line indicates the field of view boundary of AGASA.

agree within 0.5%, which is much smaller than the signal expected from the aforementioned anisotropy claims.

These claims are tested by comparing the number of events and background in the respective 'signal regions' (cf. Fig. 3) and energy ranges. As a result we find $n_{obs}/n_{exp} = 0.98 \pm 0.02$ for the AGASA region between 10^{18} and $10^{18.4}$ eV instead of 1.22 ± 0.05 [20] with an exposure five times larger than AGASA. Concerning the SUGAR excess, we have $n_{obs}/n_{exp} = 0.98 \pm 0.06$ between $10^{17.9}$ and $10^{18.5}$ eV instead of 1.85 ± 0.29 [21] with more than an order of one magnitude larger statistics.

Finally, a point-like neutron source was searched for by comparing the number of events and background within a Gaussian window corresponding the surface detector resolution around the GC. Again, no significant excess was observed, and an upper limit on the flux arriving from a point source at the GC of 0.08 km^{-2} yr^{-1} could be set at 95% C.L. Theoretical predictions [25, 26] exceed this upper bound by more than one order of magnitude, and [27] is at the level of the present Auger sensitivity.

Search for Primary Photons

As mentioned already in the introduction, non-accelerator 'top-down' models predict a significant fraction of primary photons at ultra high energies. Experimentally, photon initiated showers can be clearly distinguished from hadronic showers via their muon content or the depth of their shower maximum (X_{max}). The latter is used in the analysis presented here [28, 29], which takes advantage of the direct observation of the shower maximum by the fluorescence telescopes for a hadron/photon separation. The expected shower maxima are shown in Fig.4. Already at 'low' energies of 10^{18} eV photons

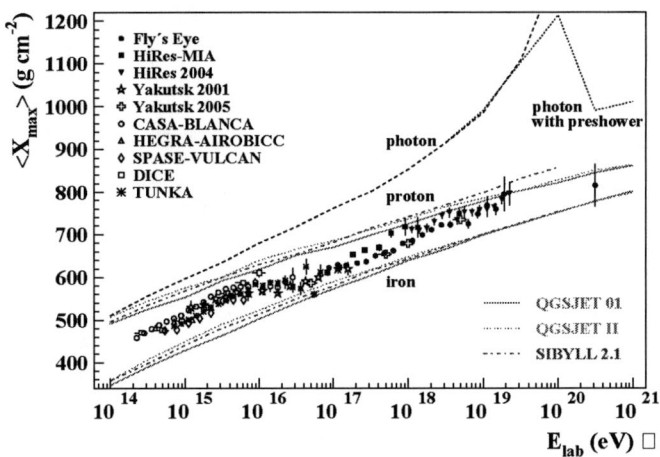

FIGURE 4. Simulated photon shower maxima compared to proton and iron showers as well as data.

penetrate deeper into the atmosphere, since in electromagnetic interactions less particles are produced than in hadronic ones and correspondingly the primary energy is dissipated slower. Above these energies radiative energy loss processes are suppressed by the LPM effect [30, 31] resulting in an even deeper X_{max} until 10^{20} eV, when the primary photons can convert within the geomagnetic field [32] and create a pre-shower outside of the atmosphere.

The data analysis is performed as follows: After applying quality and fiducial volume cuts to ensure a good X_{max} resolution as well as an equal acceptance for hadrons and photons, 29 candidate showers above 10^{19} eV remain. For each of these events 100 photons are simulated [33, 34, 35] according to the measured energies and directions. The residual between the measured and average simulated X_{max}, in units of the expected fluctuations (due to the width of the simulated distribution and the statistical uncertainty of the X_{max} measurement) yield a measure of the deviation Δ_γ of data from the photon expectation. As no photon candidate was found (Δ_γ ranges from +2.0 to +3.8), an upper limit on the photon fraction above 10^{19} eV is derived from the residuals using the method from [36]. This limit of 0.16 at 95% C.L. is compared to previous limits and predictions from 'top-down' models in Fig. 5.

First Estimate of the Primary Cosmic Ray Energy Spectrum

For the first estimate of the primary cosmic ray energy spectrum [42], 3525 surface detector events were selected from the first 17 months of data taking by requiring an energy above $3 \cdot 10^{18}$ eV and a zenith angles below $60°$. For such showers the array is fully efficient and its acceptance is purely geometric [43].

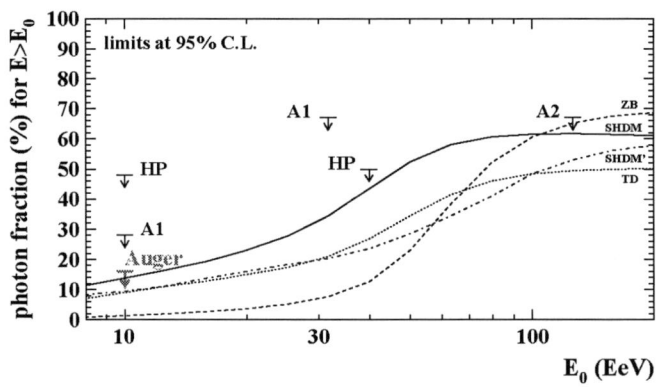

FIGURE 5. Upper limits (95% C.L.) to the cosmic ray photon fraction derived from the Auger data compared to previously obtained limits from AGASA (A1) [37], (A2) [36] and Haverah Park (HP) [38, 39] data. Expectations for non-acceleration models (ZB, SHDM, TD [40] and SHDM' [41]) are indicated as lines.

The energy assignment for the SD events works as follows: For each event the time-integrated signals of the water Cherenkov tanks are fitted with a lateral distribution function $S(r)$ [44], to obtain the shower size S_{1000} at $r = 1000$ m from the shower core. Since for larger zenith angles (θ), the ground array observes later and thus more attenuated stages of the shower development, for a given primary energy E, the measured S_{1000} decreases with the zenith angle. Assuming a constant intensity I of cosmic rays at each zenith angle, an attenuation correction function $CIC(\theta)$ can be found empirically for which $I(S_{1000}/CIC(\theta)) \equiv I(S_{38}) = \text{const}$. After this is achieved, the zenith angle corrected shower size S_{38} can be converted to an energy using the sub-set of events for which both FD and SD measurements are available to obtain a calibration curve $E(S_{38})$ as shown in the left panel of Fig. 6.

Since the fluorescence detectors provide a nearly calorimetric estimate of the primary energy, this calibration procedure yields an energy scale of the surface detector which avoids large systematic uncertainties from the modelling of air shower physics. The theoretical uncertainty on the fraction of the primary energy that does not produce fluorescence light is smaller than 3% above 10^{18} eV [45]. Instead, the energy scale uncertainty is dominated by uncertainties of the FD calibration [46], the fluorescence yield [47] and the low number of hybrid events. This experimental uncertainties are expected to shrink considerably in the near future.

The first estimate of the energy spectrum is shown in the right panel of Fig. 6. As can be seen, this analysis gives similar results to the ones obtained by HiRes, but it should be noted that both have correlated energy scale uncertainties, as they use the same fluorescence yield from [47]. Given the current surface detector energy uncertainties, indicated by the horizonal arrows, the AGASA data are compatible with this result, too. The additional Auger data collected in the last year and reduced fluorescence detector systematics, will allow for a clarification soon.

Finally, it is worth mentioning, that one event well above 10^{20} [48] was already

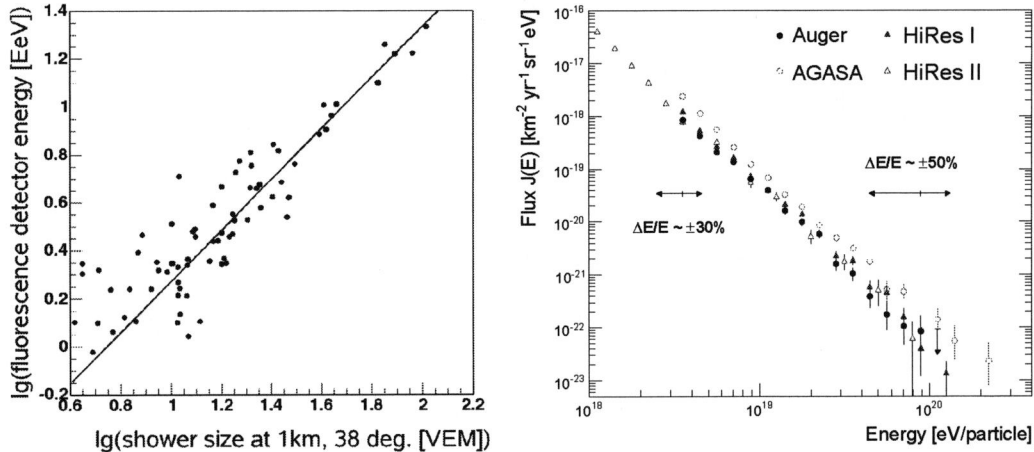

FIGURE 6. Left: Correlation of FD energy and SD shower size S_{38}. Right: First energy spectrum estimate compared to the AGASA [5] and HiRes [6] measurements. Horizontal arrows indicate the current energy scale uncertainties.

observed by the Pierre Auger Observatory. Since its shower core fell outside the array it does however not survive the fiducial volume cuts of the energy spectrum analysis.

SUMMARY AND OUTLOOK

In this presentation we reported first results from the southern Pierre Auger Observatory. Although the experiment is still under construction, the Auger exposure around the galactic centre exceeds the ones of previous experiments, and a competitive limit on a point-like neutron source from that region could be set.

Using the data of the fluorescence telescopes, an improved limit to the cosmic ray photon fraction was derived which is already close to the predictions of 'top down' models. Within a short time, the Auger photon fraction sensitivity will allow stringent tests of these models.

The energy spectrum is still limited by systematic and statistical uncertainties. Note however, that the 1.5 years of data used in this analysis correspond to only 3 months of the fully installed array. With the obvious increase of statistics, an improved understanding of the detector and dedicated laboratory measurements of the fluorescence yield an energy scale uncertainty at the 15% level should be reachable.

With more UHECR data to arrive, the rich surface and fluorescence detector data will allow for several other physics studies not covered here like the estimate of the chemical composition around the ankle, correlation studies with extra-galactic source candidates, measurement of the proton-air cross section and searches for ultra high energy neutrinos.

REFERENCES

1. J. Linsley, *Phys. Rev. Lett.* **10**, 146–148 (1963).
2. M. Nagano, and A. A. Watson, *Rev. Mod. Phys.* **72**, 689–732 (2000).
3. K. Greisen, *Phys. Rev. Lett.* **16**, 748–750 (1966).
4. G. T. Zatsepin, and V. A. Kuzmin, *JETP Lett.* **4**, 78–80 (1966).
5. M. Takeda, et al., *Astropart. Phys.* **19**, 447–462 (2003).
6. R. U. Abbasi, et al., *Phys. Lett.* **B619**, 271–280 (2005).
7. A. M. Hillas, *Ann. Rev. Astron. Astrophys.* **22**, 425–444 (1984).
8. M. Takeda, et al., *Astrophys. J.* **522**, 225–237 (1999), astro-ph/9902239.
9. S. Westerhoff, and C. B. Finley for the HiRes Collaboration, *Proc. 29th ICRC* (2005), astro-ph/0507574.
10. P. Bhattacharjee, and G. Sigl, *Phys. Rept.* **327**, 109–247 (2000), astro-ph/9811011.
11. V. Berezinsky, A. Z. Gazizov, and S. I. Grigorieva, *Phys. Lett.* **B612**, 147–153 (2005).
12. A. M. Hillas, *J. Phys.* **G31**, R95–R131 (2005).
13. D. Allard, E. Parizot, and A. V. Olinto (2005), astro-ph/0512345.
14. R. Engel, *Nucl. Phys. Proc. Suppl.* **151**, 437–461 (2006).
15. T. Pierog, R. Engel, and D. Heck, *Proc. C2RC 2005* (2006), astro-ph/0602190.
16. J. Abraham, et al., *Nucl. Instrum. Meth.* **A523**, 50–95 (2004).
17. X. Bertou for the Pierre Auger Collaboration, *Proc. 29th ICRC* (2005), astro-ph/0508466.
18. J. A. Bellido for the Pierre Auger Collaboration, *Proc. 29th ICRC* (2005), astro-ph/0507103.
19. M. Mostafa, et al., *Proc. 29th ICRC* (2005).
20. N. Hayashida, et al., *Astropart. Phys.* **10**, 303–311 (1999).
21. J. A. Bellido, R. W. Clay, B. R. Dawson, and M. Johnston-Hollitt, *Astropart. Phys.* **15**, 167–175 (2001).
22. A. Letessier-Selvon for the Pierre Auger Collaboration, *Proc. 29th ICRC* (2005), astro-ph/0507331.
23. M. Aglietta, et al. (2006), subm. to Astropart. Phys., astro-ph/0607382.
24. C. Bonifazi for the Pierre Auger Collaboration, *Proc. 29th ICRC* (2005).
25. M. Bossa, S. Mollerach, and E. Roulet, *J. Phys.* **G29**, 1409–1422 (2003).
26. F. Aharonian, and A. Neronov, *Astrophys. J.* **619**, 306–313 (2005).
27. D. Grasso, and L. Maccione, *Astropart. Phys.* **24**, 273–288 (2005).
28. M. Risse for the Pierre Auger Collaboration, *Proc. 29th ICRC* (2005), astro-ph/0507402.
29. J. Abraham, et al. (2006), subm. to Astropart. Phys., astro-ph/0606619.
30. L. D. Landau, and I. Pomeranchuk, *Dokl. Akad. Nauk Ser. Fiz.* **92**, 535–536 (1953).
31. A. B. Migdal, *Phys. Rev.* **103**, 1811–1820 (1956).
32. T. Erber, *Rev. Mod. Phys.* **38**, 626–659 (1966).
33. D. Heck, G. Schatz, T. Thouw, J. Knapp, and J. N. Capdevielle (1998), FZKA-6019.
34. N. Kalmykov et al., *Nucl. Phys.* **B (Proc. Suppl.)**, 7 (1997).
35. P. Homola, et al., *Comput. Phys. Commun.* **173**, 71 (2005), astro-ph/0311442.
36. M. Risse, et al., *Phys. Rev. Lett.* **95**, 171102 (2005).
37. K. Shinozaki, et al., *Astrophys. J.* **571**, L117–L120 (2002).
38. M. Ave, J. A. Hinton, R. A. Vazquez, A. A. Watson, and E. Zas, *Phys. Rev. Lett.* **85**, 2244–2247 (2000).
39. M. Ave, J. A. Hinton, R. A. Vazquez, A. A. Watson, and E. Zas, *Phys. Rev.* **D65**, 063007 (2002).
40. G. Gelmini, O. Kalashev, and D. Semikoz (2005), astro-ph/0506128.
41. J. R. Ellis, V. E. Mayes, and D. V. Nanopoulos (2005), astro-ph/0512303.
42. P. Sommers for the Pierre Auger Collaboration, *Proc. 29th ICRC* (2005), astro-ph/0507150.
43. D. Allard for the Pierre Auger Collaboration, *Proc. 29th ICRC* (2005), astro-ph/0511104.
44. D. Barnhill for the Pierre Auger Collaboration, *Proc. 29th ICRC* (2005), astro-ph/0507590.
45. T. Pierog, et al., *Proc. 29th ICRC* (2005).
46. P. Bauleo for the Pierre Auger Collaboration, *Proc. 29th ICRC* (2005), astro-ph/0507347.
47. M. Nagano, K. Kobayakawa, N. Sakaki, and K. Ando, *Astropart. Phys.* **22**, 235–248 (2004).
48. J. Matthews for the Pierre Auger Collaboration, *Proc. 29th ICRC* (2005).

LHC Potential for the Higgs Boson Discovery

R. Kinnunen

Helsinki Institute of Physics
P.O. Box 64, 00014 University of Helsinki, Helsinki, Finland

Abstract. The expected searches for the Higgs boson(s) of the Standard Model and its Minimal Supersymmetric extension with the CMS and ATLAS detectors at the LHC are discussed.

Keywords: Higgs boson, experimental particle physics
PACS: 14.80Bn

INTRODUCTION

The CERN LHC collider is expected to start functioning in the near future allowing, within few years, the direct search for Higgs bosons in the full expected mass range. In this report, the LHC potential for the Higgs boson discovery is discussed in the framework of the Standard Model (SM) and its Minimal Supersymmetric extension (MSSM). In the SM Higgs mechanism, the Higgs boson mass m_H is a free parameter bounded from below to $m_H > 114.4$ GeV/c^2 by the LEP results [1]. The MSSM contains five Higgs bosons: the lighter scalar h, the heavier scalar H, the pseudoscalar A and the two charged bosons $H^\pm$. The MSSM parameter space is in general presented as a function of the pseudoscalar mass m_A and the ratio $\tan\beta$ of the vacuum expectation values of the two Higgs doublets. The SUSY corrections to Higgs boson masses couplings come from the $t/\tilde{t}$ sector and at large $\tan\beta$ from the $b/\tilde{b}$ sector. The size of the correction is particularly sensitive to the Higgsino mass parameter μ. For most of the LHC studies, the m_h^{max} scenario, used in the LEP studies, has been chosen and has the following parameter values: $M_2 = 200$ GeV/c^2, $\mu = 200$ GeV/c^2, $M_{SUSY} = 1$ TeV/c^2, $M_{\tilde{g}} = 800$ GeV/c^2 and $X_t = 2M_{SUSY}$. The value of the top mass is fixed to 175 GeV/c^2. The LEP measurements yield lower bounds of 91.0, 91.9 and 78.6 GeV/c^2 for the masses of the h/H, A and $H^\pm$ bosons in the MSSM [2, 3], respectively. The excluded $\tan\beta$ regions are for $0.5 < \tan\beta < 2.4$ in the maximal m_h scenario [2].

At tree level the h(H) mass is bound to be below(above) the Z boson mass but the radiative corrections, proportional to m_{top}^4, bring the upper (lower) bound to a significantly larger value. The one loop and dominant two loop calculations, with the SUSY parameters listed above and with a top quark mass of 175 GeV/c^2, predict an upper bound of about 128 GeV/c^2 with maximal stop quark mixing [4].

This report summarizes the expected CMS and ATLAS searches for the SM and MSSM Higgs bosons with the most important discovery channels. Detailed descriptions of the CMS and ATLAS detectors can be found in Refs. [5, 6]. In CMS, the calorimeters are located between the tracker and the superconducting coil. Other features of the CMS detector [5] are a strong 4T axial magnetic field, a multilayer muon system in the return yoke and a scintillating crystal electromagnetic calorimeter. The tracker, placed

CP881, *Cairo International Conference on High Energy Physics (CICHEP II)*,
edited by S. Khalil

closest to the beam pipe, is made of fine-grained micro-strip and pixel detectors. In the ATLAS detector [6], the inner tracking system is placed inside a solenoid providing a 2T axial magnetic field. In this detector, the electromagnetic and hadron calorimeters are outside the solenoid. The muon measurements are performed with air-core-toroid muon spectrometers in the barrel and uncap regions.

The Higgs boson production and decay are discussed in Section 2. The discovery potential for the SM and MSSM Higgs bosons is discussed and presented in Sections 3-5 and conclusions are given in Section 6.

HIGGS BOSON PRODUCTION AND DECAY

In the SM, the Higgs boson production is dominated by the gluon-gluon fusion $gg \to H$ process, mediated by top and bottom quark loops, over the full expected mass range $100 \lesssim m_H \lesssim 1$ TeV/c^2. The QCD corrections for the $gg \to H$ process are large, with the next-to-leading (NLO) k factor ranging from 1.6 to 1.9 [7]. The weak gauge boson fusion (WBF) process $qq \to qqH$ has a cross sections lower by one order of magnitude than the $gg \to H$ process for light Higgs bosons but approaches that of the $gg \to H$ process for heavy Higgs bosons. In this process the energetic quaks jets, emitted in the forward directions, and the absence of significant jet activity in the central rapidities can be used to efficiently suppress the W/Z+jets and $t\bar{t}$ backgrounds at the LHC [8]. The other production processes $q\bar{q}' \to HW$, $q\bar{q} \to HZ$, $gg/q\bar{q} \to t\bar{t}H$ and $gg/q\bar{q} \to b\bar{b}H$ can be also exploited and yield a background suppression through the identification of the associated b jets and through the reconstruction of W/Z or top masses.

In the mass range of $m_H \lesssim 130$ GeV/c^2 the SM Higgs boson mainly decays to $b\bar{b}$ and $\tau^+\tau^-$ pairs with branching fractions of ~85% and ~8%, respectively. The total decay width is below 1 GeV/c^2 for $m_H \lesssim 200$ GeV/c^2, increasing to the size of the order of the mass for heavy Higgs bosons. Due to the small width, the $H \to \gamma\gamma$ and $H \to ZZ^* \to 4\ell$ decay channels with very small branching fractions can yield excellent signatures at the LHC.

In the MSSM, the lighter scalar h is SM-like for $m_A > m_h^{\max}$ (decoupling region), with production cross sections and decay partial widths close to those of the SM Higgs boson. At large $\tan\beta$, the couplings of the heavy neutral MSSM Higgs bosons to the W and Z bosons are strongly suppressed, while those to the down-type fermions are enhanced with increasing $\tan\beta$. The production of the H and A bosons proceeds mainly through the $gg \to H/A$ and $gg/q\bar{q} \to b\bar{b}H/A$ processes. At large $\tan\beta$, the $b\bar{b}H/A$ associated production dominates and presents about 90% of the total rate for $\tan\beta \gtrsim 10$ and $m_A \gtrsim 300$ GeV/c^2. If the charged Higgs bosons are light, $m_{H^\pm} < m_{top}$, they are produced in the $t\bar{t}$ events through the $t \to H^\pm b$ decay. Heavier ($m_{H^\pm} \geq m_t$) charged Higgs bosons are mainly produced in association of top quarks in the $gb \to tH^\pm$ process and in the NLO $gg \to t\bar{b}H^\pm$ process [9].

For the heavy neutral MSSM Higgs boson H and A, the branching fraction to $\tau^+\tau^-$ is about 10% and that to $\mu^+\mu^-$ about 3×10^{-4}. Light charged Higgs bosons ($m_{H^\pm} < m_{top}$) decay to $\tau\nu_\tau$ with an almost 100% branching fraction. For $m_{H^\pm} \gtrsim 200$ GeV/c^2 the $H^\pm \to tb$ decay dominates at large $\tan\beta$ while the $H^\pm \to \tau\nu_\tau$ branching fraction decreases and is about 10% for $m_{H^\pm} \gtrsim 400$ GeV/c^2. At small $\tan\beta$, the branching fractions of the neutral and charged MSSM Higgs bosons to gauginos, when kinematically allowed, dominate and suppress the branching fractions to SM particles. The branching fraction for $A/H \rightarrow t\bar{t}$, however, reaches $\sim 70\%$ at large m_A.

SEARCHES FOR THE SM HIGGS BOSON

Figure 1 shows the statistical significance for the SM Higgs boson in the most important discovery channels with an integrated luminosity of 30 fb^{-1} in the mass range of $100 \leq m_H \leq 800$ GeV/c^2 in CMS [10] and for $m_H \leq 200$ GeV/c^2 in ATLAS [11]. In the CMS analysis, NLO cross sections are used for the signal and background processes, and the significance is calculated according to Poisson statistics including the systematic uncertainties in the determination of all background components. The WBF channels play a significant role in the searches of a light SM Higgs boson. These channels have been proven particularly interesting for the Higgs boson coupling measurement and give the possibility of indirect total width measurement for $m_H < 200$ GeV/c^2 [12].

For the inclusive search of the $H \rightarrow \gamma\gamma$ channel there are large backgrounds from photon pairs produced from two initial gluons or quarks, from single photon production, where an associated jet fakes the photon signature, from multi-jet production, where two hadronic jets fake the photon signal, and from Drell-Yan production of electron pairs. The fake photon signals due to $\pi^0 \rightarrow \gamma\gamma$ decays in hadronic jets can be rejected with photon isolation and shower shape variables in the electromagnetic calorimeter. Two independent analysis have been performed for this channel in the recent CMS full simulation [10]. With a standard cut based analysis the Higgs boson can be discovered in this channel with a 5σ significance from the LEP lower limit to $m_H = 140$ GeV/c^2 with less than 30 fb^{-1} of integrated luminosity. With the optimized analysis with event by event estimation of the signal-to-background ratio this mass range can be covered with less than 16 fb^{-1} of integrated luminosity. The Higgs boson mass can be measured in the $H \rightarrow \gamma\gamma$ channel with a statistical precision of 0.1 to 0.2% already with 30 fb^{-1} of integrated luminosity. The expected mass measurement precision is shown in Fig. 2. The exclusive search of $H \rightarrow \gamma\gamma$ in the WBF and in the associated production channels $t\bar{t}H$, WH and ZH yield better signal-to-background ratios but require larger integrated luminosities, in excess of 100 fb^{-1}, for a significant signal [10].

Backgrounds for the four-lepton final state in the $H \rightarrow ZZ^*/ZZ \rightarrow \ell^+\ell^-\ell'^+\ell'^-$ signal are from the ZZ^*, $t\bar{t}$ and $Zb\bar{b}$ production and can be efficiently suppressed with lepton isolation in the tracker and in hadron calorimeter, an upper bound on the lepton impact parameter significance and cuts on the di-lepton invariant masses [10][11]. Discovery in the four-lepton channel is possible already with 10 fb^{-1} for the mass ranges of $140 < m_H < 150$ GeV/c^2 and $190 < m_H < 400$ GeV/c^2 in CMS, with similar reach in the ATLAS detector [11]. For 30 fb^{-1} the discovery range would open to $130 < m_H < 500$ GeV/c^2, apart the point around $m_H \sim 170$ GeV/c^2, which would require an integrated luminosity of 100 fb^{-1}. The Higgs boson mass can be measured in the $H \rightarrow ZZ^*/ZZ \rightarrow \ell^+\ell^-\ell'^+\ell'^-$ channel with a precision of 0.1 to 5.4%, depending on the mass, with 30 fb^{-1} of integrated luminosity [10], as shown in Fig. 2. The intrinsic Higgs boson width can be measured, for the masses greater than 190 GeV/c^2, with a precision of $\sim$35%, and the production cross section can be determined with a precision of $\sim$30%

[10]. The $H \rightarrow ZZ \rightarrow e^+e^-\mu^+\mu^-$ channel can be used to determine the CP properties of the Higgs boson [10].

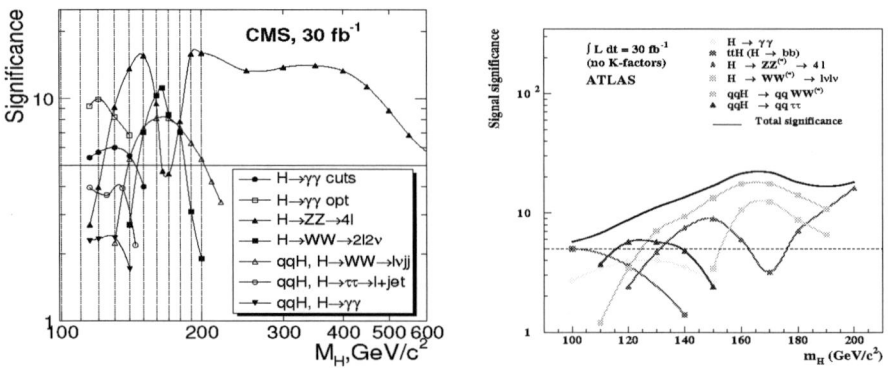

FIGURE 1. Expected statistical significance for the SM Higgs boson with 30 fb^{-1} of integrated luminosity as a function of m$_H$ in CMS (left) and for m$_H$ < 200 GeV/c^2 in ATLAS (right).

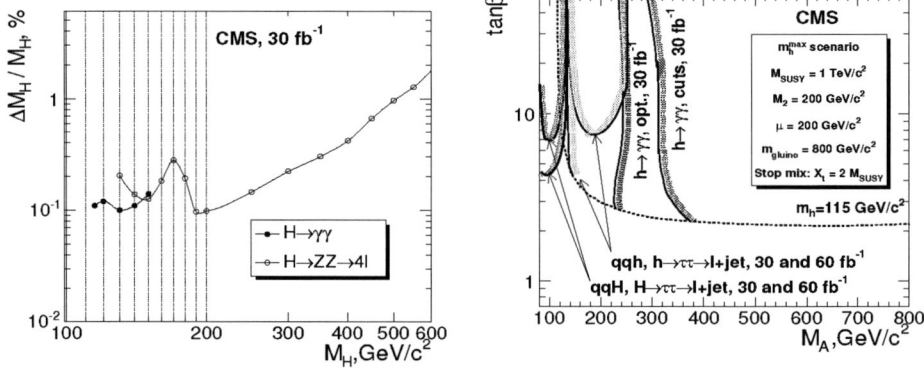

FIGURE 2. Expected precision of Higgs boson mass measurement in the $H \rightarrow \gamma\gamma$ and $H \rightarrow ZZ^*/ZZ \rightarrow \ell^+\ell^-\ell'^+\ell'^-$ channels in CMS with 30 fb^{-1} of integrated luminosity (left). The 5σ-discovery potential for the lighter scalar MSSM Higgs bosons h with the $h \rightarrow \gamma\gamma$ and $h \rightarrow \tau^+\tau^- \rightarrow \ell +$ jet decay modes as a function of m$_A$ and tanβ in the m$_h^{max}$ scenario with 30 and 60 fb^{-1} of integrated luminosity in CMS (right).

Around m$_H \sim$ 170 GeV/c^2, where the $H \rightarrow ZZ^*/ZZ$ branching fraction is smallest, the $H \rightarrow WW^*/WW$ channel can be exploited [10, 11]. For m$_H \lesssim$ 200 GeV/c^2, the backgrounds are from the processes $q\bar{q} \rightarrow WW \rightarrow 2\mu2\nu$, $gg \rightarrow t\bar{t} \rightarrow 2\mu2\nu$ and $q\bar{q} \rightarrow \gamma^*/Z \rightarrow 2\mu$. The t$\bar{t}$ background is reduced with muon isolation cuts and with a veto on central jets while a cut in the missing transverse energy and in the invariant mass of the muon pair have been used to suppress the $Z \rightarrow \mu\mu$ background. The WW background can be reduced by taking advantage of WW spin correlations, which turn into small $\ell^+\ell^-$ opening angles for the signal. Including a detailed evaluation of background uncertainties, discovery is possible in this channel for $150 < m_H < 180$ GeV/c^2 with greater than 5σ significance with less than 10 fb^{-1} of integrated luminosity. More efficient background suppression of the same decay mode, but with the $\ell\nu jj$ final state, can be obtained in the WBF production process. In this channel the Higgs boson mass can be reconstructed determining the neutrino momentum from the missing transverse energy with the W mass constraint.

In the SM, the $H \to \tau^+\tau^-$ decay mode with the subsequent $\tau_1 \to$ hadrons $+ \nu_\tau$ and $\tau_2 \to \ell + \nu_\tau\nu_\ell$ decays can be searched for in the WBF production. In addition to the forward jet tagging cuts, the methods of hadronic τ identification and Higgs boson mass reconstruction from leptonically or hadronically decaying τ's are exploited in this channel. Isolation of the narrow τ jet, mainly performed in the tracker, has been proven the most powerful method against the hadronic jet background [13]. Further suppression can be obtained from track counting in the jet, p$_T$ cuts for the tracks, impact parameter cuts for the tracks, secondary vertex reconstruction for the 3-prong τ decays and τ mass reconstruction with the calorimeter information [13]. The Higgs boson mass can be reconstructed in the $H \to \tau^+\tau^-$ channels from the missing transverse energy and the visible τ momenta exploiting the neutrino collinearity with the parent τ direction. The mass resolution improves for a decreasing opening angle between the two τ directions and is sensitive to the precision of the missing transverse energy measurement. For 60 fb^{-1} of integrated luminosity a significance greater than 5σ can be obtained for $m_H < 140$ GeV/c^2 with the $H \to \tau^+\tau^- \to \ell +$jet channel.

SEARCHES FOR THE NEUTRAL MSSM HIGGS BOSONS

Figure 2 shows the expected 5σ-discovery potential of CMS for the lighter scalar MSSM Higgs boson h in the most important discovery channels $h \to \gamma\gamma$ and $h \to \tau^+\tau^- \to \ell +$jet with 30 and 60 fb^{-1} of integrated luminosity. The results are shown in the the m_h^{max} scenario. In the parameter space outside the LEP reach the lighter scalar Higgs boson is largely SM-like and the discovery channels are closely the same as for the light SM Higgs boson. The enhancement of the Higgs coupling to down type fermions renders the $h \to \tau^+\tau^-$ decay channel in the WBF process particularly interesting in the MSSM. In the decoupling region the m_A, tanβ-plane is covered with the $h \to \tau^+\tau^- \to \ell +$jet channel while the region of small m_A is covered with the SM-like heavy scalar in $H \to \tau^+\tau^- \to \ell +$jet, also shown in Fig. 2.

At large tanβ, the coupling enhancement to down-type fermions can be exploited to search the H and A bosons in the H/A $\to \mu^+\mu^-$ and H/A $\to \tau^+\tau^-$ decay channels in the associated production gg $\to$ b$\bar{b}$H/A. In this production process, the tagging of the associated b jets suppresses efficiently the Z/γ^* and QCD multi-jet backgrounds. As the associated jets are not produced at high E$_T$ scales and are emitted largely to forward rapidities, only one jet is tagged. The signal efficiencies are typically ~20% for a hadronic jet rejection of about 100. The t$\bar{t}$ and Wt backgrounds with genuine b jets can be suppressed with a veto on an additional central jet.

Figures 3 show the expected 5σ-discovery potential of CMS [10] and ATLAS [11] for the heavy neutral MSSM Higgs bosons with the H/A $\to \tau^+\tau^-$ and H/A $\to \mu^+\mu^-$ decay modes. The CMS discovery potential is shown for 30 fb^{-1} and 60 fb^{-1} of

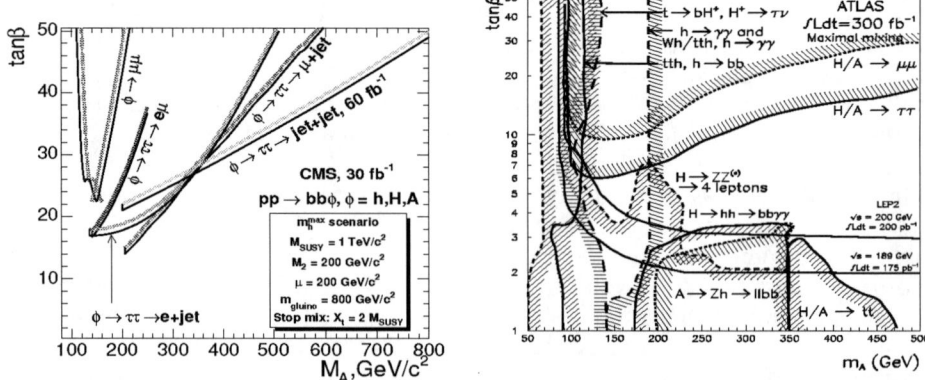

FIGURE 3. The 5σ-discovery potential for the heavy neutral MSSM Higgs bosons with the H/A → $\mu^+\mu^-$ and H/A → $\tau^+\tau^-$ decay modes as a function of m_A and $\tan\beta$ in the m_h^{max} scenario with 30 and 60 fb^{-1} of integrated luminosity in CMS (left) and with 300 fb^{-1} of integrated luminosity in ATLAS (right). The discovery potential for the A → ZH → $\ell\ell$bb, H → hh → bbγγ, H,A → t$\bar{t}$, H → ZZ* → 4lepton and for the lighter scalar Higgs boson are also shown in the ATLAS reach.

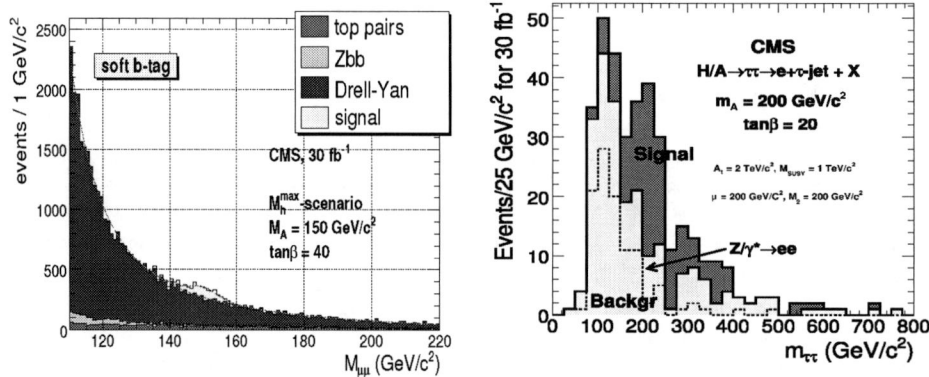

FIGURE 4. Invariant mass distribution for the H/A → $\mu^+\mu^-$ signal with $\tan\beta$ = 40 and m_A = 150 GeV/c^2, and for the total background (left), and for the H/A → $\tau^+\tau^-$ → electron + jet signal with $\tan\beta$ = 20 and m_A = 200 GeV/c^2 and for the total background with 30 fb^{-1} integrated luminosity (right).

integrated luminosity while the the ATLAS reach is given for the expected ultimate LHC luminosity. The ATLAS reach includes several other discovery channels, like A → ZH → $\ell\ell$bb, H → hh → bbγγ, H,A → t$\bar{t}$, H → ZZ* → 4lepton and the discovery channels for the lighter scalar Higgs boson.

The branching fraction for the H/A → $\mu^+\mu^-$ decay mode is only small but this channel leads to a clean final state and a good Higgs boson mass reconstruction. At large tanβ experimental mass resolution is comparable to the intrinsic Higgs boson width. Therefore the width measurement yields a constraint for the value of tanβ. For tan$\beta = 40$, for instance, the uncertainty on the tanβ measurement is from 17% to 25% for $150 \leq m_A \leq 200$ GeV/c^2, including the theoretical uncertainty in the production rate. Figure 4 (left) shows the invariant mass distribution for the H/A $\rightarrow \mu^+\mu^-$ signal with tan$\beta = 40$ and $m_A = 150$ GeV/c^2, and for the total background with 30 fb^{-1} integrated luminosity.

The H/A $\rightarrow \tau^+\tau^-$ decay channels can be searched for with the fully hadronic 2jets, electron+jet, μ+jet and two-lepton final states. The fully hadronic H/A $\rightarrow \tau^+\tau^- \rightarrow$ 2jets+X channel is particularly challenging experimentally due to the obligation to use a fully hadronic trigger and the need to suppress the very large hadronic multi-jet background [10]. The Higgs boson mass can be reconstructed and a visible signal is reached within the expected discovery range. Figure 4 (right) shows the invariant mass distribution for the H/A $\rightarrow \tau^+\tau^- \rightarrow$ electron + jet signal with tan$\beta = 20$ and $m_A = 200$ GeV/c^2, and for the total background with 30 fb^{-1} integrated luminosity. The discovery potential in Fig. 3 is shown for the m_h^{max} scenario with $\mu = 200$ GeV/c^2. Combined effect from the supersymmetric radiative corrections and decay modes into supersymmetric particles has been shown to be a a shift of the discovery contours toward lower tanβ values for negative values of μ. This shift is significant for large m_A ($\gtrsim 300$ GeV/c^2).

SEARCHES FOR THE CHARGED MSSM HIGGS BOSONS

Figures 5 show the expected 5σ-discovery potential of CMS [10] for the charged MSSM Higgs bosons with the H$^\pm \rightarrow \tau\nu_\tau$ and H$^\pm \rightarrow$ tb decay channels for a 30 fb^{-1} integrated luminosity. A similar reach has been obtained for the H$^\pm \rightarrow \tau\nu_\tau$ decay channels in the ATLAS experiment [11].

To search for the heavy charged Higgs bosons the H$^\pm \rightarrow \tau\nu_\tau$ decay channel with hadronic τ decays can be used in the t$\bar{t}$ events in the region $m_{H^\pm} < m_{top}$ and in the associated production process gb $\rightarrow$ tH$^\pm$ for $m_{H^\pm} > m_{top}$. In these channels, the t$\bar{t}$, Wt and W+3jet backgrounds with genuine τ's can be suppressed exploiting the opposite τ helicity correlations in the H$^\pm \rightarrow \tau\nu_\tau$ and the W$^\pm \rightarrow \tau\nu_\tau$ decays [14]. These correlations lead to a more energetic leading pion in the signal process from the $\tau \rightarrow \pi^\pm + \nu_\tau$ decay and from the longitudinal components of the 3-prong decay channels through ρ and a$_1$ mesons. A large background suppression can be obtained requiring at least 80% of the visible τ-jet energy to be carried by a single charged pion. For $m_{H^\pm} < m_{top}$, the H$^\pm$ signal is obtained from an excess of τ's in t$\bar{t}$ events relative to electrons and muons. This channel is expected to be triggered on a lepton from the decays of one of the top quarks. For $m_{H^\pm} > m_{top}$, in the purely hadronic events with hadronic top decays, the missing transverse energy originates mainly from the H$^\pm \rightarrow \tau\nu_\tau$ decay, making possible a reconstruction of the transverse mass from the τ jet and missing transverse energy with an endpoint at $m_{H^\pm}$ for the signal and at m_W for the backgrounds with the W$^\pm \rightarrow \tau\nu_\tau$ decay.

The dominating H$^\pm \rightarrow$ tb decay channel can be used in the associated gb $\rightarrow$ tH$^\pm$ production process, with a leptonic decay of one the top quarks. The channel is a subject to a large background from the $t\bar{t}$ production with associated standard jets or b jets. Due to systematic uncertainties in the determination of this background no discovery is expected in the CMS detector with this decay channel for low ($\leq$ 60 fb^{-1}) integrated luminosities [10]. Methods of tagging three b jets, exploiting the gb $\rightarrow$ tH$^{\pm}$ process, and tagging four b jets, selecting the NLO gg $\rightarrow$ tbH$^{\pm}$ process, were used. Figure 5 (right) shows the change in the expected discovery range when the systematic uncertainties are included for the gb $\rightarrow$ tH$^{\pm}$ production process.

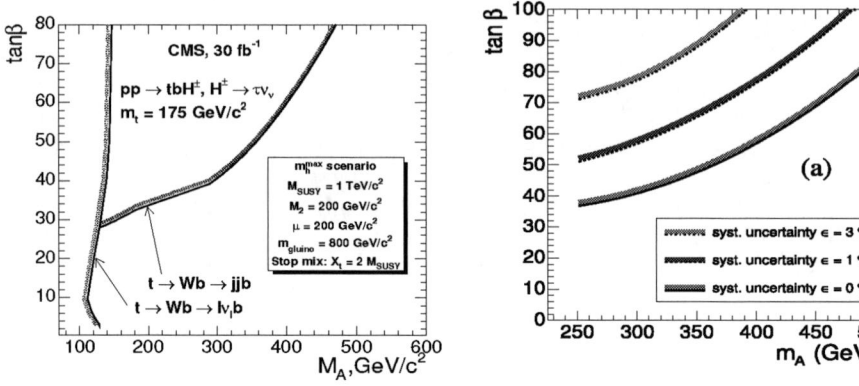

FIGURE 5. The 5σ-discovery potential of CMS for the charged MSSM Higgs bosons with the H$^{\pm}$ $\rightarrow$ $\tau\nu_{\tau}$, τ $\rightarrow$ hadrons + ν_{τ} decay modes (left) and with the H$^{\pm}$ $\rightarrow$ tb (right) in the gb $\rightarrow$ tH$^{\pm}$ production process as a function of m_A and tanβ in the m_h^{max} scenario with 30 fb^{-1} of integrated luminosity.

CONCLUSIONS

The most important channels for the searches of the Higgs bosons at the LHC in the SM and in its Supersymmetric extension MSSM were discussed. The SM Higgs boson is expected to be found at the LHC with several decay channels over the full expected mass range with the CMS and ATLAS detectors. In the region 140 GeV/c^2 $\lesssim$ m_H $\lesssim$ 400 GeV/c^2 the discovery is possible already with an integrated luminosity of 10 fb^{-1} or less with the H $\rightarrow$ WW*/WW and H $\rightarrow$ ZZ*/ZZ decay channels. The inclusive h $\rightarrow$ $\gamma\gamma$ channel can also yield a discovery already with 10 fb^{-1} of integrated luminosity from the LEP limit to m_H $\lesssim$ 150 GeV/c^2. The weak boson fusion production channels have been investigated, as a particular scope the Higgs boson coupling measurement. The H $\rightarrow$ WW* $\rightarrow$ $\ell\nu$jj decay channel leads to a 5σ signal already 30 fb^{-1} of integrated luminosity within 150 GeV/c^2 $\lesssim$ m_H $\lesssim$ 200 GeV/c^2 while the H $\rightarrow$ $\gamma\gamma$ and h $\rightarrow$ $\tau^+\tau^-$ decay channels require larger integrated luminosity in the SM. The Higgs boson mass can be measured with a precision of 0.1 to 5.4% depending on the mass. The total decay width and the production rate can be measured with the precisions of $\sim$35 and $\sim$30%, respectively.

In the MSSM, a significant fraction of the parameter space is covered with the lighter scalar Higgs boson with the inclusive h $\rightarrow$ $\gamma\gamma$ channel and with the h $\rightarrow$

$\tau^+\tau^-$ channel in the gauge boson fusion $qq \to qqh$ already with 30 fb^{-1} of integrated luminosity. For 60 fb^{-1} of integrated luminosity only a small area around 130 GeV/$c^2 \lesssim m_A \lesssim$ 140 GeV/c^2, not yet excluded by LEP, is left uncovered. The heavy neutral MSSM Higgs bosons can be found through the H/A $\to \mu^+\mu^-$ and H/A $\to \tau^+\tau^-$ decays channels at large tanβ. The two-lepton and lepton+jet final states from the H/A $\to \tau^+\tau^-$ decays cover the domain $m_A \lesssim$ 300 GeV/c^2 and tan$\beta \gtrsim$ 10 already with 30 fb^{-1}. The two-τ-jet final states extend the sensitivity up to $m_A \sim$ 800 GeV/c^2 for tanβ values larger than 25 for 60 fb^{-1} of integrated luminosity. The heavy scalar H can be discovered in the H $\to \tau^+\tau^-$ decay channel also in the gauge boson fusion for $m_A \lesssim$ 120 GeV/c^2 and tan$\beta >$ 8 already with 30 fb^{-1}. For the searches of the charged Higgs bosons the H$^\pm \to \tau\nu_\tau$ decay channel with hadronic τ decays plays a crucial rule. For $m_{H^\pm} < m_{top}$ the reach is for $m_A \lesssim$ 140 GeV/c^2 in the $t\bar{t}$ events with leptonic decay of one of the top quarks. The heavy charged Higgs bosons can be found at large tanβ ($\gtrsim$ 30) with this decay channel in the associated production with top quarks in fully hadronic final states.

ACKNOWLEDGMENTS

The author would like to thank A. Nikitenko for many helpful discussions.

REFERENCES

1. LEP Higgs Working Group, Phys. Lett. **B569** (2003) 61.
2. LEP Higgs Working Group, Searches for the Neutral Higgs Bosons of the MSSM, hep-ex/0107030.
3. LEP Higgs Working Group for Higgs Boson Searches, LHWG-Note 2004-01.
4. S. Heinemeyer and G. Weiglein, Leading Electroweak Two-Loop Corrections to Precision Observables in the MSSM, J. High Energy Phys. **10** (2002) 072.
5. CMS Collaboration, Technical Design Reports: The Tracker Project CERN/LHCC 98-6, CMS TDR 5, 199; The Hadron Calorimeter Project CERN/LHCC 97-31, CMS TDR 2, 1997; The Electromagnetic Calorimeter Project, CERN/LHCC 97-33, CMS TDR 4, 1997; CMS Collaboration, The Muon Project, CERN/LHCC 97-32, CMS TDR 3, 1997.
6. ATLAS Collaboration, Technical Design Reports: Calorimeter Performance CERN/LHCC 96-40, 1997; Inner Detector CERN/LHCC 97-16 and CERN/LHCC 97-17; Muon Spectrometer CERN/LHCC 97-22; Tile Calorimeter CERN/LHCC 96-42.
7. R. Rainwater, M. Spira and D. Zeppenfeld, Higgs Boson Production at Hadron Colliders, hep-ph/0203187.
8. D. Zeppenfeld, *Int. J. Mod. Phys.* **A16** (2001) 831; T. Plehn, D. Rainwater, D. Zeppenfeld, *Phys. Rev. Lett.* **88** (2002) 051801.
9. T. Plehn, Charged Higgs Boson Production in Bottom-Gluon Fusion, hep-ph/0206121; E. Boos and T. Plehn, Higgs Boson Production Induced by Bottom Quarks, hep-ph/0304034.
10. CMS Collaboration, CMS Physics Technical Design Report: Physics Performance CERN/LHCC 2006-021, CMS TDR 8.2, June 2006.
11. ATLAS Collaboration, ATLAS Detector and Physics Performance, Technical Design Report, ATLAS TDR 14, CERN/LHCC 99-14 and ATLAS TDR 15, CERN/LHCC 99-15.
12. D. Zeppenfeld, R. Kinnunen, A. Nikitenko and E. Richter-Was, Measuring Higgs boson couplings at the LHC, *Phys. Rev.*, **D62** (2000) 0130019/1-10.
13. CMS Collaboration, CMS Physics Technical Design Report: Volume 1, CERN/LHCC 2006-001, CMS TDR 8.1.
14. D.P. Roy, *Phys. Lett.*, **B459** 607 (1999).

Confinement Driven by Scalar Field in 4d Non Abelian Gauge Theories

Mohamed Chabab

LPHEA, Physics Department, Faculty of Science Semlalia, Cadi-Ayyad University,
40000-Marrakech, Morocco,
Email:mchabab@ucam.ac.ma

Abstract. We review some of the most recent work on confinement in 4d gauge theories with a massive scalar field (dilaton). Emphasis is put on the derivation of confining analytical solutions to the Coulomb problem versus dilaton effective couplings to gauge terms. It is shown that these effective theories can be relevant to model quark confinement and may shed some light on confinement mechanism. Moreover, the study of interquark potential, derived from Dick Model, in the heavy meson sector proves that phenomenological investigation of this mechanism is more than justified and deserves more efforts.

Keywords: dilaton, confinement, quark potential
PACS: 11.25.Mj, 12.90.+b, 12.38.Aw, 12.39.Pn

INTRODUCTION

Full Understanding of the QCD vacuum structure and color confinement mechanism are still lacking. Despite enormous amount of work performed over more than thirty years, particularly in lattice simulations of QCD, direct derivation of confinement from first principles remain still elusive, and there is no totally convincing proposal about its generating mechanism. On the other hand, it is known that the vacuum topological structure of theories with dilaton fields is drastically changed compared to the non dilatonic ones [1]. Therefore much about confinement might be learned from such theories, particularly string inspired ones. Indeed the appearance of fundamental scalars with direct coupling to gauge curvature terms in string theories offers a challenge with attractive implications in four-dimensional gauge theories. [1] Besides, color confinement can be signaled through the behavior of the interaction potential at large distances. In this context, it was suggested in [3] that an effective coupling of a massive dilaton to the 4-dimensional gauge fields may provide an interesting mechanism wich accomodate both the Coulomb and confining phases. The derivation performed in [3, 4] suggest a new scenario to generate color confinement which may be considered as a challenge to the mechanism based on monopole condensation.

The outline of this contribution is as follows. In the next section, We describe the influence of the dilaton on a low energy gauge theorie and look into the problem how dilatonic degrees of freedom modifies Coulomb potential and how transition to

---

[1] The dilaton is an hypothetical scalar particle predicted by string theory and Kaluza-Klein type theories. In string theory, its expectation value probes the strength of the gauge coupling [2].

CP881, *Cairo International Conference on High Energy Physics (CICHEP II)*,
edited by S. Khalil

a confinining phase occurs. Then, we review several recent work by presenting the corresponding effective coupling functions used. We brifly comment on the analytic solutions of the field equations and their confinement features. Also, it seems to us more than justified to dedicate some efforts to phenomenological investigations. We summarize the results obtained from study of Dick interaction potential in the heavy quarkonium systems.

THE MODEL

The imprint of dilaton on a 4d effective nonabelian gauge theory is described by a Lagrangian density:

$$\mathcal{L}(\phi,A) = -\frac{1}{4F(\phi)}G^a_{\mu\nu}G_a^{\mu\nu} + \frac{1}{2}\partial_\mu\phi\partial^\mu\phi - V(\phi) + J_a^\mu A_\mu^a \tag{1}$$

where $V(\phi)$ denotes the non perturbative dilaton potential and $G^{\mu\nu}$ is the standard field strength tensor of the theory. $F(\phi)$ is the coupling function depending on the dilaton field. Several forms of $F(\phi)$ have been proposed in literature. The most popular one $F(\phi) = e^{-k\frac{\phi}{f}}$ occured in string theory and Kaluza-Klein theories[2].

The problem of the Coulomb gauge theory augmented with dilatonic degrees of freedom in (1) is analyzed as follows:

First, we consider a point like static Coulomb source which is defined in the rest frame by the current:

$$J_a^\mu = g\delta(r)C_a v_0^\mu = \rho_a \eta_0^\mu \tag{2}$$

where C_a is the expectation value of $SU(N_c)$ generator.

The field equations emerging from the static configuration (2) are given by:

$$[D_\mu, F^{-1}(\phi)G^{\mu\nu}] = J^\nu \tag{3}$$

and

$$\partial_\mu\partial^\mu\phi = -\frac{\partial V(\phi)}{\partial\phi} - \frac{1}{4}\frac{\partial F^{-1}(\phi)}{\partial\phi}G_a^{\mu\nu}G_a^{\nu\mu} \tag{4}$$

At this stage, by seting $G_a^{0i} = E^i\chi_a = -\nabla^i\Phi_a$, after some algebra, we derive the chromo-electric field:

$$E_a = \frac{Q_{eff}^a(r)}{r^2} \tag{5}$$

where the effective charge is defined by

$$Q_{eff}^a(r) = \left(g\frac{C_a}{4\pi}\right)F(\phi(r))$$

239

. From Eq(5), we learn that it is the running of the effective charge that makes the potential stronger than the Coulomb potential. In other words, Coulomb spectrum is recovered if the effective charge did not run.

Thereby the interquark potential reads as [4],

$$U(r) = 2\tilde{\alpha}_s \int \frac{F(\phi(r))}{r^2} dr \tag{6}$$

with $\alpha_s = \frac{g^2}{4\pi}$ and $\tilde{\alpha} = \frac{\alpha_s}{8\pi}\left(\frac{N_c-1}{2N_c}\right)$

The formula in is remarkable since it provides a direct relation between the interquark potential and the coupling function $F(\phi(r))$. Moreover, it shows that exixtence of a confining phase in the theory in (1) is subject to the following requirement,

$$\lim_{r \to \infty} rF^{-1}(\phi(r)) = finite \tag{7}$$

The main objective is to solve the field equations of motion (3) and (4) and determine analytically $\phi(r)$ and $\Phi_a(r)$. For this, $F(\phi)$ and $V(\phi)$ have to be fixed. In the sequel the dilaton potential is set to $V(\phi) = \frac{1}{2}m^2\phi$. Below, we will briefly describe the main features of three recent models and present their solutions.

1. Dick Model

In this effective theory, Dick used the form: $\frac{1}{F(\phi)} = \frac{\phi^2}{f^2 + \beta\phi^2}$ where f represents a coupling scale characterising the strength of the scalar-gluon coupling. β is a parameter. Then he found for the radial dependance of the dilaton field and the interquark potential (up to a color factor) [3]:

$$\phi(r) = \pm\frac{1}{r}\sqrt{\frac{k}{m} + (y_0^2 - \frac{k}{m})exp(-2mr)}$$

$$V(r) = [\frac{\beta g^2}{4\pi r} - gf\sqrt{\frac{N_c}{2(N_c-1)}}ln[e^{2mr} - 1 + \frac{m}{k}y_0^2]$$

With the abbreviation: $k^2 = \frac{\alpha_s f^2}{8\pi}\frac{N_c-1}{N_c}$

Note that the potential $V(r)$ comes with the required behavior: a first term which accomodates the Coulomb interaction at short distances, and a second term linearly rising in the asymptotic regime with a string tension [2]

---

[2] In the massless case, $V(\phi) = 0$, solutions of the field equations reduced to: $\phi(r) = \pm\left(\frac{gf}{2\pi}\right)\sqrt{\frac{N_c-1}{N_c}}r^{\frac{-1}{2}}$,
$V(r) = \frac{g^2\beta(N_c-1)}{8\pi rN_c} - \frac{fg}{2}\sqrt{\frac{N_c-1}{N_c}}r$

$\sigma \sim gmf$ which depends on the dilatonic degrees of freedom m, f.

2. Cornwall-Soni Model

In this model, the glueballs are represented by a massive scalar field ϕ, and couple in a non mimimal way to gluons, through $\frac{1}{F(\phi)} = \frac{\phi}{f}$ [6]

Analytical Solutions were found for $r \to \infty$ [7],

$$\phi(r) = \left[\frac{\alpha_s f (N_c - 1)}{16 \pi m^2 N_c} \right]^{\frac{1}{3}} r^{-\frac{4}{3}}$$

$$V(r) = -3g \frac{N_c - 1}{2N_c} \left[\frac{g f^2 N_c m^2}{\pi (N_c - 1)} \right]^{\frac{1}{3}} r^{\frac{1}{3}}$$

These formulas show that their model provides confinement of quarks detected through an interaction potential propotional to $\tilde{r}^{1/3}$ at large distances and considered by the authors as non perturbative correction to the Coulomb potential.

3. Chabab-Sanhaji Model

The main aim in this work was to construct a low energy effective field theory from which some of the popular phenomenological potentials may emerge. For this, we used the following coupling function $F(\phi) = \left(1 - \beta \frac{\phi^2}{f^2} \right)^{-n}$ [8].

By substituting $F(\phi)$ in the field equations, they were found too complicated to integrate analytically. However, as in Cornwall-Soni Model, since the focus is on the long range behavior of the dilaton field and on how it modifies the Coulom phase, the analysis is restricted to the infrared region. Thus, the asymptotic solutions are found to be,

$$\phi = \left[\frac{f^2}{\beta} - \left(\frac{\beta}{f^2} \right)^{\frac{-n}{n+1}} \left(\frac{2n\alpha_s}{m^2} \right)^{\frac{1}{n+1}} \left(\frac{1}{r} \right)^{\frac{4}{n+1}} \right]$$

and the chromo-electric potential:

$$\Phi_a(r) = -\frac{gC_a}{4\pi} \left(\frac{2n\alpha_s}{m^2 f^2} \right)^{\frac{-4n}{n+1}} \frac{n+1}{3n-1} r^{\left(\frac{3n-1}{n+1} \right)}$$

We see that the occurrence of confinement depends on the parametre n and our effective theory can serve to model quark quark confinement when $n \in \left[\frac{1}{3}, 1 \right]$.

On the other hand, we attained the above mentioned objective: by selecting specific values of n, we reproduced the following known interquark potentials

- $n = 1 \Rightarrow$ linear term of Cornwall potential.

- $n = 11/29 \Rightarrow$ Martin's potential [13].
- $n = 3/5 \Rightarrow$ Song-Lin, or Motyka-Zalewski' potential [14].
- $n = 5/9 \Rightarrow$ Turin potential [15].

These quark potentials, which gained credibility only through their confrontation to the hadron spectrum, are now supplied with a theoretical framework since they can be derived from a low energy effective theory.

PHENOMENOLOGICAL ANALYSIS: RESULTS AND DISCUSSION

The interquark potential resulting from Dick model is quit attractive and deserves phenomenological investigations. A first study has been performed in [26]. in the heavy mesons sector. Therein, the semi-relativistic wave equation has been solved using Dick potential. This problem was addressed as in [11] where the shifted-l expansion technique is used (SLET), l is the angular momentum. This method provides a powerful analytic technique for determining the bound states of the semi-relativistic wave equation consisting of two quarks of masses m_1, m_2 and total binding meson energy M in any spherically symmetric potential. It is rapidly converging and handles highly excited states which pose problems for variational methods [12]. Moreover, relativistic corrections are included in a consistent way.

Dick interquark potential reads,

$$V_D(r) = -\frac{4}{3}\frac{\alpha_s}{r} + \frac{4}{3}gf\sqrt{\frac{N_c}{2(N_c-1)}}\ln[exp(2mr)-1] \tag{8}$$

The SLET technique used to obtain results from the theory requires us to specify several inputs, namely, m_c, m_b, m, f and α_s. In our numerical analysis, we set the charm and bottom quark masses to the values $m_c = 1.89$ GeV and $m_b = 5.19$ GeV. For the QCD coupling constant, in contrast to the Lattice potentials which use the same effective coupling in the description of heavy quarkonium, we take into account the running of α_s,

$$\alpha_s(\lambda) = \frac{\alpha_s(m_z)}{1-(11-\frac{2}{3}n_f)[\alpha_s(m_z)/2\pi]\ln(m_z/\lambda)}, \tag{9}$$

where the renormalization scale is fixed to $\lambda = 2\mu$, with μ is the reduced mass,

$$\mu = \frac{m_1 m_2}{m_1 + m_2}, \tag{10}$$

Thus, combination of the leading order formula (9) and the world experimental value $\alpha_s(m_z) = 0.118$ yields,

$$\alpha_s(charmonium) = 0.31, \qquad \alpha_s(bottomonium) = 0.20, \tag{11}$$

while $\alpha_s = 0.22$ for the $b\bar{c}$ quarkonia. On the other hand, the interquark potential parameters m and f are treated as being free in our analysis and are obtained by fitting the spin-averaged $c\bar{c}$ and $b\bar{b}$ boundstates. An excellent fit with the available experimental data can be seen to emerge when the following values are assigned,

$$m = 57 \, MeV \qquad gf\sqrt{\frac{N_c}{2(N_c - 1)}} = 430 \, MeV.^3 \tag{12}$$

TABLE 1.   Calculated mass spectra (in units of GeV) $M_{n\ell}$ of $c\bar{c}$ boundstates from Dick inerquark potential [26]

State, $n\ell$	$M_{n\ell}$, SLET	$M_{n\ell}$, Exp.	State, $n\ell$	$M_{n\ell}$, SLET	$M_{n\ell}$, Exp.
1S	3.073	3.068	1P	3.546	3.525
2S	3.662	3.663	2P	3.871	-
3S	4.027	4.028	1D	3.787	3.788

TABLE 2.   Calculated mass spectra (in units of GeV) $M_{n\ell}$ of $b\bar{b}$ from Dick interquark potential [26]

State, $n\ell$	$M_{n\ell}$, SLET	$M_{n\ell}$, Exp.	State, $n\ell$	$M_{n\ell}$, SLET	$M_{n\ell}$, Exp.
1S	9.450	9.446	1P	9.903	9.900
2S	10.014	10.013	2P	10.227	10.260
3S	10.299	10.348	1D	10.129	-

Tables (1,2) list the results of the analysis for the spin-averaged energy levels of interest. In all cases, where comparison with experiment is possible, agreement is generally very good. Next step, to check the consistency of our predictions, we estimate the bound states energies of the $b\bar{c}$ quarkonia. These states are expected to be produced at LHC and Tevatron. Moreover, they should provide an excellent test to discriminate between various techniques used to probe nonperturbative properties of hadrons. In table 3 we show our calculated spectrum. The estimate of the mass of the lowest pseudoscalar S-state of the B_c spectra is close to the experimental value reported by CDF collaboration [16]. As to the higher states masses, they compare favorably with other predictions based on QCD sum-rules [17, 18] or potential models [19]-[25]. In conclusion, Dick interquark potential (08) is tested successfully to fit the spin-averaged quarkonium spectrun. In view of these results, it is quite encouraging to pursue phenomenological appliation of $V_D(r)$ and other quark potentials emerging from such low effective gauge theory with dilaton.

GENERAL CONCLUSION

In summary, We reviewed some of the most recent work on confinement in 4d non abelian gauge theories with a massive scalar field (dilaton) and effective coupling func-

---

3 if we adopt the usual number $0.18 GeV^2$ for the string tension, the dilaton mass will be shifted to a value about $158 MeV$.

TABLE 3. Calculated mass spectra (in units of GeV) $M_{n\ell}$ of $b\bar{c}$ boundstates from Dick potential [26]

State,$n\ell$	$M_{n\ell}$, SLET	$M_{n\ell}$, Exp.	State,$n\ell$	$M_{n\ell}$, SLET	$M_{n\ell}$, Exp.
1S	6.322	6.40 $\pm$0.39$\pm$0.13	1P	6.767	-
2S	6.876	-	2P	7.072	-
3S	7.181	-	1D	6.994	-

tions to gauge fields. Analytrical solutions have been found with confinement feature in the asymptotic regime. Thus, These low energy effective theories can serve well to model quark confinement. Moreover, by using Dick interquark potential in the heavy quarkonium sector, we showed that phenomenological investigation of the confinement generating mechanism suggested by these models is more than justified. Indeed, the obtained results for charmoniun and bottomonium fit well experimental data when the dilaton mass is given a value about 57 MeV. Also, for B_c system, we found that the S-state energy level is close to the value reported by CDF collaboration, while those of excited states agree favorably with predictions of other theoretical works. On the other hand, This analysis allows a test to the physics beyond the standard model in relation to hadron spectroscopy. Indeed, as a by-product, the estimate of the dilaton mass lies in the range of values proposed in [27, 28]. This determination may shed some light on the search of the dilaton since, as suggested in [29, 30], the possibility to identify this hypotetical particle to a fundamental scalar invisible to present day experiments should not be exluded.

ACKNOWLEDGMENTS

The author thanks the CICHEPII organizers for the invitation to this nice Conference.This Work is partially supported by the government research program PROTARS III, contract number D16/04.

REFERENCES

1. M. Cvetic, A.A. Tseytlin, *Nucl. Phys.* **B 416**, 137 (1983).
2. M. Green, J. Schwartz, E. Witten, Superstring Theory, (Cambridge University Press, Cambridge 1987)
3. Dick R 1999 *Eur. Phys. J.* **C6** 701; 1997 *Phys. lett.* **B 397** 193; *Phys. lett.* **B 409** 321, (1997).
4. M. Chabab, R. Markazi, E. H. Saidi, *Eur. Phys. J.* **C 13**, 543 (2000).
5. Bali G S 2001 *Phys. Rep.* **343** 1; Petreczky P and Petrov K *ArXiv:* hep-lat/0405009.
6. J. M. Cornwall, A. Soni, *Phys. Rev.* **D 29**, 1424 (1984).
7. R. Dick, L. P. Fulcher, *Eur. Phys. J.* **C 9**, 271 (1999).
8. M. Chabab, L. Sanhaji, *Int. J. Mod. Phys.* **A 20** 1863 (2005); **hep-th/0311096**.
9. A. Galperin, E. Ivanov, V. Ogievetsky, P.T. Towsend, *Class. Quantum Gravity* **1**, 469 (1985).
10. E. Eichten et al.,*Phys. Rev. Lett.* **34**, 369 (1975).
11. Barakat T, *Int. J. Mod. Phys.* **A 16** 2195 (2001).
12. Sung Hwang D and Hee Kim G, *Phys. Rev.* **D 53** 3659 (1996); Sung Hwang D et al., *Phys. Rev.* **D 53**, 4951 (1996).
13. Martin A, *Phys. Lett.* **B 100** 511 (1981).
14. Motyka L and Zalewski K, *Z. Phys.* **C 69** 342 (1996); Song X, Lin H, *Z. Phys.* **C 34**, 223 (1987).

15. Lichtenberg D B et al., *Z. Phys.* **C 41** 615 (1989) 107.
16. Albe F et al. (CDF Collaboration), *Phys. Rev. Lett.* **81** 2432 (1998); *Phys. Rev.* **D 58** 112004 (1998).
17. Kiselev V V, *Int. J. Mod. Phys.* **A 11** 3689 (1996).
18. Chabab M, *Phys. Lett.* **B 325** 205 (1994); *7th Inter. Conf. Hadon Spectroscopy*, AIP Conference Proceedings **432**, Upton, New York, 1997, 856.
19. Eichten E and Quigg C, *Phys. Rev.* **D 49** 584 (1994).
20. Gershtein S S, Kiselev V V, Likhoded A K and Tkabladze A V, *Phys. Rev.* **D 51**, 3613 (1995).
21. Godfrey S and Isgur N, *Phys. Rev.* **D 32**, 189 (1985); Godfrey S, *Phys. Rev.* **D 70**, 054017 (2004).
22. Zhang J, Van Orden J W and Roberts W, *Phys. Rev.* **D 52**, 5229 (1995).
23. Gupta S N and Johnson J N, *Phys. Rev.* **D 53**, 074006 (1996).
24. Ebert D, Faustov R N and Galkin V O, *Phys. Rev.* **D 67**, 014027 (2003).
25. Brambilla N et al., *CERN Yellow Report on Heavy Quarkonium, hep-ph/0412158* (2004).
26. Barakat T, Chabab M, *hep-ph/0101056.*
27. Gasperini M, *Phys. Lett.* **B 327** 214 (1994).
28. Cho Y M and Keum Y Y, *Mod. Phys. Lett.* **A 3** 108 (1998).
29. Bando M, Matumoto K I and Yamawaki K, *Phys. Lett.* **B 178**, 308 (1986).
30. Halyo E, *Phys. Lett.* **B 271**, 415 (1991).

Multivariate Search of the Standard Model Higgs Boson at LHC

Mostafa Mjahed [a]

Ecole Royale de l'Air, Maths and Systems Department,
40000 Marrakech, Morocco
E-mail: mmjahed@hotmail.com
[a] *also at LPTN, Faculty of Sciences Semlalia, 40000 Marrakech, Morocco.*

Abstract: We present an attempt to identify the SM Higgs boson at LHC in the channel ($p\bar{p} \to HX \to W^+W^-X \to l^+\nu l^-\nu X$). We use a multivariate processing of data as a tool for a better discrimination between signal and background (via Principal Components Analysis, Genetic Algorithms and Neural Network). Events were produced at LHC energies ($M_H = 140 - 200$ GeV), using the Lund Monte Carlo generator PYTHIA 6.1. Higgs boson events ($p\bar{p} \to HX \to W^+W^-X \to l^+\nu l^-\nu X$) and the most relevant background are considered.

Keywords: Higgs boson, LHC, PCA, neural network, classification

PACS: 07.05 Mh; 07.05 Kf

INTRODUCTION

In the Standard Model (SM) of electro-weak and strong interactions, there are four types of gauge vector bosons (*gluon, photon, W* and *Z*) and twelve types of fermions (*six quarks* and *six leptons*) [1, 2]. These particles have been observed experimentally. The SM also predicts the existence of one scalar boson, the *Higgs boson* [3]. The observation of the *Higgs boson* remains one of the major cornerstones of the SM. This is a primary focus of the LHC experiments.

Several mechanisms contribute to the production of SM *Higgs bosons* in proton collisions. The dominant mechanism is the gluon fusion process, $pp \to gg \to H$, which provides the largest production rate for the entire *Higgs* mass range of interest. For large *Higgs* masses, the fusion process $qq \to WW, ZZ \to H$ becomes competitive, while for *Higgs* particles in the intermediate mass range $M_Z < M_H < 2M_Z$ the Higgs-strahlung off top quarks and *W; Z* gauge bosons are additional important production processes.

CP881, *Cairo International Conference on High Energy Physics (CICHEP II),*
edited by S. Khalil
© 2007 American Institute of Physics 978-0-7354-0382-6/07/$23.00

It has been established by many studies that a SM *Higgs boson* can be discovered with high significance over the full mass range of interest, from the lower limit set by the LEP experiments of *114.1 GeV/c²* up to *1 TeV/c²* [4].

In previous works, multivariate analysis methods, as *neural networks, discriminant analysis* [5] and *genetic algorithms* [6] have been used to identify *Higgs boson* events at LHC. Several attempts have been made to combine different gender classifiers containing complementary information to improve the classification accuracy. Experiments show that the combined classifiers generally outperform individual classifiers [6, 7, 17].

In this paper, we shall not consider all the possible *Higgs decay* channels, but we shall limit this analysis to some specific case studies. We mainly focus on the detection of the *Higgs boson* in the channel $p\bar{p} \to HX \to W^+W^-X \to l^+ \nu l^- \nu X$. We seek improved classification performance using a coupling of *principal components analysis* (PCA), *neural networks* (NN) and *genetic algorithms* (GA).

SIGNAL AND BACKGROUNDS

As introduced above, we will identify the SM *Higgs boson* in the channel $p\bar{p} \to HX \to W^+W^-X \to l^+\nu l^-\nu X$. The decay channel chosen is $H \to W^+W^- \to e^+\mu^-\nu\nu$, $e^-\mu^+\nu\nu$, $e^+e^-\nu\nu$, $\mu^+\mu^-\nu\nu$. The basic signature of this process is:

Two charged oppositely leptons with large transverse momentum P_T.

Two energetic jets in the forward detectors.

Large missing transverse momentum P'_T.

A number of backgrounds are relevant to the considered channel:

a) $t\bar{t}$ *production*, with $t \to Wb \to l\nu j$. In this process a pair of W and a pair of jets (j) are produced.

b) *QCD W^+W^- +jets production:* This is due to QCD emissions to the production of W^+W^-.

The physical observables used for the separation between signal and backgrounds are:

$\Delta\eta_{ll}$, $\Delta\phi_{ll}$: *the pseudo-rapidity and the azimuthal angle differences between the two leptons,*

$\Delta\eta_{jj}$, $\Delta\phi_{jj}$: *the pseudo-rapidity and the azimuthal angle differences between the two jets,*

M_{ll}, M_{jj}: *the invariant mass of the two leptons and jets,*

T_{nm} ($n,m=1,2,3$) *some rapidity weighted transverse momentum,*

$$T_{nm} = \sum_{i \in event} \eta_i^{n} \cdot p_{iT}^{m} \quad n,m=1,2,3, \ldots \tag{1}$$

where η_i is the rapidity of the leptons or jets, p_{iT} their transverse momentums.

The production of signal and background processes has been modelled with PYTHIA6.1 [8], in the *Higgs* mass range, *115< M_H < 200 GeV/c²*. To achieve this analysis, some selection cuts are made to the generated Monte Carlo events.

METHODOLOGY

Multivariate techniques have been extensively used in physics analyses. The common purpose of these methods, for classification or identification tasks, is to construct a set of rules called classifier. The computation of such rules requires a first learning phase where the respective properties of the events are previously known. The validation of this classification is then done with a test events set. There are various combination strategies that can used for combining classifiers. In this work, we aim to couple three multivariate methods: *principal components analysis, genetic algorithms* and *neural networks*.

Principal Components Analysis

The objective of the *principal components analysis* (PCA) is to linearly transform the initial attributes in order to concentrate the maximum amount of information in a minimum number of transformed attributes [9].

PCA aims at finding a set of *m* orthogonal vectors in input data space that account for as much as possible of the data's variance. Thus, the differences in classes of events originally present in the data set may still be identified by projecting data of the original *p*-dimensional input space onto the *m*-dimensional subspace spanned by these vectors ($m < p$). This would perform a dimensionality reduction with the preservation of the information spread around the full input space.

The components are the eigenvectors of the autocorrelation matrix of the random process under consideration, sorted by the decreasing order of the corresponding eigenvalues.

Genetic Algorithms

Genetic algorithms (GA) [10,11] are stochastic global search and optimization methods that mimic the metaphor of natural biological evolution and which are developed based on the Darwinian theory of "survival of fittest". GA have found numerous applications in a number of problem domains where a randomized local search of the parameter space is applicable. Some examples of such applications are various pattern detection and recognition problems [12], training the weights of neural networks [13] and signal enhancement in High Energy Physics (HEP) [14].

Simple GA has three basic operators: selection, crossover and mutation. A GA starts iteration with an initial population. Each member in this population is evaluated and assigned a fitness value. In the selection procedure, some selection criterion is applied to select a certain number of strings, namely parents, from this population according to their fitness values. Strings with higher fitness values have more opportunities to be selected for reproduction in next step. Next, in crossover procedure, selected strings from old population are randomly paired to mate. Crossover usually results in two new strings, namely, two children that are expected to combine the best characters of their parents. Mutation simply changes one bit 0 to 1 and vice versa, at a position determined by some rules. Mutation is simple but still important in evolution because it further increases the diversity of the population members and enables the optimization to get out of local optima. After mutation, a new generation is created, and thus becomes the parents for next generation. This process is iterated until convergence is achieved or a near optimal.

GA operate on a population (a number of potential solutions). The population at time t is represented by the time-dependent variable $P(t)$, with the initial population of random estimates being $P(0)$. The basic algorithm structure is shown on Fig. 1.

```
begin
      t = 0
      Initialize P(t)
      Evaluate P(t) (fitness)
      While not finished do
              begin
              t = t + 1
              Select P(t) from P(t - 1)
              Alter (cross and mutation) P(t)
              Evaluate P(t) (fitness)
              end
end
```

FIGURE 1. A simple Genetic Algorithm

Because the GA is a stochastic search method, it is difficult to formally specify convergence criteria. A common practice is to terminate the GA after a pre-specified number of generations and then test the quality of the best members of the population against the problem definition. If no acceptable solutions are found, the GA may be restarted or a fresh search initiated.

ANALYSIS

To identify Higgs boson signature, we define two classes: the *Higgs boson* process, (signal, denoted C_{Higgs} : $p\overline{p} \rightarrow HX \rightarrow W^+WX \rightarrow l^+ \nu l \nu X$) and the *background* events (C_{Back} : $t\overline{t}$ and QCD W^+W^- +*jets* production).

After the selected cuts made to the PYTHIA6.1 generated events, the samples retained amounted to *40000* events, with *20000* learning events and *20000* test events (*10000* samples for each class).

The efficiency β_i, the purity η_i and the errors ε_i corresponding to the methods are computed from the *confusion matrix A (A_{ij})*, (A_{ij} being the value of events of genuine *class C_i* classified as *class C_j*). For each class C_i we have:

$$\beta_i = \frac{A_{ii}}{\sum_l A_{il}}, \qquad \eta_i = \frac{A_{ii}}{\sum_l A_{li}}, \qquad \varepsilon_i = 1 - \beta_i \tag{2}$$

In the following, we present the three used approaches in this analysis. First a *back-propagation neural network* (NN) is optimized. Then, the NN inputs are processed by using PCA and the corresponding network (PCANN) is re-optimized. In the third application, and to reach better performance, we tune PCANN parameters via GA. A comparison between these approaches is given and the overall results are discussed.

Higgs Boson Identification using NN (NN)

Neural networks are broadly used in pattern recognition problems and functions approximation tools. There are many types of artificial neural networks whose differ in architecture, in the type of implemented transfer functions and strategy of learning.

The identification of *Higgs boson* process C_{Higgs} was done by a "*three layer*" *neural network*. Its architecture is (*10, 10, 2*) as shown in Fig. 2 .

Each process p described with the above defined *variables, x= ($\Delta\eta_{11}$, $\Delta\phi_{11}$, $\Delta\eta_{jj}$, $\Delta\phi_{jj}$, M_{11}, M_{jj}, T_{11}, T_{21}, T_{31}, T_{41}),* was presented to the input layer, which in turn feeds to a *hidden layer h*, feeding finally to *two output* neurons.

The rules for calculating the outputs o_1 and o_2 are:

$$o_j = \sum_i w_{ij}^{ho} h_i , \qquad h_i = f(\sum_j w_{ji}^{xh} x_j - \theta_i) \tag{3}$$

$$f(x) = \frac{1}{1 + e^{-2x}} \tag{4}$$

$$E = \sum_p \sum_j (r_j^p - o_j^p)^2 \tag{5}$$

where w_{ij}^{kl} are synaptic weights of the connection between *neuron i (of layer k)* and *neuron j (of layer l)*, and θ_i thresholds. These parameters are adjusted in such a way that the error function E is minimized using the *back-propagation algorithm* [15].

The desired outputs (r_1, r_2) for the *2* classes C_{Higgs} and C_{Back} are respectively: (*+1,+1*) *and (-1, -1)*.

The classification of a test process x_0 is straightforward obtained according to the rule:

$$if\ o_1(x_0) > 0\ and\ o_2(x_0) > 0\ then\ x_0 \in C_{Higgs}\ else\ x_0 \in C_{Back} \tag{6}$$

TABLE1. *Classification matrix* (validation) using 2×10000 test processes performed by the *three neural networks* NN, PCANN and GAPCANN.

Method	NN		PCANN		GAPCANN	
	C_{Higgs}	C_{Back}	C_{Higgs}	C_{Back}	C_{Higgs}	C_{Back}
C_{Higgs}: *10000*	*6841*	*3159*	*7121*	*2879*	*7365*	*2635*
C_{Back}: *10000*	*3111*	*6889*	*2865*	*7135*	*2611*	*7389*

The classification of the test processes are given in *Table 1* and the deduced efficiencies and purities are translated in *Table 2*.

Higgs Boson Identification using PCA based NN (PCANN)

In the second use of neural technique, the above defined inputs of NN are replaced by some extracted PCA.

As presented above, the principal components are those linear combinations of the p original variables ($p=10$) which maximize the variance of the linear combination and which have zero covariance (and hence zero correlation) with the previous principal components. Generally, there are exactly p such linear combinations. However, typically, the first few of them explain most of the variance in the original data.

So instead of working with all the original *10* variables $\Delta\eta_{ll}$, $\Delta\phi_{ll}$, $\Delta\eta_{ij}$, $\Delta\phi_{ij}$, M_{ll}, M_{ij}, T_{11}, T_{21}, T_{31}, T_{41}, we perform PCA and then we only select first three principal components: v_1, v_2 and v_3.

In the neural optimization stage, we proceed as above. Each event is represented by the new inputs: $x= (v_1, v_2, v_3)$. The same rules for calculating the outputs o_1 and o_2 and for classifying the processes are used. The architecture of the optimal obtained neural network (PCANN) is (*3, 10, 2*) (as illustrated by Fig. 2).

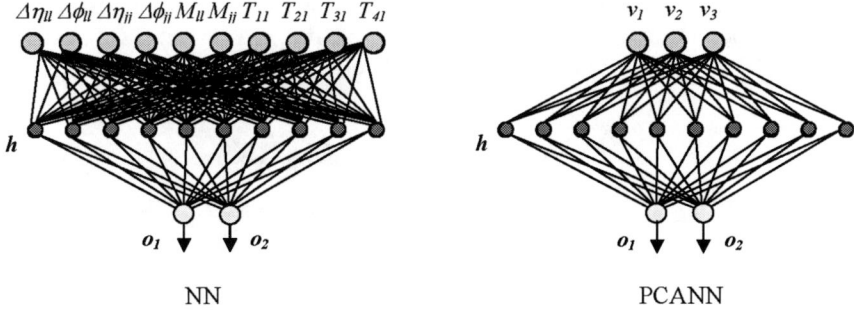

FIGURE 2. Topology of the two neural networks NN and PCANN

TABLE 2. *Efficiencies* and *purities* of classifications using 2×10000 test processes performed by the *three neural networks* NN, PCANN and GAPCANN.

Method	Efficiency (%)			Purity (%)		
	C_{Higgs}	C_{Back}	All	C_{Higgs}	C_{Back}	All
NN	68.41	68.89	68.65	68.74	68.56	68.65
PCANN	71.21	71.35	71.28	71.30	71.25	71.27
GAPCANN	73.65	73.89	73.77	73.82	73.71	73.76

The validation results are summarized in the *confusion matrix* (*Table 1*) and the corresponding efficiencies are reported in *Table 2*.

Higgs Boson Identification using GA based PCANN (GAPCANN)

In order to get a better identification of the SM *Higgs* events in the channel $p\bar{p} \rightarrow HX \rightarrow W^+WX \rightarrow l^+ v \, l \, v X$, a GA based optimization approach for PCANN weights is performed.

The design parameters to be optimized are the real components of populations $P(t)$. $P(t) = (P_i(t)) = (P_{ij}(t))$; $(i = 1...N_{ind}; j = 1...N_{par})$, $(N_{ind}$ being the population size and N_{par} the number of parameters to be optimized).

Initialized from the given parameter ranges by genetic algorithm, these components are then evolved generation by generation and new solutions are obtained. After a number of generations (N_{gen}), the GA is terminated. The best solution correspond to the minimal value of the fitness function (corresponding here to the total misclassification rate ε)..

For the *(3, 10, 2)*.architecture (Fig. 2), the number of parameters to seek rises to 60. The obtained *back-propagation weights* of PCANN are retained in the initial population: $P_1(0) = (W^{h0}_{ij} ; (i=1,...,10 ; j=1,2); W^{xh}_{kl} (k =1,...,3; l=1,...,10))$. At each step of the GA procedure (Fig.1) with N_{gen} = 1000, new NN weights are reproduced. The corresponding NN are validated and their fitness functions (ε) are then compared. We report the optimal classification results of the GA based PCANN approach (GAPCANN) in *Table 1* and the corresponding efficiencies and purities in *Table 2*.

CONCLUDING REMARKS

In this work, we were aiming to combine several multivariate methods for a better identification of *Higgs boson* process in $p\bar{p} \rightarrow HX \rightarrow W^+WX \rightarrow l^+ v \, l \, v X$ channel. The idea developed is that of using principal components analysis to select more discriminant variables and genetic algorithms to optimize the weights of neural networks.

Concerning PCA, the reduction of dimensionality is practicable if the selected new axes account for approximately *75%* or more of the variance. The cumulative percentage of variance explained by the principal axes is consulted in order to make this choice. No theoretical results exist to define the optimal number of PCA components to be used in a regression; it entirely depends on the problem to be solved. Experience with the NN technique shows that, if the problem is well regularized, once sufficient information is provided as input, adding more PCA components to the inputs does not have a large impact on the retrieved results.

With respect to GA, it can be observed that GA has been able to find optimal parameters. These results are encouraging and suggest that GA can be easily used in other complex and realistic designs often encountered in HEP. GA allows to minimize the classification error and to improve efficiencies and purities of classifications.

Tables 1 and 2 show the difference between the three applied approaches. If *neural network* methods have proven themselves to be more efficient classifiers than the linear methods as discriminant analysis [5], it should be noted that PCA based neural network (PCANN) is here higher, which confirms the importance of the extracted principal components. The third neural network (GAPCANN) is better because of its ability to handle complex relationships between the attributes and the physical processes considered in this work. The efficiencies and purities achieved with the optimized *neural network* (GAPCANN) are in average *2.4* to *5.3 %* higher than those obtained with NN and PCANN. This result described otherwise in Fig. 3, through the distribution of the 2 considered classes with respect to GAPCANN outputs, illustrate the effectiveness of the third proposed approach. The total GAPCANN efficiency and purity are respectively of *73.77 %* and *73.76 %*.

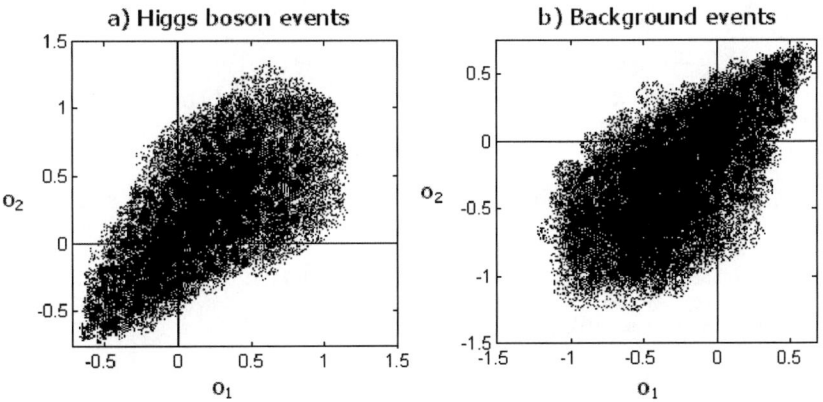

FIGURE 3. First and second GAPCANN output distributions for the 2 classes: a) *Higgs boson events*, b) *Background events*

Notice that, the computed classifications and their efficiencies are made according to the rule: $o_1 > 0$ and $o_2 > 0$ for *Higgs process* ($o_1 < 0$ and $o_2 < 0$ for *background*). Other cut's values on the neural network outputs give other efficiencies and purities. We can reach values of purity permitting to identify the C_{Higgs} *processes* more purely [16]. For $o_1 > 0.71$ and $o_2 > 0.72$, more than *52%* of *Higgs process* may be identified with higher value of purity (about *99%*).

We are conscious of the limits of the present analysis. In fact, in spite of the complexity of this tool, further work should include additional processes and variables and an accurate computation of detector effects.

ACKNOWLEDGMENTS

I am very grateful to the Moroccan Administration of Defence which supported in part this work. I would like to thank the CICHEP II Organizing Committee and ICTP for their support. Particular thanks are due to Dr K. Shaaban, Dr H. El-Sharkawy, Dr M. Chabab and Dr N. Belalaoui.

REFERENCES

1. S.L. Glashow, *Nucl. Physics*. B22 (1961) 579.
2. S.L. Glashow, J. Iliopoulos and L. Maiani, *Phys. Rev.* D2 (1970) 1285.
3. P.W. Higgs, *Phys. Letters*. 12 (1964) 132.
4. N. Knauer et al, *Phys. Letters*. B503 (2001) 113.
5. M. Mjahed, *Nucl. Physics* B Vol 140, (2005) 799.
6. M. Mjahed, *Nucl. Instrum. and Methods* A559 (2006)172.
7. M. Mjahed, *Nucl. Instrum. and Methods*. A 481 (1-3) (2002) 601.
   M. Mjahed, *Nucl. Physics* B Vol 106-107C, (2002) 1094.
8. T. Sjostrand et al., "High-Energy-Physics Event Generation with PYTHIA 6.1", *Comp. Phys. Comm.* 135 (2001) 238.
9. M. S. Srivastava, E. M. Carter, *Applied multivariate statistics*, North Holland Amsterdam, 1983.
10. D.E. Goldberg, *Genetic Algorithms in Search, Optimization, and Machine Learning*, Addison Wesley, Reading, MA, 1989.
11. Z. Michalewicz, *Genetic Algorithms + Data Structures = Evolution Programs*, Springer Verlag, New York, 1994.
12. S.H. Pal, P.P. Wang , *Genetic Algorithms for Pattern Recognition*, CRC Press, 1996.
13. A.J.F. van Rooij et al, "Neural Network Training Using Genetic Algorithms", *Machine Perception & Artificial Intelligence*, vol. 26, 1996.
14. L. Karimova et al, *Nuc. Instrum. and Methods* A 534, 1-2 (2004) 170.
15. F. L. Luo et al, *Applied neural networks for signal processing*, Cambridge Univ. press, 1997.
   J. A. Freeman, M. Skapura, Neural Networks, Addison - Wesley, 1991.
16. M. Mjahed, *Nucl. Physics* B 119C (2003) 1027.
17. J. Kittler et al, *IEEE Trans. on Pattern Analysis and Machine Intelligence*, Vol. 20, No. 3, March 1998.

Quasifree pion photoproduction on the deuteron: Single spin asymmetries[1]

Eed M. Darwish[2]

Physics Department, Faculty of Science, Sohag University, Sohag 82524, Egypt

Abstract. Quasifree single-pion photoproduction on the deuteron via the exclusive reaction $d(\gamma, \pi^0 n)p$ is studied in the $\Delta(1232)$-resonance region including all *leading* πNN effects. Special emphasizes are given to single spin observables. The elementary pion photoproduction amplitude from free nucleons is taken from the dynamical model of Sato and Lee (SL model) [Phys. Rev. C54 (1996) 2660]. The interactions in the final two-body subsystems are taken in separable form. Final-state interaction effects are found to be significant and lead to an improved agreement with the most recent experimental data from Brookhaven National Laboratory (LEGS-exp.L3b).

Keywords: Meson production; Photoproduction reactions; Spin observables; Deuteron.
PACS: 13.60.Le; 25.20.Lj; 14.20.-c; 25.45.De.

1. INTRODUCTION

Since the advent of high duty-factor accelerators, such as MAMI in Mainz and ELSA in Bonn (Germany), JLab in Newport News and LEGS in Brookhaven (USA) or Max-Lab in Lund (Sweden), the study of single pion production reaction in intermediate energy nuclear physics has been getting more and more attention in recent years (see Refs. [1, 2, 3, 4, 5] where also references to earlier work can be found) with respect to the study of hadron structure in the non-perturbative domain of Quantum Chromodynamics (QCD) and therefore the nature of strong interactions.

Rescattering effects in pion photoproduction on the deuteron were treated approximately by including complete hadronic rescattering in the final NN- and πN-subsystems [1] and applied to the spin asymmetry with respect to circular photon polarization [2], which determines the Gerasimov-Drell-Hearn (GDH) sum rule. During the work of the present paper, the approach presented in Refs. [1, 2] was improved in Refs. [3, 4], in which a better elementary production operator from the MAID model [6] was taken and the role of final-state interaction (FSI) effects on cross sections and polarization observables has been studied. Incoherent pion photoproduction from the deuteron has been also studied in the $\Delta(1232)$-resonance region [5] using the elementary production operator from the SAID [7] analysis and including NN- and πN-FSI effects.

The main goal of the present paper is to report on a theoretical prediction for spin-dependent observables of the exclusive reaction $d(\gamma, \pi^0 n)p$ in the $\Delta(1232)$-resonance

---

[1] Talk given at the 2nd Cairo International Conference on High Energy Physics (CICHEP II), German University in Cairo (GUC), January 14-17, 2006, Cairo, Egypt.
[2] *E-mail address:* eeddarwish@yahoo.com.

CP881, *Cairo International Conference on High Energy Physics (CICHEP II)*,
edited by S. Khalil

region. We are going to extend the work in the preceding paper [1] in order to under-stand the dynamics of pion photoproduction amplitude. The extention includes the fol-lowing important aspects: (i) A more realistic elementary pion photoproduction operator from Ref. [8] is used. (ii) The influence of FSI effects on spin-dependent observables is quantitatively studied. We consider besides the pure impulse approximation (IA) com-plete rescattering in the final two-body subsystems, i.e. in the NN- and πN-subsystems. (iii) Comparison with recent experimental data is discussed. The calculation is of high theoretical interest, because it provides an important test of our understanding of the πNN dynamics. That understanding is a prerequisite for reliable extraction of the pion photoproduction amplitude on the neutron.

2. THE MODEL

In the deuteron rest frame, the differential cross section for an initial spin states (λ, M_d) of $\vec{d}(\vec{\gamma}, \pi)NN$ is given by [9]

$$
\begin{aligned}
d\sigma(\lambda, M_d) &= \frac{(2\pi)^4}{v_{rel}} \vec{dk} d\vec{p}_1 d\vec{p}_2 \delta(q + m_d - E_\pi(k) - E_N(p_1) - E_N(p_2)) \\
&\times \delta(\vec{q} - \vec{k} - \vec{p}_1 - \vec{p}_2) \sum_{m_{s1} m_{s2}} |T_{fi}|^2
\end{aligned}
\tag{1}
$$

where the relative velocity $v_{rel} = 1$ in the Laboratory frame and the other notations are obvious. Choose

$$
\vec{P}_c = \vec{p}_1 + \vec{p}_2, \qquad \vec{p} = \frac{1}{2}(\vec{p}_1 - \vec{p}_2)
\tag{2}
$$

we then have

$$
E_N(p_1) = E_N(\frac{\vec{P}_c}{2} + \vec{p}), \qquad E_N(p_2) = E_N(\frac{\vec{P}_c}{2} - \vec{p})
\tag{3}
$$

Momentum conservation leads to $\vec{P}_c = \vec{q} - \vec{k}$. We then have the differential cross section of pions

$$
\frac{d\sigma(\lambda, M_d)}{dk d\Omega_k} = (2\pi)^4 \int d\Omega_p F \sum_{m_{s1} m_{s2}} |T_{fi}|^2
\tag{4}
$$

where

$$
F = \left| \frac{k^2 p^2 E_N(\frac{\vec{P}_c}{2} + \vec{p}) E_N(\frac{\vec{P}_c}{2} - \vec{p})}{\frac{\vec{P}_c \cdot \vec{p}}{2}(E_N(\frac{\vec{P}_c}{2} - \vec{p}) - E_N(\frac{\vec{P}_c}{2} + \vec{p})) + p(E_N(\frac{\vec{P}_c}{2} + \vec{p}) + E_N(\frac{\vec{P}_c}{2} - \vec{p}))} \right|
\tag{5}
$$

and p is the solution of

$$
q + m_d = E_\pi(k) + E_N(\frac{\vec{P}_c}{2} + \vec{p}) + E_N(\frac{\vec{P}_c}{2} - \vec{p})
\tag{6}
$$

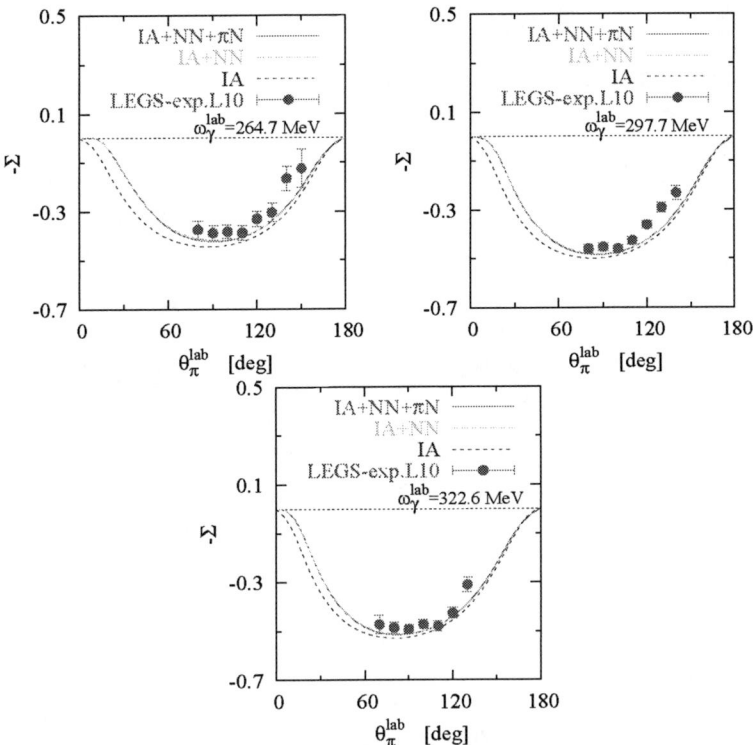

FIGURE 1. Linear photon asymmetry Σ for the reaction $d(\vec{\gamma}, \pi^0 n)p$ in comparison with the experimental data from LEGS-collaboration (LEGS-exp.L3b) [10]. Notation of curves: dashed: IA; dotted: IA+NN-rescattering; solid: IA+NN- and πN-rescattering.

All observables are determined by the photoproduction amplitude T_{fi}. In principle, the full treatment of all interaction effects requires a full unitary πNN three-body calculation. However, the treatment of the scattering amplitude in this work is completely analogous to our previous work [1], to which the reader is referred for formal details. With respect to polarization observables, we consider in this paper the linear photon asymmetry Σ, the vector and tensor deuteron target asymmetries T_{11} and T_{2M} (M=0,1,2), respectively. The explicit expressions for these asymmetries are given in Refs. [4, 5].

3. RESULTS

For the calculation of the NN-rescattering contribution we have taken the separable representation of the realistic Paris potential from Ref. [11] and included all partial waves up to $^3D_3$. Also, the deuteron wave function was calculated using this potential. Similarly, πN-rescattering is evaluated using a realistic separable representation of the

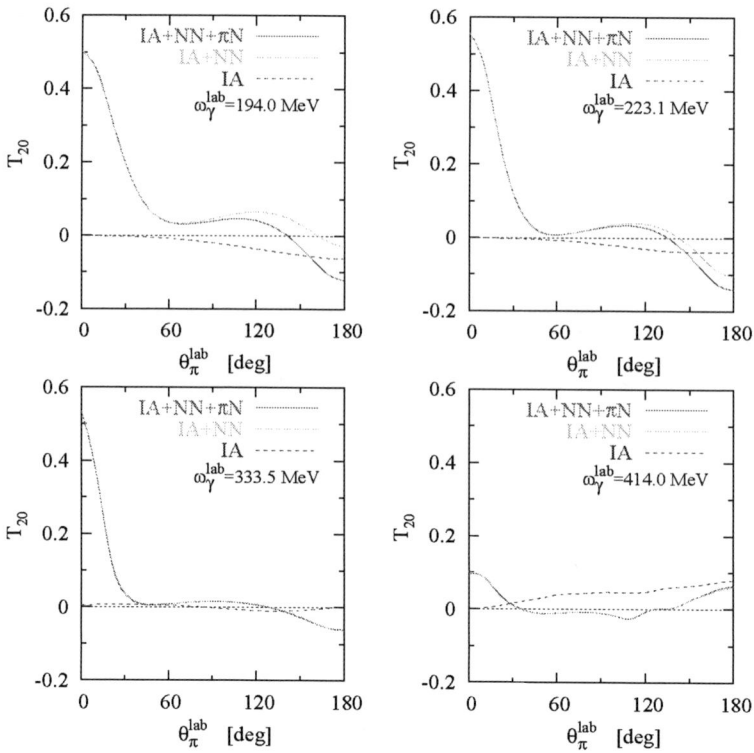

FIGURE 2. Target asymmetry T_{20} for oriented deuteron for the exclusive reaction $\vec{d}(\gamma, \pi^0 n)p$. Notation as in Fig. 1.

πN-interaction from Ref. [12] and taking into account all partial waves up to $l = 2$. I have evaluated the various polarization asymmetries in impulse approximation alone and with inclusion of NN- and πN-rescattering. As already mentioned, the elementary pion photoproduction amplitude is taken from the Sato-Lee (SL) model [8].

We begin the discussion of the results with the beam asymmetry Σ for linearly polarized photons for the reaction $d(\vec{\gamma}, \pi^0 n)p$ displayed in Fig. 1 as a function of the pion angle in the laboratory system at various photon lab-energies in comparison with the recent experimental data from LEGS Brookhaven National Laboratory (LEGS-exp-L3b) [10]. In general, one finds for the Σ-asymmetry a broad distribution with a maximum around $80°$ to $90°$. In all of these energies one notes a relatively small influence from FSI effects. The influence from FSI is noticeable at the lowest energy and forward pion angles. With increasing photon lab-energy the effect of FSI becomes smaller although not negligible even at the highest energy. The relatively largest influence appears at the lowest energy of $\omega_\gamma^{lab} = 264.7$ MeV and at forward pion angles. A comparison to recent data from the LEGS-collaboration (LEGS-exp.L3b) [10] on the linear photon asymmetry Σ for the reaction $d(\vec{\gamma}, \pi^0 n)p$ is exhibited also in Fig. 1 for three photon

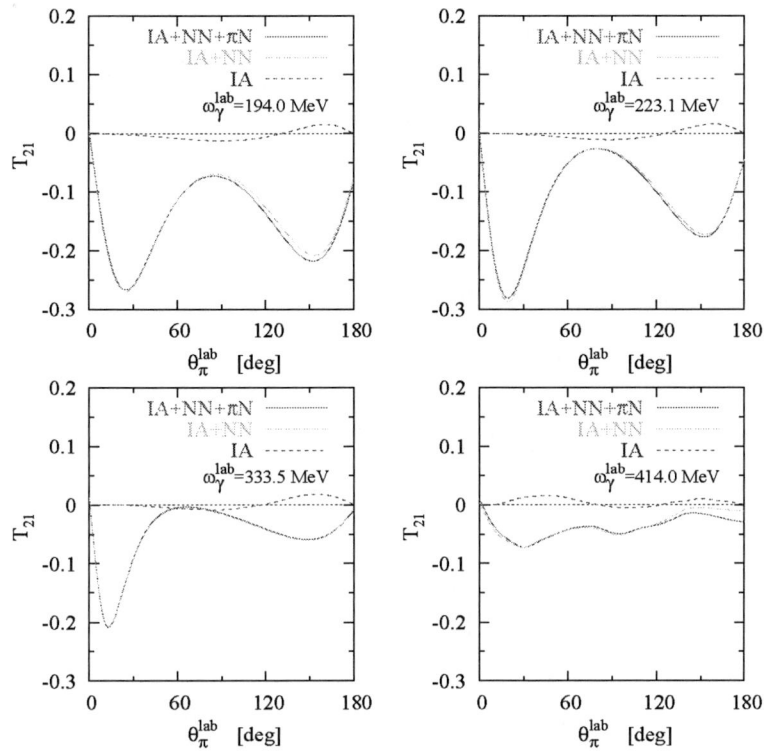

FIGURE 3. Target asymmetry T_{21} for oriented deuteron for $\vec{d}(\gamma, \pi^0 n)p$. Notation as in Fig. 1.

lab-energies. One notes quite a perfect agreement of the theoretical description with the data.

As next we discuss the tensor target asymmetries T_{2M} displayed in Figs. 2 through 4 for $M = 0$, 1, and 2, respectively, as a function of the emission pion angle. In general NN-FSI effects are sizeable, whereas πN-rescattering is negligible. The tensor target asymmetry T_{20} exhibits a pronounced sharp peak at $0°$ and a rapid fall-off with increasing angles, remaining quite small above $30°$. It is small in absolute size both in IA and IA+FSI. The FSI effect is very large at forward angles. The tensor target asymmetry T_{21} shows drastic FSI influence. The resulting absolute size of T_{21} is small. T_{21} peaks at small angles around $20°$ for small energies. For high energies this peak disappears and becoming a broader distribution with a considerably smaller size. FSI shows some notable influence. Finally, the tensor target asymmetry T_{22} exhibits a prominent peak in forward direction which becomes sharper and moves towards smaller angles with increasing photon lab-energy. This asymmetry is very small in IA. Its absolute size is less than 0.02 in the kinematic region under consideration. FSI manifests itself in a pronounced peak around $30°$ although even in the center of the peak the asymmetry T_{22} is still small. The tensor asymmetries are much more sensitive to FSI. This is particularly

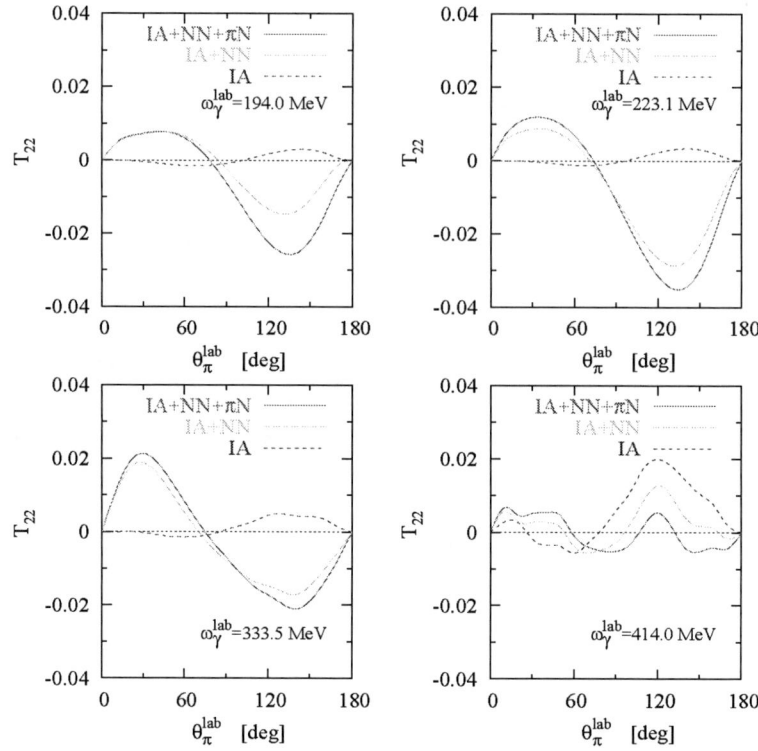

FIGURE 4. Target asymmetry T_{22} for oriented deuteron for $\vec{d}(\gamma, \pi^0 n)p$. Notation as in Fig. 1.

apparent in T_{20} exhibiting a forward negative minimum in IA which turns into a positive forward peak when FSI is switched on. Also T_{21} shows such a drastic influence from FSI. T_{22} is much smaller and shows an oscillatory behavior. FSI effects are noticeable again.

The vector target asymmetry T_{11} for the reaction $\vec{d}(\gamma, \pi^0 n)p$ at four selected energies of $\omega_\gamma^{lab} = 194$, 223.1, 333.5, and 414 MeV is observed in Fig. 5 as a function of the emission pion angle. The T_{11}-asymmetry shows a broad distribution over the whole angular range. We see that the structure of T_{11} changes significantly with energy. While at $\omega_\gamma^{lab} = 194$ MeV one finds a maximum around $150°$, one notes a negative minimum at backward pion angles and a positive maximum at forward pion angles at $\omega_\gamma^{lab} = 414$ MeV.

4. CONCLUSION

Quasifree single-pion photoproduction from the deuteron via the exclusive reaction $d(\gamma, \pi^0 n)p$ has been studied in the $\Delta(1232)$-resonance region with particular emphasis

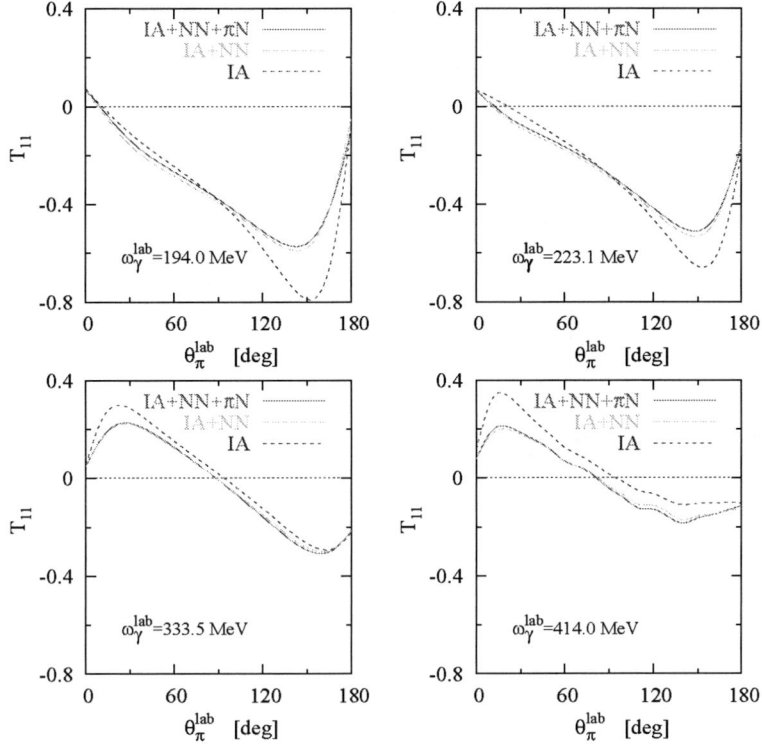

FIGURE 5. Vector target asymmetry T_{11} for $\vec{d}(\gamma, \pi^0 n)p$. Notation as in Fig. 1.

on the influence of final state interaction in the NN- and πN-subsystems of the final state. As polarization observables we have considered the linear photon asymmetry, tensor and vector target asymmetries of the semi-exclusive differential cross section. Many of these asymmetries are found to be quite sizeable, in particular the linear photon asymmetry Σ and the vector target asymmetry T_{11}. The tensor target asymmetries are in general considerably smaller. They are often quite insensitive to final state rescattering. A comparison of our results for the photon asymmetry Σ to recent experimental data from the LEGS-collaboration (LEGS-exp.L3b) [10] has been achieved and a satisfactory agreement has been obtained.

ACKNOWLEDGMENTS

I would like to express my gratitude to the organizing committee and the chairman of the CICHEP II Conference for providing the pleasant environment for useful discussions. I am indebted to Profs. T.-S. H. Lee, T. Sato and A. Sandorfi for fruitful discussions and valuable information.

REFERENCES

1. E. M. Darwish, H. Arenhövel, and M. Schwamb, *Eur. Phys. J. A* **16**, 111 (2003).
2. E. M. Darwish, H. Arenhövel, and M. Schwamb, *Eur. Phys. J. A* **17**, 513 (2003).
3. H. Arenhövel, A. Fix, and M. Schwamb, *Phys. Rev. Lett.* **93**, 202301 (2004).
4. H. Arenhövel and A. Fix, *Phys. Rev. C* **72**, 064004 (2005);
   A. Fix and H. Arenhövel, *Phys. Rev. C* **72**, 064005 (2005).
5. M. I. Levchuk *et al.*, *Phys. Rev. C* **74**, 014004 (2006).
6. D. Drechsel, O. Hanstein, S. Kamalov, and L. Tiator, *Nucl. Phys. A* **645**, 145 (1999); http://www.kph.uni-mainz.de/MAID/maid2003/.
7. R. A. Arndt, W. J. Briscoe, I. I. Strakovsky, and R. L. Workman, *Phys. Rev. C* **66**, 055213 (2002), SAID database, http://gwdac.phys.gwu.edu
8. T. Sato and T.-S. H. Lee, *Phys. Rev. C* **54**, 2660 (1996).
9. T.-S. H. Lee and T. Sato, private communication.
10. A. Sandorfi, private communication.
11. J. Haidenbauer and W. Plessas, *Phys. Rev. C* **30**, 1822 (1984);
    J. Haidenbauer and W. Plessas, *Phys. Rev. C* **32**, 1424 (1985).
12. S. Nozawa, B. Blankleider, and T.-S. H. Lee, *Nucl. Phys. A* **513**, 459 (1990).

Forward-Backward Charge Fluctuations at RHIC Energies

Stephane Haussler[a], Mohamed Abdel-Aziz[b], Marcus Bleicher[b].

[a] Frankfurt Institute for Advanced Studies (FIAS), J.W. Goethe Universität,
Max von Laue Straße 1, 60438 Frankfurt am Main, Germany

[b] Institut für Theoretische Physik, J.W. Goethe Universität,
Max von Laue Straße 1, 60438 Frankfurt am Main, Germany

We use the ultra-relativistic quantum molecular dynamic UrQMD version 2.2 to study forward backward fluctuations and compare our results with the published data by the PHOBOS.

1. Introduction

One of the main goals of the relativistic heavy ion program is to understand the nature of the hadron production mechanism (e.g. parton coalescence, string fragmentation or cluster decay). Recent RHIC data suggested the formation of a quark gluon plasma (QGP) during the collision of two heavy gold nuclei at center of mass energy $\sqrt{s_{NN}} = 200$ GeV. Using correlations and fluctuations to probe the nature of the created QCD matter has been proposed by many authors, see for example[1]. Recently, the PHOBOS experiment performed a similar analysis to the UA5 experiment [3] for Au+Au reactions at $\sqrt{s_{NN}} = 200$ GeV [4]. In ref [2], a simple model was introduced to extract the effective cluster multiplicity K_{eff} from PHOBOS data and $K_{\text{eff}} \sim 2.7$ was found for peripheral collisions and $K_{\text{eff}} \sim 2.2$ for central collisions. The value of K_{eff} in central collisions is close to the pp value reported by UA5 [3]. Note that all measured cluster multiplicities K_{eff} are larger than the one that is computed for a hadron resonance gas ($K_{\text{HG}} = 1.5$) [5], indicating that the measured charge correlations can not be described by simple statistical models based on hadronic degrees of freedom. Our goal in the current study is to get a baseline estimate for forward-backward fluctuations based on the microscopic hadronic transport model UrQMD. For a complete review of the model see [6]. In this paper we analyze 5×10^5 pp and minimum bias Au+Au events at $\sqrt{s_{NN}} = 200$ GeV.

2. Forward-Backward Fluctuations

In this section, we introduce the variable C that measures the asymmetry between the forward and backward charges. We define two symmetric rapidity regions at $\pm\eta$ with equal width $\Delta\eta$. The number of charged particles in the forward rapidity interval $\eta \pm \Delta\eta/2$ is N_F while the corresponding number in the backward hemisphere $-\eta \pm \Delta\eta/2$ is given by N_B. We define the asymmetry variable $C = (N_F - N_B)/\sqrt{N_F + N_B}$, in each event.

CP881, *Cairo International Conference on High Energy Physics (CICHEP II)*,
edited by S. Khalil

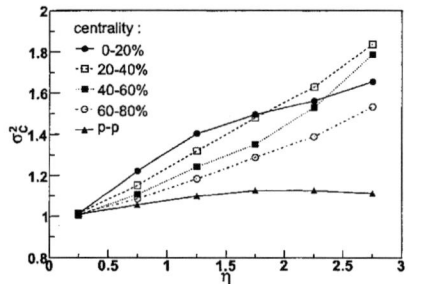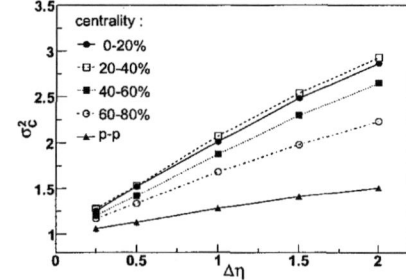

Figure 1. UrQMD results of dynamic fluctuations as a function of η with $\Delta\eta=0.5$ (left) and as a function of $\Delta\eta$ with $\eta=2$ (right).

The variance of the charged particle multiplicity in the forward hemisphere is given by $D_{FF} = \langle N_F^2 \rangle - \langle N_F \rangle^2$ and similarly for the backward hemisphere $D_{BB} = \langle N_B^2 \rangle - \langle N_B \rangle^2$. We also introduce the covariance of charged particles in both hemispheres by $D_{FB} = \langle N_F N_B \rangle - \langle N_F \rangle \langle N_B \rangle$, where $\langle .... \rangle$ stands for the average over all events. The PHOBOS measure of the dynamical fluctuations σ_C^2 can be written as

$$\sigma_C^2 \approx \frac{D_{FF} + D_{BB} - 2D_{FB}}{\langle N_F + N_B \rangle}. \tag{1}$$

Recently STAR [7] reported a preliminary results of the so called correlation strength parameter $b = D_{FB}/D_{FF}$. The effective cluster multiplicity K_{eff} is propotional to σ_C^2, such that if b=0, then the covariance D_{FB} vanishes. In this case we have $\sigma_C^2 = K_{\text{eff}}$. We emphasize that K_{eff} should be understood as a product of the true cluster multiplicity times a leakage factor ξ that takes into account the limited observation window $\Delta\eta$. The event by event fluctuations of the asymmetry parameter (variance) σ_C^2 in the absence of any correlations among the produced particles will be $\sigma_C^2 = 1$.

3. Results

In Fig. 1(left) we show σ_C^2 as a function of η at $\sqrt{s_{NN}} = 200$ GeV computed from UrQMD. We find that for pp collisions $\sigma_C^2 \approx 1$ when $\eta = 0.25$, then σ_C^2 increases to 1.1 for $\eta = 2.75$. In Fig. 1(right) we plot σ_C^2 as a function of $\Delta\eta$, we notice that σ_C^2 reaches approximately 1.6. This value is close to the resonance gas value $K_{\text{eff}} \sim 1.5$ mentioned in [5]. In Fig. 1, we plot UrQMD results for Au+Au at different centralities. UrQMD shows that σ_C^2 as a function of η and $\Delta\eta$ has a clear centrality dependance, such that σ_C^2 increases with centrality and then starts to decrease with more centrality cuts. In [2] we predicted that, for PHOBOS data, with more centrality cut, σ_C^2 may be reduce to 1.9, and this reduction in $\sigma_C^2 \sim K_{\text{eff}}$ may be regarded as an indication for cluster melting at RHIC. In Fig. 2 we show our UrQMD with PHOBOS data [4] and their HIJING results for $0 - 20\%$ central (left) and $40 - 60\%$ peripheral (right) Au+Au collisions. In both figures, we keep $\Delta\eta = 0.5$. From Fig. 2 we find that for both centralities σ_C^2 increases with increasing η. This behavior exists in both the experimental data and HIJING. For both

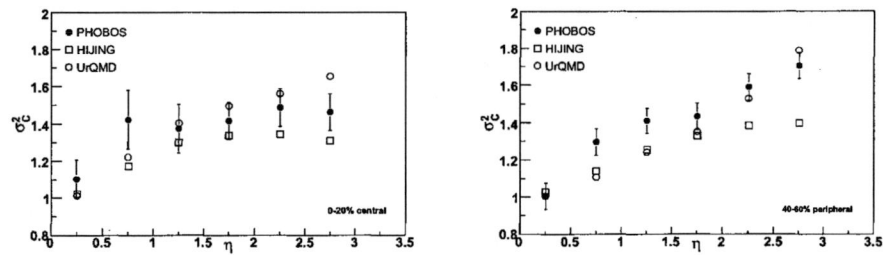

Figure 2. σ_C^2 as a function of η at $\sqrt{s_{NN}} = 200$ GeV, with $\Delta\eta$=0.5. (Left) $0-20\%$ central Au+Au, (right) $40-60\%$ peripheral Au+Au collisions. Black circles are PHOBOS [4] data, open squares HIJING and open circles are UrQMD results.

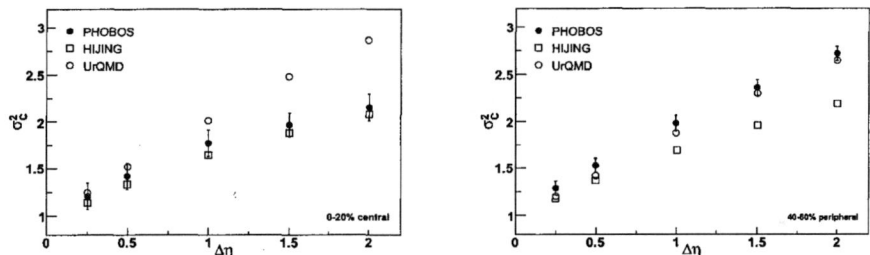

Figure 3. σ_C^2 as a function of $\Delta\eta$ at $\sqrt{s_{NN}} = 200$ GeV, with η=2. (Left) $0-20\%$ central Au+Au, (right) $40-60\%$ peripheral Au+Au collisions. Black circles are PHOBOS data [4], open squares HIJING and open circles are UrQMD results.

centralities we find that $\sigma_C^2 \approx 1$ when $\eta = 0.25$. This is because the competition between long and short range correlations almost cancels. In $0-20\%$ central collisions Fig. 2, UrQMD and HIJING roughly reproduce the experimental data within 1.5σ. In $40-60\%$ peripheral collisions Fig. 2(right), UrQMD can roughly produce the PHOBOS data while HIJING deviates by more than two σ for the large rapidity gaps. In ref [2], we show that by varying the observation window $\Delta\eta$ one can see the whole cluster structure. We keep the center of the observation window at $\eta = \pm 2$ while we allow for the observation window to change. In Fig. 3 we show the PHOBOS data [4] in addition to HIJING and our UrQMD analysis. We see that the measured σ_C^2 increases up to 2.2 and 2.8 for central and peripheral collisions respectively. In contrast to HIJING which gives the same value for both centralities. UrQMD shows a centrality dependance as shown in Fig 3. For $0-20\%$ cental collisions, HIJING can reproduce the data. HIJING fails to reproduce the the peripheral data. UrQMD over estimates σ_C^2 in the central collisions while it succeeded to reproduce the peripheral collisions. The failure of UrQMD to reproduce the the central data indicated that the cluster structure in UrQMD can survive, because the hadronic rscattering is not strong enough to destroy such clusters.

4. Summary and Conclusion

In this paper we computed forward-backward fluctuations and compared UrQMD calculations results to the available experimental data measured by PHOBOS. We started by studying proton-proton collisions and we find that long range correlation persists over a wide rapidity gap between the two rapidity hemispheres. The variance of the asymmetry parameter C was found to increase with increasing $\Delta\eta$ such that σ_C^2 changes from $\sigma_C^2(\eta = 2, \Delta\eta = 0.25) \approx 1$ to $\sigma_C^2(\eta = 2, \Delta\eta = 3) \approx 1.6$, this can be due the saturation of the leakage factor $\xi \to 1$. For Au+Au collisions, we find that for both centrality bins $0-20\%$ and $40-60\%$, $\sigma_C^2 \approx 1$ for small η. This can be seen as a cancellation between short and long range fluctuations. By increasing η we see that σ_C^2 also increases and approaches 1.6 and 1.8 for $0-20\%$ and $40-60\%$ respectively. This increase can be attributed to the decrease in the long range correlations. This will be true if the particle production mechanism does not change with η. To see the whole cluster structure, we fix the center of the observation window at 2 and allow $\Delta\eta$ to increase. We find that UrQMD can reproduce the peripheral data while it overestimates the experimental results for central collisions. HIJING gives the opposite behavior to UrQMD. HIJING produces the central data while it fails to reproduce the peripheral data as shown by PHOBOS. Also we find that UrQMD shows a smaller centrality dependance than the data. The discrepancies between HIJING, UrQMD and the data encourage more theoretical study to be done in order to clarify the correlation between produced particles in high energy nuclear collisions. The next step in this work is to measure b and σ_C^2 consistently to extract the cluster multiplicity K_{eff} and test for the survival or melting of such clusters.

Acknowledgements

This work is supported by BMBF and GSI.

REFERENCES

1. S. Haussler, H. Stoecker and M. Bleicher, Phys. Rev. C **73**, 021901 (2006); S. Jeon, L. Shi and M. Bleicher, Phys. Rev. C **73**, 014905 (2006); V. Koch, M. Bleicher and S. Jeon, Nucl. Phys. A **698**, 261 (2002) [Nucl. Phys. A **702**, 291 (2002)]; M. Bleicher, S. Jeon and V. Koch, Phys. Rev. C **62**, 061902 (2000); M. Bleicher *et al.*, Phys. Lett. B **435**, 9 (1998); N. Armesto, L. McLerran and C. Pajares, arXiv:hep-ph/0607345; L. Cunqueiro, E. G. Ferreiro, F. del Moral and C. Pajares, Phys. Rev. C **72**, 024907 (2005).
2. M. Abdel-Aziz and M. Bleicher, arXiv:nucl-th/0605072.
3. K. Alpgard *et al.* [UA5 Collaboration], Phys. Lett. B **123**, 361 (1983); R. E. Ansorge *et al.* [UA5 Collaboration], Z. Phys. C **37**, 191 (1988).
4. B. B. Back *et al.* [PHOBOS Collaboration], Phys. Rev. C **74**, 011901 (2006).
5. M. A. Stephanov, K. Rajagopal and E. V. Shuryak, Phys. Rev. D **60**, 114028 (1999).
6. S. A. Bass *et al.*, Prog. Part. Nucl. Phys. **41**, 225 (1998), M. Bleicher *et al.*, J. Phys. G **25** (1999) 1859.
7. T. J. Tarnowsky [STAR Collaboration], arXiv:nucl-ex/0606018.

Heterotic parastrings

N. Belaloui and L. Khodja

Département de physique, Faculté des Sciences Exactes, Université Mentouri, Constantine, Algeria

Abstract. We investigate a parabose parafermi version of the heterotic strings. When we impose the modular invariance of the one-loop amplitude, we find an other possibility of the heterotic strings based on the group E_8. In this case, a consistent analysis of the spectrum with respect to the partition function is done, and the SUSY generators algebra which correspond to the algebra of the SUSY Quantum Mechanics is constructed.

Keywords: strings, supersymmetry, parastatistics
PACS: 11.30.Pb, 11.25.-w, 11.25.Hf

INTRODUCTION

Paraquantum bosonic and superstring were constructed in various critical dimensions depending on the order of the paraquantization. The first study of the parabosonic and paraspinning string theories was done by F. Ardalan and F. Mansouri [1] .This study is based on the particular manner in which the center of mass variables of the string are to be handled. In this hypothesis, they find that the resulting parabosonic (resp paraspinning) string theories are consistent if the dimension D (resp. D') of the spacetime and the order Q of the paraquantization are related by the expressions $D = 2 + \frac{24}{Q}$ (resp $D' = 2 + \frac{8}{Q}$).

A second study of these two cases without Ardalan and Mansouri hypothesis on the center of mass variables of the string is done by Bennacer and Belaloui [2,3]. Like in Ardalan and Mansouri work [1], D (resp. D') is again given as a function of the paraquantization order through the relation $D = 2 + \frac{24}{Q}$ (resp $D' = 2 + \frac{8}{Q}$).

In particular, one can have paraspinning strings with critical dimensions $D' = 10, 6, 4, 3$ (respectively in orders $Q = 1, 2, 4, 8$). This coincides with the dimensions in which the classical superstrings can be formulated.

Another work is done by L. Khodja , N. Belaloui and H. Bennacer [4,5] which consists to investigate the existence possibilities of the $D = 3, 4, 6$ parasuperstrings.

As superstrings can be viewed as $Q = 1$ parasuperstrings, this theory presented some anomalies vanished for the two gauge groups $SO(32)$ and $E_8 \otimes E_8$.

In this work, we investigate the consequences of the study of [4,5] on the heterotic parastring, in this case, three points are developped

- Modular invariance of the one-loop amplitude for external massless string states which imposes some conditions on the lattice, this imply that, in addition to the ordinary case $(D, D') = (26, 10)$,where the two theories exist ($SO(32)$ and $E_8 \otimes E_8$), the heterotic parastring can only survives in the case (14,6) which is constructed from the only possible group E_8.

CP881, *Cairo International Conference on High Energy Physics (CICHEP II)*,
edited by S. Khalil
© 2007 American Institute of Physics 978-0-7354-0382-6/07/$23.00

- The susy generators algebra. We found that : a parabosonic-parafermionic susy system conducts to the algebra of the susy Qantum Mechanic .
- The spectrum: we found a common chord between the number of degenerations of the states and the one given by the partition function.

HETEROTIC PARASTRINGS

As heterotic string can be viewed as $Q = 1$ heterotic parastring, one might be tempted to seek what hybrid combination one can construct from closed parabosonic string and parasuperstrings .

We consider the light-cone gauge. The right-moving sector consists of the right-modes of the oriented parasuperstring that is $\frac{8}{Q}$ transverse parabosonic coordinates $X^i(\tau - \sigma)$ $(i = 1, \overline{\frac{8}{Q}})$ and a Majorana and (or) Weyl parafermionic coordinates $S^a(\tau - \sigma)$ where $(a = 1, \overline{2^{\frac{D'}{2}}})$. The left-moving sector consists of the left-modes of the oriented parabosonic string that is $\frac{8}{Q}$ transverse coordinates $X^i(\tau + \sigma)$ with $(i = 1, \overline{\frac{8}{Q}})$ .The remaining $\frac{16}{Q}$ coordinates $X^I(\tau + \sigma)$ with $(I = 1, \overline{\frac{16}{Q}})$ are compactified on a lattice which properties will be determined.

The light cone gauge action describing the dynamics of the free heterotic strings is given by (see for example [6])

$$S = -\frac{1}{4\pi\alpha'} \int d\sigma d\tau \left(\partial_a X^i \partial^a X^i + \sum_{I=1}^{D-D'} \partial_a X^I \partial^a X^I + i\bar{S}\gamma^-(\partial_\tau + \partial_\sigma)S \right) \qquad (1)$$

where in addition, the constraints $(\partial_\tau - \partial_\sigma)X^i = 0$ and $\gamma^+ S^a = 0$ have to be imposed.

The solutions of the equations of motion are given by the following oscillators expansions:

$$X^i(\tau - \sigma) = \frac{1}{2}x^i + \frac{1}{2}p^i(\tau - \sigma) + \frac{1}{2}i\sum_{n=1}^{\infty}\frac{1}{n}\alpha_n^i \exp[-2in(\tau - \sigma)] \qquad (2)$$

$$X^i(\tau + \sigma) = \frac{1}{2}x^i + \frac{1}{2}p^i(\tau + \sigma) + \frac{1}{2}i\sum_{n=1}^{\infty}\frac{1}{n}\tilde{\alpha}_n^i \exp[-2in(\tau + \sigma)] \qquad (3)$$

$$S^a(\tau - \sigma) = \sum_{n=-\infty}^{\infty} S_n^a \exp[-2in(\tau - \sigma)] \qquad (4)$$

$$X^I(\tau + \sigma) = x^I + p^I(\tau + \sigma) + \frac{1}{2}i\sum_{n=1}^{\infty}\frac{1}{n}\tilde{\alpha}_n^I \exp[-2in(\tau + \sigma)] \qquad (5)$$

In this gauge, the paraquantum operators $x^-, p^+, x^i, p^i, \alpha_n^i, x^I, p^I; \tilde{\alpha}_n^I$ and s_n^a verify the trilinear relations

$$\left[x^i, \left[p^j, p^k\right]_+\right] = 2i\left(\delta^{ij}p^k + \delta^{ik}p^j\right) \qquad (6)$$

268

$$\left[x^i,\left[p^j,A\right]_+\right] = 2\iota\delta^{ij}A \tag{7}$$

$$\left[\alpha_n^i,\left[\alpha_m^j,\alpha_l^k\right]_+\right] = 2(n\delta^{ij}\delta_{n+m,0}\alpha_l^k + n\delta^{ik}\delta_{n+l,0}\alpha_m^j) \tag{8}$$

$$\left[\alpha_n^i,\left[\alpha_m^j,B\right]_+\right] = 2\delta^{ij}n\delta_{n+m}B \tag{9}$$

$$\left[\tilde{\alpha}_n^I,\left[\tilde{\alpha}_m^J,\tilde{\alpha}_l^K\right]_+\right] = 2(n\delta^{IJ}\delta_{n+m,0}\tilde{\alpha}_l^K + n\delta^{IK}\delta_{n+l,0}\tilde{\alpha}_m^J) \tag{10}$$

$$\left[\tilde{\alpha}_n^I,\left[\tilde{\alpha}_m^J,C\right]_+\right] = 2\delta^{IJ}n\delta_{n+m}C \tag{11}$$

$$\left[s_n^a,\left[\bar{s}_m^b,\bar{s}_l^c\right]_-\right] = 2\left[(\gamma^+h)^{ab}\delta_{n+m}\bar{s}_l^c - (\gamma^+h)^{ac}\delta_{n+l}\bar{s}_m^b\right] \tag{12}$$

$$\left[s_n^a,\left[\bar{s}_m^b,D\right]_+\right]_+ = 2(\gamma^+h)^{ab}\delta_{n+m}D \tag{13}$$

$$\left[x^I,\left[p^J,p^K\right]_+\right] = i(\delta^{IJ}p^K + \delta^{IK}p^J) \tag{14}$$

$$\left[x^I,\left[p^J,E\right]_+\right] = \iota\delta^{IJ}E \tag{15}$$

$$i,j,k\ldots\ldots = \overline{1,D'-2} \tag{16}$$

$$a,b,c\ldots\ldots = \overline{1,D'-2} \tag{17}$$

$$and\ I,J,K\ldots = \overline{1,D-D'=1,\frac{16}{Q}} \tag{18}$$

and all the others are null. Here A, B, C, D and E represent the following operators
$A = x^k,\alpha_n^k,\tilde{\alpha}_n^I,s_n^a,x^I,p^I$
$B = x^k,p^k,\tilde{\alpha}_n^I,s_n^a,x^I,p^I$
$C = x^i,p^i,\alpha_n^i,s_n^a,x^K,p^K$
$D = x^i,p^i,\alpha_n^i,\tilde{\alpha}_n^I,x^I,p^J$
$E = x^i,p^i,\alpha_n^i,\tilde{\alpha}_n^K,s_n^a,x^K$
applying the Green decomposition [7], the set of the precedent trilinear commutation relations is equivalent to the following anomalous bilinear relations:

$$\left[x^{i(\alpha)},p^{j(\alpha)}\right] = \iota\delta^{ij} \quad ; \quad \left[x^{i(\alpha)},p^{j(\beta)}\right]_+ = 0 \quad \alpha \neq \beta \tag{19}$$

$$\left[\alpha_n^{i(\alpha)},\alpha_m^{j(\alpha)}\right] = n\delta^{ij}\delta_{n+m,0} \quad ; \quad \left[\alpha_n^{i(\alpha)},\alpha_m^{j(\beta)}\right]_+ = 0 \quad \alpha \neq \beta \tag{20}$$

$$\left[\tilde{\alpha}_n^{I(\alpha)},\tilde{\alpha}_m^{J(\alpha)}\right] = n\delta^{IJ}\delta_{n+m,0} \quad ; \quad \left[\tilde{\alpha}_n^{I(\alpha)},\tilde{\alpha}_m^{J(\beta)}\right]_+ = 0 \quad \alpha \neq \beta \tag{21}$$

$$\left[s_n^{a(\alpha)},\bar{s}_m^{b(\alpha)}\right]_+ = \delta^{ab}\delta_{n+m,0} \quad ; \quad \left[s_n^{a(\alpha)},\bar{s}_m^{b(\beta)}\right] = 0 \quad \alpha \neq \beta \tag{22}$$

$$\left[\alpha_n^{i(\alpha)},s_m^{a(\alpha)}\right] = 0 \quad ; \quad \left[\alpha_n^{i(\alpha)},s_m^{a(\beta)}\right]_+ = 0 \quad \alpha \neq \beta \tag{23}$$

$$\left[x^{I(\alpha)},p^{J(\alpha)}\right] = \frac{1}{2}\iota\delta^{IJ} \quad ; \quad \left[x^{I(\alpha)},p^{J(\beta)}\right]_+ = 0 \quad \alpha \neq \beta \tag{24}$$

and all the other commutators (and anticommutators) of the type $[A^{(\alpha)},B^{(\alpha)}] = 0$ (and $[A^{(\alpha)},B^{(\beta)}]_+ = 0$), for $\alpha \neq \beta$.

MODULAR INVARIANCE

In order to have a consistent theory, one has to deal with left-moving and right-moving coordinates with the same order Q. this imposes contraintes on Q

$$
\begin{aligned}
Q &= 1 \longrightarrow (D,D') = (26,10) \\
Q &= 2 \longrightarrow (D,D') = (14,6) \\
Q &= 4 \longrightarrow (D,D') = (8,4) \\
Q &= 8 \longrightarrow (D,D') = (5,3)
\end{aligned}
$$

Before developping this point, one can recall that the finiteness of the one-loop amplitude for external massless string states is subject to their modular invariance.

By analogy with the ordinary case, one can impose to the one-loop amplitude in the paraquantum case to be modular invariante.

From the expressions of the one-loop amplitude of the parabosonic string established by Ardalan and Mansouri [8] and the one of the heterotic string given by Gross et al [9,10]. In the paraquantum case, This latter turned out that it is equivalent to the one of the ordinary case with the following substitutions

$$
\begin{aligned}
D &\rightarrow Q(D-2)+2 \\
D' &\rightarrow Q(D'-2)+2
\end{aligned}
$$

so that, it takes the form

$$
A_{Loop} \sim \int \prod_{i=1}^{4} d^2 z_i \, |\omega|^{-2} \left[-\frac{4\pi}{\ln|\omega|} \right]^{\frac{Q(D'-2)+2}{2}} \times \prod_{1 \leq i \leq 4} [\chi(C_{JI},\omega)]^{\frac{1}{2}k_i k_j} B(\overline{\omega},\overline{z}_i,K_i) \quad (25)
$$

where

$$
B(\overline{\omega},\overline{z}_i,K_i) = \overline{\omega}^{-1} f(\overline{\omega})^{-Q(D-2)} \left[\psi(\overline{C}_{JI},\overline{\omega}) \right]^{K_i K_j} L(\overline{\omega},\overline{z}_i,K_i) \quad (26)
$$

and

$$
\chi(z,\omega) = \exp\left[\frac{\ln^2 |z|}{2\ln\overline{\omega}} \right] \left| \frac{1-z}{\sqrt{z}} \prod_{m=1}^{\infty} \frac{(1-\omega^m z)\left(1-\frac{\omega^m}{z}\right)}{(1-\omega^m)^2} \right| \quad (27)
$$

$$
\psi(\overline{z},\overline{\omega}) = \exp\left[\frac{\ln^2 \overline{z}}{2\ln\overline{\omega}} \right] \frac{1-\overline{z}}{\sqrt{\overline{z}}} \prod_{m=1}^{\infty} \left(\frac{(1-\overline{\omega}^m \overline{z})\left(1-\frac{\overline{\omega}^m}{\overline{z}}\right)}{(1-\overline{\omega}^m)^2} \right) \quad (28)
$$

$$
L(\overline{\omega},\overline{z}_i,K_i) = \sum_{P \in \Lambda} \exp\left[\frac{1}{2}\ln\overline{\omega}\left(P - \sum_{i=1}^{4} \frac{\ln\overline{z}_i}{\ln\overline{\omega}} Q_i \right)^2 \right] \quad (29)
$$

$$
Q_i = \sum_{j=1}^{i-1} K_i \quad (30)
$$

$$v_i = \sum_{j=1}^{i} \frac{\ln z_i}{2\pi i} \tag{31}$$

$$\tau = \frac{\ln \omega}{2\pi i} \tag{32}$$

$$C_{ji} = z_i z_{i+1} \ldots \ldots z_j \tag{33}$$

$$f(\omega) = \prod_{m=1}^{\infty} (1 - \omega^m) \tag{34}$$

here, the sum $\sum_{P \in \Lambda}$ runs over the points P of a $(D - D')$ dimensional lattice Λ and the notations are the same as in [10], in particular:

$$Q_i = \sum_{j=1}^{i-1} j, \quad v_i = \sum_{j=1}^{i} \frac{\ln z_j}{2\pi i}, \quad \tau = \frac{\ln \omega}{2\pi i} \tag{35}$$

$$C_{ji} = z_i z_{i+1} \ldots \ldots z_j, \quad f(\omega) = \prod_{m=1}^{\infty} (1 - \omega^m) \tag{36}$$

MODULAR TRANSFORMATION

Now, we consider the modular transformation:

$$\tau = \frac{\ln \omega}{2\pi i} \to \tau' = -\frac{1}{\tau} = -\frac{2\pi i}{\ln \omega'} \tag{37}$$

and the jacobian:

$$\prod_{i=1}^{4} d^2 z_i |\omega|^{-2} \sim d^2 \tau \prod_{i=1}^{3} d^2 v_i \tag{38}$$

from the transformations (37), we can write

$$\prod_{i=1}^{4} d^2 z_i |\omega|^{-2} \left[-\frac{4\pi}{\ln|\omega|} \right]^{\frac{Q(D'-2)+2}{2}}$$

$$\to \frac{1}{|\tau|^4} d^2 \tau \frac{1}{|\tau|^6} \prod_{i=1}^{3} d^2 v_i \left[-\frac{4\pi}{\ln|\omega|} \right]^{\frac{Q(D'-2)+2}{2}} |\tau|^2 \frac{Q(D'-2)+2}{2} \tag{39}$$

and

$$\overline{\omega}^{-1} f(\overline{\omega})^{-Q(D-2)} \prod_{1 \leq i < j \leq 4} [\chi(C_{JI}, \omega)]^{\frac{1}{2} k_i k_j} \left[\psi \left(\overline{v}_{ji}, \tau \right) \right]^{K_i K_j} L(\tau, v_i)$$

$$\to \overline{\omega}'^{-1} f(\overline{\omega}')^{-Q(D-2)} \overline{\tau}^{\frac{Q(D-2)}{2}} \times \prod_{1 \leq i < j \leq 4} \left[\psi \left(\overline{v}'_{ij}, \tau' \right) \overline{\tau} \exp \left(\frac{i\pi \overline{v}_{ij}^2}{\overline{\tau}} \right) \right]^{K_i K_j}$$

271

$$\times L^*(\tau', v_i')\overline{\tau}^{-\frac{Q(D-D')}{2}} \exp\left(-\frac{i\pi}{\overline{\tau}}\left(\sum_{i=1}^{N} Q_i \overline{v}_i\right)^2\right) \tag{40}$$

where $L^* \equiv L(\sum_{P \in \Lambda^*})$ and $\sum_{1 \le i < j \le 4} k_i k_j = 0$
Λ^* is the dual lattice of Λ
by the use of the identities

$$\prod_{1 \le i < j \le 4} \overline{\tau}^{K_i K_j} = \overline{\tau}^{-4} \text{ and } \sum_{1 \le i < j \le 4} \overline{v}_{ij}^2 K_i K_j = \left(\sum_{i=1}^{4} Q_i \overline{v}_i\right)^2 \tag{41}$$

the modular invariance is satisfied provided that $D' = 2 + \frac{8}{Q}$ and $\Lambda = \Lambda^*$ (the lattice Λ is self-dual)

Finally, the fact that the string is closed imposes a self dual even lattice.

Let us recall that the self-dual-even lattices exist only in eight n dimensions, i.e $D - D' = 8n$ and for $n = 1$, there exists only one such lattice which is the E_8 lattice This imply that in addition to the ordinary case $(Q = 1)$ where $(D, D') = (26, 10)$ based on the groups $E_8 \otimes E_8$ or $SO(32)$, there is a second possibility for $Q = 2$ where $(D, D') = (14, 6)$ based on the only semi-laced group E_8!

Notice that, in $D' = 6$, the spinors are Weyl and the number of internal coordinates is eight $(I = \overline{1,8})$

SUPERALGEBRA

Like in the parasuperstrings theory [4,5], in this case, we define the parasupercharges in the same way as follows :

$$Q^a = \frac{i}{2}\left[(p^+)^{\frac{1}{2}}, (\gamma^+ s_0)^a\right]_+ + i(p^+)^{\frac{-1}{2}} \sum_{n=-\infty}^{\infty} \left[(\gamma_i s_{-n})^a \alpha_n^i\right]_+ \tag{42}$$

these generators obey to an ordinary supersymmetric quantum mechanic algebra, i.e to the following ordinary anticommutation relations

$$\left[Q^a, \overline{Q}^b\right]_+ = -2\left(h\gamma_\mu p^\mu\right)^{ab} \tag{43}$$

This result is in accordance with what we find in the litterature about the parasuper-symmetric quantum mechanics systems.

PARTITION FUNCTION

In this section, the spectrum is analysed through the development of the partition function. We introduce the mass operator

$$\frac{1}{4}M^2 = N + \left(\tilde{N} - 1\right) + \frac{1}{2}\sum_{I=1}^{8}(p^I)^2 \tag{44}$$

272

where N is the number operator for the right-sector and $\tilde{N}$ for the left-sector. p^I are the momentum of the internal space.

$$N = \sum_{n=1}^{\infty} \sum_{\alpha=1}^{Q} \left(\alpha_{-n}^{i(\alpha)} \alpha_n^{i(\alpha)} + \frac{1}{2} \bar{s}_{-n}^{(\alpha)} \gamma^- s_n^{(\alpha)} \right) \tag{45}$$

$$\tilde{N} = \sum_{n=1}^{\infty} \sum_{\alpha=1}^{Q} \left(\tilde{\alpha}_{-n}^{i(\alpha)} \tilde{\alpha}_n^{i(\alpha)} + \tilde{\alpha}_{-n}^{I(\alpha)} \tilde{\alpha}_n^{I(\alpha)} \right) \tag{46}$$

The closure of the strings imposes the constraint:

$$\tilde{N} + \frac{1}{2} p_L^2 - 1 = N \tag{47}$$

where the left hand side $\left(\tilde{N} + \frac{1}{2} p_L^2 - 1 \right)$ corresponds to the left sector and the right

hand side (N) to the right sector, with $p_L^2 = \sum_{I=1}^{8} (p^I)^2$.

Let us define the partition functions F_L for the left-sector, F_R for the Right-sector and F for the heterotic parastring by the folowing relations:

$$F_L(x) = \sum_{N=-1}^{+\infty} d_L(N) x^N \tag{48}$$

$$F_R(x) = \sum_{N=1}^{+\infty} d_R(N) x^N \tag{49}$$

$$F(x) = \sum_N d(N) x^N \tag{50}$$

where :

$d_L(N)$ represents the degeneracy of the parabosonic left-moving modes at the Nth mass level compactified on an eight dimensional lattice Λ_{E_8}.

$d_R(N)$: degeneracy of parasuperstring right-moving modes at the Nth mass level

$d(N)$: the number of the heterotic parastring states at the Nth mass level

One can then write $F(x) = F_L(x) \otimes F_R(x)$ where the notation $\otimes$ means $d(N) = d_L(N).d_R(N)$

and where

$$F_L(x) = \frac{1}{x} P_{E_8}.F_L \left(\tilde{\alpha}_n^i, \tilde{\alpha}_n^I \right) \tag{51}$$

P_{E_8} is the partition function for the lattice Λ_{E_8} given by the relation (see for example [11]):

$$P_{E_8} = \sum_w e^{i\pi\tau|w|^2} = \sum_w q^{\frac{1}{2}|w|^2}$$

$$\equiv 1 + 240 \sum_{n=1}^{\infty} \sigma_3(n) q^n$$

$$= 1 + 240x + 2160x^2 + 6720x^3 + 17520x^4 + 30240x^5 + ... \tag{52}$$

with

$$\sigma_\alpha(n) = \sum_{d/n} d^\alpha \quad (d/n \text{ integer divisors of } n)$$

and $F_L\left(\tilde{\alpha}^i_n, \tilde{\alpha}^I_n\right)$ is the partition function for $(14-2)$ dimensions parabosonic string given by:

$$
\begin{aligned}
F_L\left(\tilde{\alpha}^i_n, \tilde{\alpha}^I_n\right) &= \sum_n d(n)x^n = Tr x^{\tilde{N}} \\
&= Tr x^{\sum_{n=1}^\infty \sum_{\alpha=1}^Q \left(\tilde{\alpha}^{i(\alpha)}_{-n} \tilde{\alpha}^{i(\alpha)}_n + \tilde{\alpha}^{I(\alpha)}_{-n} \tilde{\alpha}^{I(\alpha)}_n\right)} \\
&= \prod_{n=1}^\infty \left(\frac{1}{1-x^n}\right)^{12} \\
&= 1 + 12x + 90x^2 + 520x^3 + 2535x^4 + 10908x^5 + 42614x^6 + \dots \quad (53)
\end{aligned}
$$

then

$$
\begin{aligned}
F_L(x) &= \sum_N^\infty d_L(N)x^N \\
&= 252 + 5130x + 54760x^2 + 419895x^3 + 2587788x^4 + \dots\dots
\end{aligned}
$$

in the same way:

$$
\begin{aligned}
F_R(x) &= \sum_N d_R(N)x^N = Tr x^{\frac{1}{2}\sum_{n=1}^\infty \left([\alpha^i_{-n},\alpha^i_n]_+ + n[s^a_{-n},s^a_n]_-\right)} \\
&= 8\prod_{n=1}^\infty \left(\frac{1+x^n+x^{2n}}{1-x^n}\right)^4 \\
&= 8\Big[1 + 8x + 44x^2 + 188x^3 + 694x^4 + 1640x^5 + 5688x^6 \\
&\quad + 12224x^7 + 33542x^8 + 71188x^9 + O(x^{10})\Big] \quad (54)
\end{aligned}
$$

Finally

$$
\begin{aligned}
F(x) &= F_L(x) \otimes F_R(x) \\
&= \left(252 + 5130x + 54760x^2 + \dots\right) \otimes \left(8 + 64x + 352x^2 + \dots\right) \\
&= (252 \times 8) + (5130 \times 64)x + \dots. \quad (55)
\end{aligned}
$$

SPECTRUM

Let us now, describe the spectrum . The zero mass level is given by the set of the following states:

$M^2 = 0$

Left	Right
$\widetilde{\alpha}^i_{-1}\lvert 0\rangle_L \to 4$	$\lvert i\rangle \to 4\,parabosons$
$\widetilde{\alpha}^I_{-1}\lvert 0\rangle_L \to 8$	$\lvert a\rangle \to 4\,parafermions$
$\left\lvert p^I; (p^I)^2 = 2\right\rangle \to 240$	

where

- $\lvert i\rangle$: represents the four physical transverse polarizations of a massless vector field
- $\lvert a\rangle$: represent the four components spinorial partner
- In the left-sector we use i for the space time modes and I for the internal modes. The states $\left\lvert p^I; (p^I)^2 = even\right\rangle$ are the lattice vectors. In the right-sector, we use i for the parabosonic modes and a for the parafermionic modes

The total number of the states in the lowest level is $252 \times 8 = 2016$ states

As in the ordinary case, the breakdown of these states describes the (para)supergravity (PSG) and the(para)super Yang-Mills theories.

Indeed, the states $\widetilde{\alpha}^i_{-1}\lvert 0\rangle_L \times \lvert j\,or\,a\rangle_R$ describe what we will call the $D' = 6$ PSG, where their break down into distinct states gives

$$\widetilde{\alpha}^i_{-1}\lvert 0\rangle_L \times \lvert j\rangle_R + \widetilde{\alpha}^j_{-1}\lvert 0\rangle_L \times \lvert i\rangle_R \quad \to \frac{4\times 3}{2} + 4 = 10 \quad \to \text{graviton}$$

$$\widetilde{\alpha}^i_{-1}\lvert 0\rangle_L \times \lvert j\rangle_R - \widetilde{\alpha}^j_{-1}\lvert 0\rangle_L \times \lvert i\rangle_R \quad \to \frac{4\times 3}{2} = 6 \quad \to \text{antisym. tensor}$$

$$\widetilde{\alpha}^i_{-1}\lvert 0\rangle_L \times \lvert a\rangle_R \quad \to 4\times 4 = 16 \quad \to \text{gravitino}$$

The remaining states belong to what we will call the PSYM theory defined on E_8 and represented by the (992×2) states

$$\left[\widetilde{\alpha}^I_{-1}\lvert 0\rangle_L + \left\lvert p^I; (p^I)^2 = 2\right\rangle\right] \quad \times \quad \lvert i\,or\,a\rangle_R$$

One can again write all the physical states in the first and the second levels as follows:

1st level

$M^2 = 8$

	Left		Right	
	State	**Number**	**State**	**Number**
	$\frac{1}{2}\left[\tilde{\alpha}^I_{-1}, \tilde{\alpha}^J_{-1}\right]_+ |0\rangle_L$	$8 + \frac{8\times 7}{2} = 36$	$\alpha^i_{-1}|0\rangle_R$	32
	$\tilde{\alpha}^I_{-2}|0\rangle_L$	8	$s^a_{-1}|0\rangle_R$	32
	$\left| p^I ; (p^I)^2 = 4 \right\rangle$	2160		
	$\tilde{\alpha}^I_{-1}\left| p^J ; (p^J)^2 = 2 \right\rangle$	$8 \times 240 = 1920$		
	$\tilde{\alpha}^i_{-1}\left| p^J ; (p^J)^2 = 2 \right\rangle$	$4 \times 240 = 960$		
	$\tilde{\alpha}^i_{-2}|0\rangle_L$	4		
	$\frac{1}{2}\left[\tilde{\alpha}^i_{-1}, \tilde{\alpha}^I_{-1}\right]_+ |0\rangle_L$	$4 \times 8 = 32$		
	$\frac{1}{2}\left[\tilde{\alpha}^i_{-1}, \tilde{\alpha}^j_{-1}\right]_+ |0\rangle_L$	$4 + \frac{4\times 3}{2} = 10$		

2nd level

$M^2 = 16$

Left		Right	
State	**Number**	**State**	**Number**
$\frac{1}{2}\tilde{\alpha}^I_{-1}\left[\tilde{\alpha}^J_{-1},\tilde{\alpha}^K_{-1}\right]_+\lvert 0\rangle_L$	120	$\left\langle s^a_{-1},s^b_{-1}\right\rangle\lvert 0\rangle_R$	80
$\frac{1}{2}\left[\tilde{\alpha}^I_{-1},\tilde{\alpha}^J_{-2}\right]_+\lvert 0\rangle_L$	64	$\frac{1}{2}\left[\alpha^i_{-1},\alpha^j_{-1}\right]_+\lvert 0\rangle_R$	128
$\tilde{\alpha}^I_{-3}\lvert 0\rangle_L$	8	$\frac{1}{2}\left[\alpha^i_{-1},s^a_{-1}\right]_+\lvert 0\rangle_R$	80
$\tilde{\alpha}^I_{-1}\left\lvert p^I;(p^I)^2=4\right\rangle$	17280	$\alpha^i_{-2}\lvert 0\rangle_R$	32
$\frac{1}{2}\left[\tilde{\alpha}^I_{-1},\tilde{\alpha}^J_{-1}\right]_+\left\lvert p^I;(p^I)^2=2\right\rangle$	8640	$s^a_{-2}\lvert 0\rangle_R$	32
$\tilde{\alpha}^I_{-2}\left\lvert p^I;(p^I)^2=2\right\rangle$	1920		
$\frac{1}{2}\tilde{\alpha}^i_{-1}\left[\tilde{\alpha}^j_{-1},\tilde{\alpha}^k_{-1}\right]_+\lvert 0\rangle_L$	2		
$\frac{1}{2}\left[\tilde{\alpha}^i_{-1},\tilde{\alpha}^j_{-2}\right]_+\lvert 0\rangle_L$	16		
$\tilde{\alpha}^i_{-3}\lvert 0\rangle_L$	4		
$\tilde{\alpha}^i_{-3}\left\lvert p^I;(p^I)^2=4\right\rangle$	8640		
$\frac{1}{2}\left[\tilde{\alpha}^i_{-1},\tilde{\alpha}^j_{-1}\right]_+\left\lvert p^I;(p^I)^2=2\right\rangle$	2400		
$\tilde{\alpha}^i_{-2}\left\lvert p^I;(p^I)^2=2\right\rangle$	960		
$\frac{1}{2}\tilde{\alpha}^I_{-1}\left[\tilde{\alpha}^i_{-1},\tilde{\alpha}^j_{-1}\right]_+\lvert 0\rangle_L$	80		
$\frac{1}{2}\tilde{\alpha}^i_{-1}\left[\tilde{\alpha}^I_{-1},\tilde{\alpha}^J_{-1}\right]_+\lvert 0\rangle_L$	144		
$\left\lvert p^I;(p^I)^2=6\right\rangle$	6720		
$\frac{1}{2}\left[\tilde{\alpha}^i_{-1},\tilde{\alpha}^I_{-1}\right]_+\left\lvert p^I;(p^I)^2=2\right\rangle$	7680		
$\frac{1}{2}\left[\tilde{\alpha}^i_{-1},\tilde{\alpha}^I_{-2}\right]_+\lvert 0\rangle_L$	32		
$\frac{1}{2}\left[\tilde{\alpha}^i_{-2},\tilde{\alpha}^I_{-1}\right]_+\lvert 0\rangle_L$	32		

where $\lvert 0\rangle_R$ means $\lvert i\rangle$ or $\lvert a\rangle$ and where, by considering the trilinear commutation relations, we use these general forms (in the precedent tables) to describe all the different possibilities for the states, in particular, we introduce the following notation $\left\langle s^a_{-1},s^b_{-1}\right\rangle \rightarrow \left\{\frac{1}{2}\left[s^a_{-1},s^b_{-1}\right]_-,(s^a_{-1})^2\right\}$ to describe the product of the parafermionic operators s^a_n, where $(s^a_n)^3=0$ describes the second order of the paraquantization.

Notice here that, in the ordinary case $(Q=1)$, $(s^a_n)^2=0$ and it is clear that the last notation is equivalent to the ordinary product $s^a_{-1}s^b_{-1}$.

One can then determine the degeneration of the first three levels and recapitulate all the results in the following table as well as the polynomials describing the left-moving sector partition function and the ones of the right-moving sector.

	Left moving sector	**Right moving sector**
Partition Fct.	$252 + 5130x + 54760x^2 + \ldots$	$8 + 64x + 352x^2 + \ldots$
Fund.state	252	8
1st *level*	5130	64
2nd *level*	54760	352
3rd *level*	419895	1504

SUMMARY

In this work, we are interested in the construction of an heterotic string theory from the combination of a closed parabosonic srting and parasuperstring with the same order Q.
A number of results are obtained:

- The modular invariance imposed that, in addition to the ordinary case $(D, D') = (26, 10)$, only survives the case $(14, 6)$ for the order $Q = 2$, based on the only semi-laced group E_8.
- The SUSY generators algebra is equivalent to an ordinary SUSYQM algebra.
- The spectrum is analysed and the total partition function is derived. A coherence between the coefficients of the development of the total partition function and the number of the degenerated states is obtained

ACKNOWLEDGMENTS

One of the authors (N.B) would like to thank the organizers of this conference particularly prof. Shaaban Khalil. This work is supported by the Algerian Ministry of the Higher Education and Scientific Research under contract: No.D1602/04/2004

REFERENCES

1. F. Ardalan and F. Mansouri : Phys. Rev. D9 (1974) 3341.
2. N. Belaloui and H. Bennacer: Czech. J. Phys. 53 (2003) 769.
3. N. Belaloui and H. Bennacer: Czech. J. Phys. 54 (2004) 621.
4. N. Belaloui, L. Khodja and H. Bennacer: JINR, Dubna, 2004 ISBN 5-9530-0069-3.
5. N. Belaloui and L. Khodja: proceedings of the International Conference on High Energy and Mathematical Physics, Marrakech, 04-07 April 2005 (to appear)
6. M.Kaku, *Introduction to Superstings*, Spinger-Verlag 1990.
7. H.S Green, Phys.Rev. 90, (1953) 270.
8. F. Ardalan and F. Mansouri : Phys. Rev. Lett. 56 (1986) 2456.
9. D.J. Gross, J.A.Harvey, E.Martinec and R.Rohni, Nucl.Phys.B256 (1985) 253.
10. D.J.Gross, J.A.Harvey, E.Martinec and R.Rohni; Nucl.Phys. B267(1986)75-124.
11. D. Lüst and S. Theisen : Lectures on string theory (Springer, Berlin Heidelberg 1989).

Supercharge Operator of Hidden Symmetry in the Dirac Equation

Tamari T. Khachidze and Anzor A. Khelashvili

Department of Theoretical Physics, Ivane Javakhishvili Tbilisi State University,
I.Chavchavadze ave. 3, 0128, Tbilisi, Georgia *

Abstract. We require that there exist some additional symmetry of the Dirac Hamiltonian in an arbitrary central potential which describes the degeneracy of spectrum with respect to the two signs of eigenvalues of famous Dirac's K operator. It is proved that this happens only for the Coulomb potential. In this particular case symmetry operator reduces to the well-known Johnson-Lippmann operator, which plays a role of supercharge in the Dirac-Coulomb problem and generates a Witten's superalgebra.

Key words: Dirac operator, supersymmetry, commutativity, central potential, Witten's algebra.

PACS: 11.30.-j, 11.30.Pb

Supersymmetry (SUSY) of the Hydrogen atom is rather old and well-studied problem. We have in mind the usual cases: SUSY in non-relativistic quantum mechanics and inclusion of spin degrees of freedom by Pauli method as well. In this last case the projection of well-known Laplace-Runge-Lenz vector onto the electron spin direction plays the role of supercharge [1].

In case of the Dirac electron the so-called radial SUSY was demonstrated a long-time ago [2]. As for 3-dimensional case, it was shown [3–6], that the supercharge operator is the one, introduced by Johnson and Lippmann in 1950 in the form of a brief abstract [7]. As regards to the more detailed derivation, to the best our knowledge, is not published in scientific literature. Moreover as far as commutativity of the Johnson-Lippmann operator with the Dirac Hamiltonian is concerned, it is usually mentioned that it can be proved by "rather tedious calculations." [5]

Below we show that among all central potentials the Coulomb potential is a distinguished one. The additional symmetry takes place only for this potential. Then we show that the operator responsible for that symmetry reduces to the Johnson-Lippmann one in case of Coulomb potential. Simultaneously we obtain this operator in a simple and transparent manner and demonstrate its commutativity with the Dirac Hamiltonian.

So, let us consider the Dirac Hamiltonian for arbitrary central potential $V(r)$:

$$H = \vec{\alpha} \cdot \vec{p} + \beta m + V(r) \tag{1}$$

In this form $V(r)$ is a fourth component of a Lorentz 4-vector. We mention this fact here because the pure Lorentz -scalar potential is also often considered [8].

Let us introduce the so-called Dirac operator [9],

$$K = \beta \left(\vec{\Sigma} \cdot \vec{l} + 1 \right) = \beta \left(\vec{J}^2 - \vec{l}^2 - \vec{S}^2 + 1 \right) \equiv \beta \kappa. \tag{2}$$

where $\vec{l}$ is the angular momentum vector, $\vec{\alpha}$ and β are the usual Dirac matrices, $\vec{\Sigma}$ is the electron spin matrix

---

*khelash@ictsu.tsu.edu.ge

CP881, *Cairo International Conference on High Energy Physics (CICHEP II)*,
edited by S. Khalil
2007 American Institute of Physics 978-0-7354-0382-6/07/$23.00

$$\vec{\Sigma} = \begin{pmatrix} \vec{\sigma} & 0 \\ 0 & \vec{\sigma} \end{pmatrix} \tag{3}$$

and $\vec{J} = \vec{l} + \frac{1}{2}\vec{\Sigma}$ - total momentum. It is easy to show that

$$[K, H] = 0 \tag{4}$$

for arbitrary central potential, $V(r)$.

Therefore the spectrum of Dirac equation depends on the eigenvalues of the K -operator as well. As a rule, this leads to a degeneracy with regard to the signs of κ, $(\pm\kappa)$ at least for Coulomb potential [10, 11]

We can find an operator A, which could connect these two signs. Naturally, such an operator should anticommute with K,

$$\{A, K\} = AK + KA = 0 \tag{5}$$

If at the same time this operator commutes with Hamiltonian H, it'll generate the symmetry of the Dirac equation.

Therefore, we are looking for an operator A with the following properties

$$\{A, K\} = 0, \ [A, H] = 0 \tag{6}$$

After that we will be able to construct supercharges as follows [3–6]

$$Q_1 = A, \qquad Q_2 = i\frac{AK}{|\kappa|} \tag{7}$$

Then it is obvious that

$$\{Q_1, Q_2\} = 0 \tag{8}$$

and we can construct Witten's superalgebra, where $Q_1^2 = Q_2^2 \equiv h$ is a Witten's Hamiltonian.

Now our goal is a construction of the A operator. For this purpose at first we recall the generalization of one theorem [1, 12], known from Pauli equation. For the Dirac case this theorem may be formulated as follows:

Theorem: *If $\vec{V}$ is a vector with respect to the angular momentum $\vec{l}$, and this vector is perpendicular to $\vec{l}$*

$$\left(\vec{l} \cdot \vec{V}\right) = \left(\vec{V} \cdot \vec{l}\right) = 0,$$

then K anticommutes with operator $(\vec{\Sigma} \cdot \vec{V})$, which is scalar with respect to the total $\vec{J}$ momentum.

The proof of this theorem is almost trivial [13] - it is sufficient to consider the product $(\vec{\Sigma} \cdot \vec{l})(\vec{\Sigma} \cdot \vec{V})$ in this and in reversed order and make use of definition of K. Then it follows that

$$\left\{K, \vec{\Sigma} \cdot \vec{V}\right\} = K\left(\vec{\Sigma} \cdot \vec{V}\right) + \left(\vec{\Sigma} \cdot \vec{V}\right)K = 0 \tag{9}$$

It is evident that the class of operators anticommuting with K (so-called, K-odd operators) is not restricted by these operators only. Any operator of the form $\hat{O}(\vec{\Sigma} \cdot \vec{V})$, where $\hat{O}$ commutes with K, but otherwise arbitrary, also is a K - odd.

Let us remark for the further application, that the following useful relation holds in the framework of conditions of the above theorem

$$K\left(\vec{\Sigma} \cdot \vec{V}\right) = -i\beta\left(\vec{\Sigma} \cdot \frac{1}{2}\left[\vec{V} \times \vec{l} - \vec{l} \times \vec{V}\right]\right) \tag{10}$$

Now one can proceed to the second stage of our problem - we wish to construct the K-odd operator A, which commutes with H. It is clear that there remains large freedom according to the above mentioned remark about $\hat{O}$ operator - one can take $\hat{O}$ into account or ignore it.

We have the following physically interesting vectors at hand which obey the requirements of our theorem. They are:

$\hat{\vec{r}}$ - unit radius vector and $\vec{p}$ - linear momentum vector.

Both of them are perpendicular to $\vec{l}$ [14]. Therefore, we choose the following K-odd terms:

$$(\vec{\Sigma} \cdot \hat{\vec{r}}) \quad \text{and} \quad K(\vec{\Sigma} \cdot \vec{p}) \tag{11}$$

As it turns out inclusion of K into the second term of (11) is necessary for obtaining our final result self-consistently. Let's remark that both operators in (11) are diagonal matrices, while the Hamiltonian (1) is nondiagonal. Therefore, in commuting of (11) with nondiagonal terms appear as well. For instance,

$$\left[\vec{\Sigma} \cdot \hat{\vec{r}}, H\right] = \frac{2i}{r}\beta K\gamma^5 \tag{12}$$

Therefore, we probe the following operator

$$A = x_1 \left(\vec{\Sigma} \cdot \hat{\vec{r}}\right) + ix_2 K \left(\vec{\Sigma} \cdot \vec{p}\right) + ix_3 K\gamma^5 f(r) \tag{13}$$

Here the coefficients x_i $(i = 1, 2, 3)$ are chosen in such a way that A is Hermitian operator for arbitrary real numbers and $f(r)$ is an arbitrary scalar function to be determined later.

The commutator of A with H can be calculated straightforwardly, the result is

$$[A, H] = x_1 \frac{2i}{r}\beta K\gamma^5 + x_2 K \left(\vec{\Sigma} \cdot \hat{\vec{r}}\right) V'(r) -$$
$$- x_3 K \left(\vec{\Sigma} \cdot \hat{\vec{r}}\right) f'(r) - ix_3 2m\beta K\gamma^5 f(r)$$

Equating the above expression to zero, i.e. requiring commutativity of our operator with the Dirac Hamiltonian, we find

$$K \left(\vec{\Sigma} \cdot \hat{\vec{r}}\right) [x_2 V'(r) - x_3 f'(r)] + \tag{14}$$
$$+2i\beta K\gamma^5 \left[\frac{x_1}{r} - mx_3 f(r)\right] = 0$$

Here terms are grouped in a way that we have a diagonal matrix in the first row, while the anti-diagonal matrix is in the second row. Therefore, the two equations follow:

$$x_2 V'(r) = x_3 f'(r) \tag{15}$$

$$x_3 mf(r) = \frac{x_1}{r} \tag{16}$$

Integration of the Eq. (15) with the requirement that functions $f(r)$ and $V(r)$ tend to zero when $r \rightarrow \infty$, yields

$$x_2 V(r) = x_3 f(r) \tag{17}$$

while the equation (16) gives

$$f(r) = \frac{x_1}{x_3} \frac{1}{mr} \tag{18}$$

281

Substituting Eq. (18) into Eq (17) results in the following potential

$$V(r) = \frac{x_1}{x_2} \frac{1}{mr} \tag{19}$$

Hence, in the very general framework we have shown that the only central potential for which the Dirac Hamiltonian would have an additional symmetry (in the above mentioned sense) is a **Coulomb potential.**

Meanwhile, the relative signs of coefficients x_1 and x_2 may be arbitrary. Therefore we have a symmetry both for attraction and repulsion. If we take the Coulomb potential in the usual attraction form

$$V_c(r) = -\frac{a}{r} \tag{20}$$

where $a = Ze^2 = Z\alpha$, it follows

$$x_2 = -\frac{1}{ma} x_1 \tag{21}$$

In this case our symmetry operator (13) becomes

$$A = x_1 \left\{ \left(\vec{\Sigma} \cdot \hat{\vec{r}} \right) - \frac{i}{ma} K \left(\vec{\Sigma} \cdot \vec{p} \right) + \frac{i}{mr} K \gamma^5 \right\} \tag{22}$$

Number x_1, as an unessential common factor may be omitted.

We mention here that if the potential in the Dirac Hamiltonian was a Lorentz-scalar (which means the change $V \to \beta V$) then, while K still commutes with H, obtained operator A does not commute anymore with H even for Coulomb potential and must be generalized [15].

What the real physical picture is standing behind this? Remark that taking into account the Eq. (10) for $\vec{V} = \vec{p}$, one can recast our operator for the Coulomb potential in the following form

$$A = \vec{\Sigma} \cdot \left(\hat{\vec{r}} - \frac{i}{2ma} \beta \left[\vec{p} \times \vec{l} - \vec{l} \times \vec{p} \right] \right) + \frac{i}{mr} K \gamma^5 \tag{23}$$

One can see that in the non-relativistic limit, i.e. when $\beta \to 1$ and $\gamma^5 \to 0$, our operator reduces to

$$A \to A_{NR} = \vec{\sigma} \cdot \left(\hat{\vec{r}} - \frac{i}{2ma} \left[\vec{p} \times \vec{l} - \vec{l} \times \vec{p} \right] \right) \tag{24}$$

Note the Laplace-Runge-Lenz vector in the parenthesis of Eq. (24). Therefore, relativistic supercharge reduces to the projection of the Laplace-Runge-Lenz vector on the electron spin direction. Precisely this operator, Eq.(24) was used in the case of Pauli electron [1].

Therefore, we see that there is a deep relation between supersymmetry of the Dirac Hamiltonian and the symmetry related to the Laplace-Runge-Lenz vector, which appeared already in classical mechanics and provides the closeness of celestial orbits.

It is noticeable that the hidden symmetry, familiar to Coulomb potential, manifests here itself unexpectedly only at the end of calculations.

We can conclude that the hidden symmetry associated to the Laplace-Runge-Lenz vector governs very wide range of physical phenomena from planetary motion to fine and hyperfine structure of atomic spectra. As for the Lamb shift, which is pure quantum field theory effect, its Hamiltonian, derived by radiative corrections to a photon propagator and photon-electron vertex function, does not commute with A operator and therefore spoils the above mentioned symmetry. In other words, symmetry of the A operator controls the absence of the Lamb shift in the Dirac theory.

Remark that our operator (22) is not new. If we make transition to the usual Dirac $\vec{\alpha}$ matrices according to the relation $\vec{\Sigma} = \gamma^5 \vec{\alpha}$, then operator A can be reduced to the form

$$A = \gamma^5 \left\{ \vec{\alpha} \cdot \hat{\vec{r}} - \frac{i}{ma} K \gamma^5 (H - \beta m) \right\} \tag{25}$$

282

which coincides precisely with the Johnson-Lippmann operator [7].

One must also remember that the form of obtained symmetry operator is not unique. One can always replace $A \rightarrow \hat{O}A$, where $[\hat{O}, H] = 0$ and $[\hat{O}, K] = 0$. One can take, for example, $\hat{O} = f(K, H)$ - arbitrary regular matrix function of $K(H)$. This arbitrariness can be used,for example, in generalization of symmetry operator to the case of Lorentz scalar-Coulomb potential.It is known [15] that the JL operator can be generalized in case of combination of vector- and scalar-Coulomb potentials. For this case relevant symmetry operator has the form [15]

$$ B = -\frac{\imath}{ma} K \gamma^5 (H - \beta m) + \vec{\Sigma} \cdot \hat{\vec{r}} \{\alpha_V + \alpha_S \frac{H}{m}\} \tag{26} $$

where α_V and α_S are Coulomb potential strengths in either cases. We see that it differs from (25) by amount $const \vec{\Sigma} \cdot \hat{\vec{r}} H$, which is permissible according our theorem above.

Moreover, SUSY in specific and mostly exotic models of Dirac equation (such as 2+1 dimensions [16], non-minimal or anomalous magnetic moment coupling [17], squared equation [18, 19], monopole [20] etc.) are not excluded by our above consideration. We think that it will be interesting to consider so-called Dirac oscillator [21] from above point of view, as the relevant Hamiltonian $im\omega\beta\vec{\alpha} \cdot \vec{r}$ commutes with K and the corresponding degeneracy still takes place. It will be reported in a forthcoming paper.

Acknowledgements

We would like to thank Professor Dr. S. Khalil for hospitality at the Cairo Int. Conference on High Energy Physics CICHEP II. This work was supported in part by the NATO Reintegration Grant No. FEL REG. 980767.

[1] R. G. Tangerman, J. A. Tjon. Phys. Rev.**A48**,1089 (1998);

[2] C. V. Sukumar, J.Phys.A: Math.Gen.**18**,L697 (1985);

[3] J. P. Dahl, T. Jorgensen. Int.J.Quantum Chemistry.**53**,161 (1995).

[4] A. A. Stahlhofen, Helv.Phys.Acta.**70**, 372 (1997);

[5] H. Katsura, H. Aoki, J. Math. Phys. **47**, 032302 (2006); see also ArXiv: quant-ph/0410174 (2004);

[6] T. T. Khachidze, A. A. Khelashvili, Mod. Phys. Lett.**A20**, 2277 (2005);

[7] M. H. Johnson, B. A. Lippmann, Phys.Rev. **78**, 329(A), (1950);

[8] C. L. Critchfield, J. Math. Phys. **17**, 261 (1976);

[9] P.A. Dirac, The Principles of Quantum Mechanics, 4th ed. (Oxford Univ.Press, London, 1958). Chap.11;

[10] J. J. Sakurai, Advanced Quantum Mechanics (Addison-Wesley, MA,1967),Chap.3;

[11] W.Greiner, B.Muller, J.Rafelski, Quantum Electrodynamics of Strong Fields (Springer-Verlag,Berlin, 1985), Chap.3;

[12] L. Biederharn, L. Louk, in Encyclopedia of Mathematics and Its Application (Addison- Wesley Reading, MA, 1981);

[13] T. T. Khachidze, A. A. Khelashvili, Am. J. Phys.**74**, 628, (2006); see also arXiv: quant-ph/0508122.

[14] Constraints of this theorem are also satisfied by Laplace-Runge-Lenz vector $\vec{A} = \hat{\vec{r}} - \frac{i}{2ma}[\vec{p} \times \vec{l} - \vec{l} \times \vec{p}]$, but this vector is associated to the Coulomb potential. Hence, we abstain from its consideration for now. Moreover $(\vec{\Sigma} \cdot \vec{A})$ is not an independent structure. It is expressible by two other structures, e.g. $(\vec{\Sigma} \cdot \vec{A}) = (\vec{\Sigma} \cdot \hat{\vec{r}}) + \frac{i}{ma}\beta K (\vec{\Sigma} \cdot \vec{p})$.

[15] A.Leviatan, Phys.Rev.Lett.**92**, 202501 (2004).

[16] H. Ui, Progr.Theor.Phys. **72**, 192 (1984);

[17] V.V.Semenov.J.Phys. A: Math. Gen.**23**,L721 (1990);

[18] F.De Jonghe et al. Phys.Lett.**B359**, 114 (1995);

[19] P.A.Horvathy et al. arXiv: hep-th/0006118 (2000).

[20] M. S. Plyushchay, Phys. Lett. **B485**, 187 (2000).

[21] J. Benitez et al. , Phys. Rev. Lett. **64**, 1643 (1990). R.P. Martinez-y-Romero, arXiv: quant-ph/9908069.

AUTHOR INDEX